Light Scattering and Absorption by Particles

The Q-space approach

Online at: https://doi.org/10.1088/978-0-7503-5157-7

Light Scattering and Absorption by Particles

The Q-space approach

Christopher M Sorensen

Department of Physics, Kansas State University, Manhattan, KS, USA

IOP Publishing, Bristol, UK

ISBN 978-0-7503-5157-7 (ebook)
ISBN 978-0-7503-5155-3 (print)
ISBN 978-0-7503-5158-4 (myPrint)
ISBN 978-0-7503-5156-0 (mobi)

DOI 10.1088/978-0-7503-5157-7

Version: 20221201

IOP ebooks

British Library Cataloguing-in-Publication Data: A catalogue record for this book is available from the British Library.

Published by IOP Publishing, wholly owned by The Institute of Physics, London

IOP Publishing, No.2 The Distillery, Glassfields, Avon Street, Bristol, BS2 0GR, UK

US Office: IOP Publishing, Inc., 190 North Independence Mall West, Suite 601, Philadelphia, PA 19106, USA

This book is dedicated to the memory of my parents Lyla Fern and Nels Anton Sorensen who set me on my way.

Contents

Preface

This book describes the scattering and absorption of light by particles, a topic well presented in previous books on this subject that are 'neither few nor small'. However, while the topic is the same, the approach here is significantly different. Here light is first and foremost a wave. This wave diffracts when it encounters an object. This diffraction can be altered when the electromagnetic aspect of light interacts electromagnetically with the object depending on the strength of the interaction. The nature of this altering is revealed here.

Waves diffract. Diffraction from an object scatters energy from the incident wave. The mathematics of diffraction is the spatial Fourier transform of the object into reciprocal space, the variable for which we will designate by q; hence diffraction resides in 'Q-space'. This mathematical variable finds physical reality in the lab as it is set by the experimenter via application of a source with a given wavelength, and then, with more facility, variation of the scattering angle. The square of the Fourier transform is the structure factor of the object. Importantly, inverse q has dimensions of length. In this work this scattering length scale will be compared to length scales of the scattering object or object ensembles (aggregates, aerosols, colloids and condensed matter systems) in a 'scaling approach' from which a physical understanding of diffraction and the structure factor will emerge.

With this foundation of diffraction, we turn to the issue of how the electromagnetic nature of light augments and alters the scattering of light. Unlike previous treatises, our study of electromagnetic light scattering will not start with Maxwell's equations. No complex mathematical machinations will ensue. Instead we will start with the electromagnetic phenomenon of Rayleigh scattering, a non-wave, non-diffractive limit. Between the antipodes of electromagnetic free diffraction and wave-less Rayleigh scattering, lies the entire realm of light scattering.

To explore this realm use is made of known electromagnetic formulations, like the Mie equations, for an empirical study of scattering. We will devote our energies not to the difficulties of obtaining the solution, but rather to digging deeply into the solutions themselves. Calculated results using systematic variation of particle properties, such as size and refractive index, when viewed from the diffractive Q-space perspective, will uncover functionalities and aspects of light scattering previously unseen and undescribed. Two new controlling parameters, the internal coupling parameter and the relative skin depth, will be discovered and applied. These analyses will yield a new, universal perspective for light scattering by all particles. The imprint of diffraction and Rayleigh scattering will remain. Q-space analysis will demonstrate for particles of any size and morphology that as the electromagnetic nature of light comes to the fore, the scattering evolves away from the foundations of the three-dimensional diffraction limit and Rayleigh scattering in an orderly fashion. The result will be a universal description of light scattering by particles of all sizes, shapes and refractive indices.

Acknowledgements

My endeavor to write this book is based on 50 years of experience in the field of light scattering in one manner or another. Along the way there have been many colleagues and students with whom I have collaborated, and I owe a debt of gratitude to them all. It began in graduate school scattering light from fluids near the critical point with my PhD advisors Professors William J O'Sullivan and Richard C Mockler, who patiently gave me free reign to do what I wanted to do. Some of my light scattering students made significant contributions to the ideas presented in this book; they are Chunghoon Oh, Matt Berg, Will and Yuli Wang Heinson (Heinson number 2 and 1, respectively), Justin Maughan and Prakash Gautam. In a similar manner I have also enjoyed close and substantial collaboration with my good friends Amit Chakrabarty and Hans Moosmuller. And I am certainly very grateful to have been well funded (but not funded well enough!) by NSF and other agencies for this work. I want to thank John Navas of IOP Publishing for his guidance and patience through the many years I spent writing this book. Finally, I acknowledge, with love, my wife Georgia who knew that beyond her, I had a second passion, that for science.

Author biography

Christopher M Sorensen

Chris is the Cortelyou-Rust University Distinguished Professor and University Distinguished Teaching Scholar, Emeritus, in the departments of physics and chemistry at Kansas State University. He received a BS from the University of Nebraska, served in Vietnam, and then a PhD from the University of Colorado. He has published over 330 papers that have accumulated over 21 000 citations in the areas of particulate systems and synthesis, condensed matter physics, and light scattering. He holds and has pending 12 patents. He is a fellow of the American Physical Society, the American Association for the Advancement of Science, and the American Association for Aerosol Research. He is the founding inventor of a startup company to produce graphene materials. He has won awards for his science and his teaching. In 2007 he was named the CASE/Carnegie National Professor of the Year.

Chapter 1

Waves

Light is a self-propagating, transverse wave disturbance in the electromagnetic field over a narrow range of wavelengths closely associated with the sensitivity of the human eye. It is a specific example of electromagnetic waves that cover all wavelength scales and are a mathematical consequence of the Maxwell equations. These waves carry energy at the speed of light, which is independent of the frame of reference. Light, and other electromagnetic waves, is emitted and detected as quanta, particle-like entities called photons, yet they propagate as waves. For the most part, light scattering and absorption is a consequence of the wave nature of light; hence we begin our work by studying waves.

1.1 Wave concepts

Waves 'wave' as a function of *phase*. Phase, ϕ, has the units of radians, which are dimensionless. The 'wavey' part of the wave can be represented mathematically with a sine or cosine function, the so-called 'harmonic functions'. Since we are concerned with light, which is an electromagnetic wave, we will use E to represent the electric field of the wave and write

$$E = E_0 \sin \phi \tag{1.1}$$

In equation (1.1) E_0 is the amplitude of the wave. The wave repeats itself, i.e. it is periodic, and the period in phase space is 2π radians.

In the physical world, waves wave with both a functionality in space, represented by x and measured in meters, and time, represented by t and measured in seconds. The period in space is the *wavelength*, λ, the 'spatial period', and the *period* in time is T, the 'temporal period' or simply the period. To replace the phase functionality by either the space or time functionality, x or t, these variables need coefficients equal to the ratio of the phase period to either the space or time period. These ratio coefficients are the *wavenumber*

doi:10.1088/978-0-7503-5157-7ch1

$$k = 2\pi/\lambda \tag{1.2}$$

and the angular frequency

$$\omega = 2\pi/T \tag{1.3}$$

We then write the wave in space as

$$E = E_0 \sin kx \tag{1.4}$$

and the wave in time as

$$E = E_0 \sin \omega t \tag{1.5}$$

The coefficients convert space and time into phase. Sketches of these three waves are given in figures 1.1(a)–(c).

A propagating wave will be a function of both x and t. If the wave's shape remains constant, the functionality on x and t is the same. At time $t = 0$, the functionality is $f(x)$. Since the wave moves to larger x at a wave velocity of v, the argument of the functionality must be reduced by vt to obtain the same functionality. Thus, at time t, the functionality is $f(x - vt)$. This was for a wave propagating in the positive x-direction; for the opposite directions, one has $f(x + vt)$. Thus, in general, a propagating wave has the form $f(k[x \mp \nu t]) = f(kx \mp \omega t)$. Note that this yields the wave velocity magnitude as

$$v = \omega/k \tag{1.6}$$

with a direction along the positive x-axis when the $\mp$ sign is negative. Equation (1.6) also follows from a dimensional analysis.

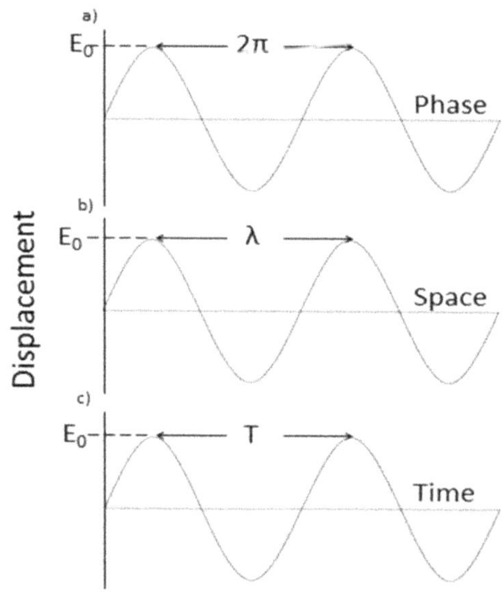

Figure 1.1. A sinusoidal wave of amplitude E_0 in (a) phase space with period 2π radians, (b) 'real' space with period λ meters, and (c) time space with period T seconds.

As an example, a harmonic wave propagating along the positive z-axis can be written as

$$E = E_0 \sin(kz - \omega t) \tag{1.7}$$

The phase of the wave (radians) is

$$\phi = kz - \omega t \tag{1.8}$$

From equations (1.2), (1.3) and (1.6) we find

$$v = \lambda/T = f\lambda \tag{1.9}$$

In equation (1.9), we define the *frequency* f as

$$f = 1/T = \omega/2\pi \tag{1.10}$$

with units of inverse seconds or Hz.

Generalization to three dimensions follows by making the wave number a *wave vector* with the magnitude given by equation (1.2) and direction along the propagation direction specified by the unit vector $\hat{\boldsymbol{k}}$. Thus, $\boldsymbol{k} = (2\pi/\lambda)\hat{\boldsymbol{k}}$. Rather than x the position is specified by the position vector $\vec{r}$. Then the equation for a propagating harmonic wave of equation (1.7) becomes

$$E = E_0 \sin(\boldsymbol{k} \cdot \boldsymbol{r} \mp \omega t) \tag{1.11}$$

Note that we are considering only scalar waves, i.e. the electric field is written as a scalar. It is, of course, a vector that gives it polarization. However, for now a scalar representation is sufficient and polarization will be described latter. The vector dot product in equation (1.11) ensures that when $\vec{r}$ changes in a direction perpendicular to the incident direction, there will be no change in phase. Thus, equation (1.11) describes plane waves of constant phase perpendicular to $\hat{k}$, the direction of propagation.

1.2 Energy transport

Propagating waves transport energy in the propagation direction. This transport is a flux of energy, which is energy per unit time per unit area, or power per unit area. Power per unit area is called *intensity*. One sometimes sees the term *irradiance* used for this same quantity (see e.g. Hecht 2002).

An electromagnetic wave such as light is composed of both electric and magnetic vector field waves coupled together in phase via the laws of electromagnetism, the Maxwell equations. These fields, $\boldsymbol{E}$ and $\boldsymbol{B}$, respectively, have magnitudes related by $E = cB$, where c is the speed of light, and they are always perpendicular to each other. A drawing of such a wave is given in figure 1.2.

The vector cross product of the fields gives the direction of propagation as described by the *Poynting vector* (Griffiths 1999) given by

$$\boldsymbol{S} = \varepsilon_0 c^2 \boldsymbol{E} \times \boldsymbol{B} \tag{1.12}$$

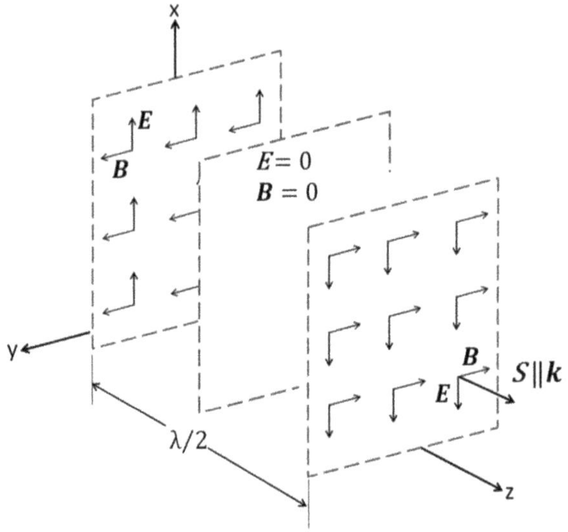

Figure 1.2. Part of a planar electromagnetic wave, frozen in time, propagating along the positive z-axis with polarization, the direction of the electric field, along the x-axis. The Poynting vector $\boldsymbol{S}$ and the wave vector $\boldsymbol{k}$ are parallel to each other and the direction of propagation. Planes of constant phase perpendicular to these vectors are drawn.

where ε_0 is the permittivity of free space. The Poynting vector describes the flux of energy, hence the intensity, for an electromagnetic wave. Given the field conditions above, it can be shown that

$$I = \varepsilon_0 c E_0^2/2 \tag{1.13}$$

The factor of 1/2 in equation (1.13) comes from the time average of the square of the sinusoidal time functionality of the wave. Note the important fact that the intensity is proportional to the wave amplitude squared. This is true for all types of waves. For most of our purposes here, we need only the proportionality to field squared, which in complex notation (see below) is

$$I \propto E^* E \tag{1.14}$$

1.3 Complex notation

The mathematics of waves is the mathematics of the harmonic trig functions sine and cosine. This is true for all shapes of waves as we will learn below when we study Fourier analysis. This will cause us to add and subtract the harmonic functions, which can be somewhat cumbersome. Here, we describe how the trig functions can be converted to exponential functions if we leave the world of real numbers and allow for complex numbers involving both real and imaginary parts.

An imaginary number involves the square root of minus one represented as $i = \sqrt{-1}$. A *complex number* has both real and imaginary parts and hence can be expressed as the sum $z = x + iy$. This can be represented as a vector, or *phasor*, in

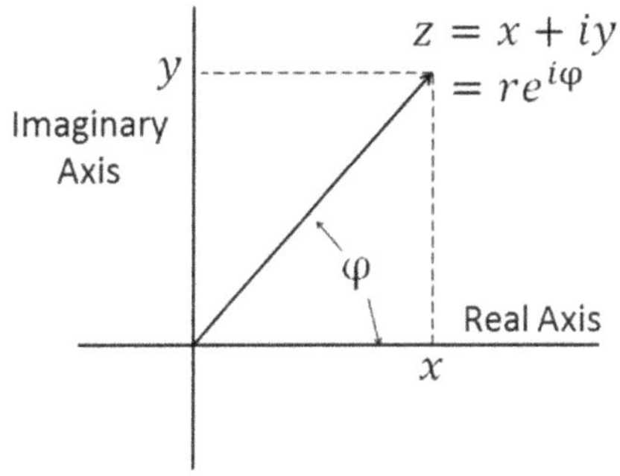

Figure 1.3. The complex number $z = x + iy = re^{i\phi}$ in the complex plane with real and imaginary axes. Note that this is not to be confused with the z-axis in a Cartesian coordinate system.

the complex plane as shown in figure 1.3. From this figure it can be seen that $x = r\sin\phi$ and $y = r\cos\phi$. The magnitude (or *modulus*) of the phasor is $r = |z| = \sqrt{zz^*}$ where $z^* = x - iy$ is the *complex conjugate* of z. To convert this to an exponential we make use of *Euler's formula*

$$e^{i\phi} = \cos\phi + i\sin\phi \tag{1.15}$$

Then any complex number can be written as

$$z = re^{i\phi} \tag{1.16}$$

The complex number z can represent a wave because it has the sine and cosine functions in it. A useful way to visualize how z works is to imagine you are looking at z from the side of the complex plane, e.g. looking in along the real or x-axis. Then as the phase ϕ advances, the visible projection of the rotating phasor z would appear to oscillate up and down, representing a wave.

Experience has taught us that calculation of wave situations with exponentials is easier than with the trig functions. Moreover, the phasor nature of the complex number and the fact that phasors add like vectors will allow for visualization of wave addition, which will enhance physical understanding. Henceforth, we will write waves in complex notation as

$$E = E_0 e^{i(kz \mp \omega t)} \tag{1.17}$$

in one dimension, or

$$E = E_0 e^{i(\mathbf{k}\cdot\mathbf{r} \mp \omega t)} \tag{1.18}$$

in three dimensions. Since measurable quantities of the physical world are real, we will always take the real part of the results of any calculation we might perform to return to reality.

1.4 Fourier analysis

The goal of this book is to describe light scattering and absorption. Since light is a wave, that goal will lead us to adding waves together. The mathematics of adding waves together is well described by *Fourier analysis*. The harmonic functions, sine and cosine, comprise a complete set of functions. 'Complete' means that any function can be represented as a sum of (or integral over) the harmonic functions. There are two distinct cases where this property is applied.

1.4.1 Fourier transformation of periodic functions

Any periodic function with period λ can be represented as

$$f(x) = \frac{A_0}{2} + \sum_{m=1}^{\infty}[A_m \cos mkx + B_m \sin mkx] \tag{1.19}$$

where $k = 2\pi/\lambda$, which foretells the connection to light waves. This is called a *Fourier series*. The *Fourier coefficients* A_m and B_m are found via the *orthogonality conditions*

$$\frac{2}{\lambda}\int_0^{\lambda} \sin mkx \sin nkx \, dx = \delta_{m,n} \tag{1.20}$$

$$\frac{2}{\lambda}\int_0^{\lambda} \cos mkx \cos nkx \, dx = \delta_{m,n} \tag{1.21}$$

$$\frac{2}{\lambda}\int_0^{\lambda} \cos mkx \sin nkx \, dx = 0 \tag{1.22}$$

where $\delta_{m,n}$ is the Kroneker delta function:

$$\delta_{m,n} = \begin{cases} 1 & m = n \\ 0 & m \neq n \end{cases} \tag{1.23}$$

It quickly follows that

$$A_m = \frac{2}{\lambda}\int_0^{\lambda} f(x)\cos mkx \, dx \tag{1.24}$$

and

$$B_m = \frac{2}{\lambda}\int_0^{\lambda} f(x)\sin mkx \, dx \tag{1.25}$$

Application of equations (1.19) through (1.25) is called *Fourier analysis*.

A good example of Fourier analysis is to apply it to the square wave. In doing so, we will lay an intuitive foundation for diffraction, which is at the heart of scattering. The first complete period of the square wave can be expressed as

$$f(x) = \begin{cases} 1 & 0 \leqslant x \leqslant \lambda/2 \\ -1 & \lambda/2 \leqslant x \leqslant \lambda \end{cases} \tag{1.26}$$

This is then repeated for all x. The square wave is drawn in figure 1.4. Application of equations (1.19)–(1.25) yields the Fourier series representation of the square wave as

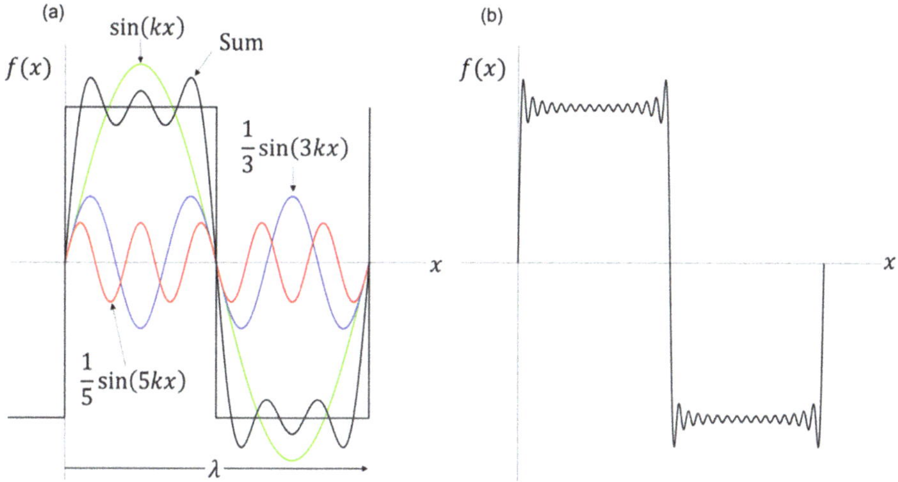

Figure 1.4. The square wave function with (a) the first three terms of its Fourier series, and (b) the first 10 terms.

$$f(x) = \frac{4}{\pi}\left(\sin kx + \frac{1}{3}\sin 3kx + \frac{1}{5}\sin 5kx + \dots\right) \tag{1.27}$$

In figure 1.4, notice how the wave takes shape as more components of the series are included. Also, notice that the sharp boundaries of the square wave depend on the bigger mk values. These correspond to shorter wavelength, λ/m, i.e. higher spatial frequencies.

The Fourier coefficients, A_m and B_m, represent the *spectrum* of the function. Their overt functionality is on m, but closer inspection of equations (1.24) and (1.25) shows that they are functions of mk, which has dimensions of inverse length. The dimensions of the variable of the spectrum function are always reciprocal to the independent variable of the function. One typically thinks of spectra, a spectrum of light for example, in terms of frequency, but frequency is just inverse time and a frequency spectrum would be simply the Fourier components of a time varying function. Since the square wave of equation (1.26) and figure 1.4 is a function of x with dimensions of length, its spectrum is a function of k with dimensions of inverse length. The wavenumber k is an angular frequency in space; it describes 2π times how often per unit length the wave repeats, see equation (1.2). The spectrum for a periodic function of infinite extent is always discrete, as for the square wave. Figure 1.5 shows the spectrum of the general square wave in figure 1.4.

1.4.2 Fourier transformation of any function

The magnificent power of Fourier analysis comes to fruition when it is applied to any function $f(x)$, not just those that are periodic. Then the sum over m in equation (1.19), which is in essence a sum over mk, will become an integral over k. The

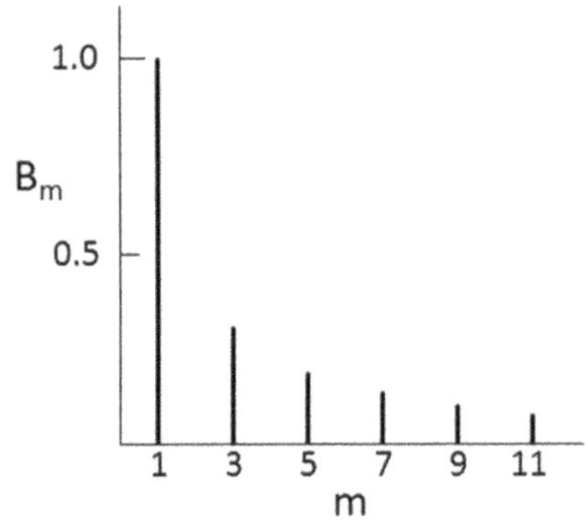

Figure 1.5. Spectrum of the infinite square wave shown in figure 1.4. Here m is the coefficient of the angular spatial frequency k. The $m = 1$ component is the *fundamental*; the $m = 3, 5, 7, \ldots$components are the *overtones*.

discrete Fourier coefficients, A_m and B_m, will be replaced by a continuous function of k, $F(k)$. The fact that the integrals we anticipate involve sines and cosines opens the opportunity to use the complex notation or representation introduced above. Complex representation allows us to express the sine and cosine in a short-hand, exponential functionality and write the *Fourier transformation* as

$$f(x) = 1/\sqrt{2\pi} \int_{-\infty}^{\infty} F(k)e^{-ikx}dk \tag{1.28}$$

The function $F(k)$ is called the *Fourier transform* of $f(x)$. Comparison to the preceding section indicates that $F(k)$ is the spectrum of $f(x)$; spectra are Fourier transforms. This applies to functions $f(x)$ that are either periodic or not. Thus, the variable k is not always related to a period, as in equation (1.2), for example. The variables x and k have dimensions that are reciprocal with each other. Thus, if the units of x are meters, the units of k are inverse meters. If one often refers to k-space as *reciprocal space*, then x is in 'real' space. Fourier transforms transform from reciprocal space k to real space x and vice versa. Also, another common Fourier parameter pair is t and ω, time and angular frequency.

Inspection of equation (1.28), while keeping in mind the Shakespearian wisdom that '...a rose by any other name ...', suggests that x and k are interchangeable as are $f(x)$ and $F(k)$. The symmetric, reciprocal nature of the transforms implies no special status of one space over the other. Indeed, the power of mathematics confirms this suggestion to yield the *inverse transform*

$$F(k) = 1/\sqrt{2\pi} \int_{-\infty}^{\infty} f(x)e^{ikx}dx \tag{1.29}$$

The interchange is not perfectly symmetrical; notice the positive exponent for the e^{ikx} term in equation (1.29) versus the negative exponent in equation (1.28). All I will say here regarding the reason for this is that 'it's in the math' and leave it to the reader to consult a math text. The same can be said for the factors of $\sqrt{2\pi}$, which can be found in different positions depending on the text one reads[1]. I do not explain these issues because I have no need to.

In case you haven't noticed, Fourier transforms are adding waves. Equation (1.28) is essentially a continuous sum of waves with wavelengths $\lambda = 2\pi/k$ and amplitudes $F(k)$. Light scattering will involve the summation of waves from various regions of a given particle or from an ensemble of particles and this will lead us to use and understand Fourier transforms.

Next we will study two very useful examples of Fourier transforms, the square pulse and the Gaussian function.

1.4.3 Fourier transform of the square pulse

The square pulse will be the foundation of our understanding of wave diffraction to be described later in this book. Consider the square pulse of width w and height h specified in equation (1.30) and drawn in figure 1.6(a).

$$\begin{aligned} f(x) &= h, \; -w/2 \leqslant x \leqslant w/2 \\ &= 0, \; x < -w/2 \text{ or } x > w/2 \end{aligned} \tag{1.30}$$

Substitution into equation (1.29) yields

$$F(k) = h/\sqrt{2\pi} \int_{-w/2}^{w/2} e^{ikx} dx \tag{1.31}$$

Through integration and application of Euler's formula, equation (1.15), and some algebra (in which, nicely, the imaginary numbers cancel out), one obtains

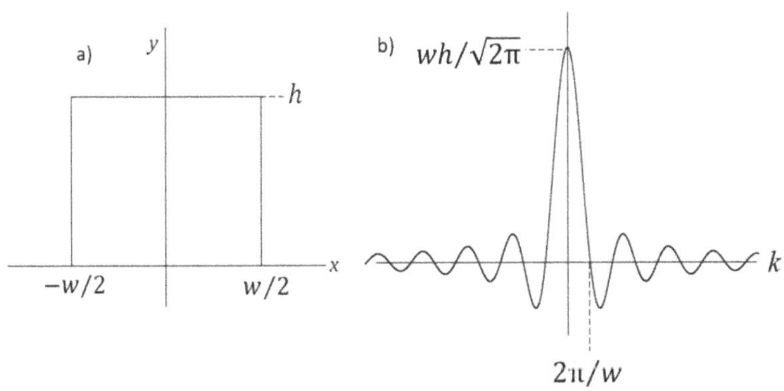

Figure 1.6. (a) The square pulse of width w and height h in x-space, (b) the sinc function with first zero at $2\pi/w$ in k-space. These two functions are Fourier transform pairs.

[1] There appears to be no universal convention for the placement of $\sqrt{2\pi}$ and the designation 'inverse transform'.

$$F(k) = 1/\sqrt{2\pi}(wh\ \sin(kw/2)/(kw/2))$$
$$= 1/\sqrt{2\pi}(A\ \sin u/u) \quad (1.32)$$
$$= 1/\sqrt{2\pi}A\ \mathrm{sinc}(u)$$

Here, an area of the square pulse has been defined as $A = hw$. Also defined are the dimensionless variable

$$u = kw/2 \quad (1.33)$$

and the sinc function

$$\mathrm{sinc}(u) = \sin u/u \quad (1.34)$$

Figure 1.6 contains plots of both the square pulse and its Fourier transform, the sinc function (figure 1.6(b)). The sinc function is a damped oscillation with zeros at $u = m\pi$ hence $k = m(2\pi/w)$. Thus, as the pulse gets wider, the sinc function gets narrower and vice versa. This is how it has to be because of the inverse nature of the dimensions of x and k; small meters transforms into big inverse meters. This inverse relationship between k and x is very important[2].

The sinc function also has a height proportional to the area of the pulse. Note, however, that the inverse is true. The sinc has a height proportional to $A = hw$ and a width proportional to the position of its first minimum at $k = 2\pi/w$. The product of this height and width, which is proportional to the area of the sinc function is proportional to the height h of the square pulse.

Finally, don't fail to remember that since the sinc function is the Fourier transform of the square pulse, then the square pulse is the Fourier transform of the sinc function; they are *Fourier transform pairs*[3].

1.4.4 Fourier transform of the Gaussian function

Consider the Gaussian function given by

$$f(x) = \frac{1}{\sqrt{2\pi}\sigma_x} e^{-x^2/2\sigma_x^2} \quad (1.35)$$

and drawn in figure 1.7. The function in equation (1.35) is normalized to unity. The quantity σ_x is the standard deviation of the Gaussian, a well-known measure of the width of the Gaussian, see figure 1.7. The subscript x is used to designate x-space. The Fourier transform of (1.35) is, via equation (1.29),

$$F(k) = \frac{1}{\sqrt{2\pi}} e^{-k^2\sigma_x^2/2} \quad (1.36)$$

[2] I remark that light scattering from small particles spreads widely, whereas scattering from big particles has a narrow spread.

[3] Precisely speaking one is the Fourier transform, the other is the *inverse* Fourier transform.

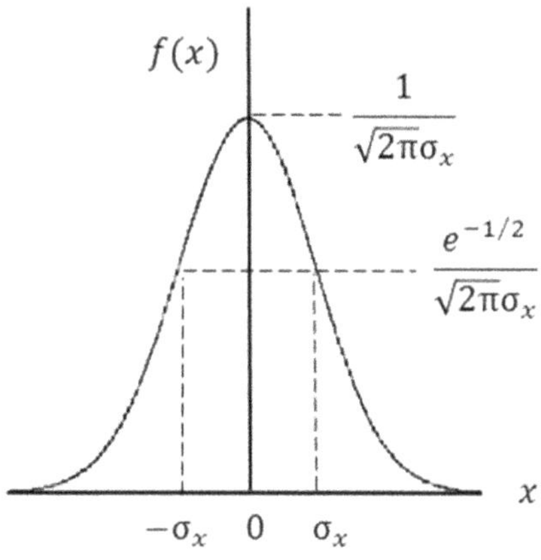

Figure 1.7. The Gaussian function in x-space area normalized to unity. σ_x is the standard deviation. 68% of the area lies within $\pm\sigma_x$.

That is, the Fourier transform of a Gaussian is another Gaussian; the Gaussian shape transforms into itself! This is the only function with this property.

A unity normalized Gaussian in k-space, in analogy to equation (1.35) should read

$$F(k) = \frac{1}{\sqrt{2\pi}\,\sigma_k} e^{-k^2/2\sigma_k^2} \tag{1.37}$$

Comparison to (1.36) yields the following relationship for the standard deviation widths in x- and k-space[4]

$$\sigma_x\sigma_k = 1 \tag{1.38}$$

This simple relationship expresses, once again, the inverse relationship between the two spaces. As one Gaussian grows wider, its Fourier transform grows narrower, and vice versa, yet the shape remains the same.

Comparison of the square pulse/sinc and the Gaussian/Gaussian Fourier pairs illustrates another important characteristic of Fourier transforms, one which we anticipated in our description of the Fourier series expression of the square wave, and that is that sharp edges transform to oscillations and soft (not sharp) edges transform to soft edges.

A Gaussian in the limit of infinitely narrow width, i.e. the limit as $\sigma \to 0$, becomes a Dirac delta function, $\delta(x)$. That is

$$\lim_{\sigma\to 0} \frac{1}{\sqrt{2\pi}\,\sigma} e^{-x^2/2\sigma} = \delta(x) \tag{1.39}$$

[4] Equation (1.38) is the foundation of the Heisenberg uncertainty principle of quantum (wave) mechanics.

This handy function can also, and more properly, be obtained from consecutive applications of the Fourier transform and its inverse, equations (1.28) and (1.29), see appendix A. Simply said, the Dirac delta function can be represented as an infinitely narrow spike that is infinitely tall such that its area is unity. Since $\sigma_x \to 0$ for $\delta(x)$, by (1.38) its Fourier transform must have $\sigma_k \to \infty$, i.e. it is infinitely wide; the function becomes a constant. Thus, infinitely wide (i.e. constant) and infinity narrow functions are Fourier transform pairs.

1.4.5 The convolution and the convolution theorem

A *convolution* is a blending together (or a folding together, hence the German synonym *faltung*) of two functions with a shift x between them. The convolution $C(x)$ of two functions $f(u)$ and $g(u)$ is given by

$$C(x) = f(u) \otimes g(u) = 1/\sqrt{2\pi} \int_{-\infty}^{\infty} f(x-u)g(u)du \tag{1.40}$$

The symbol $\otimes$ designates convolution and the factor $1/\sqrt{2\pi}$ is consistent with the Fourier transformation formulae above. A physical example of a convolution will be given below.

An important and useful theorem of Fourier analysis is the *convolution theorem*. The convolution theorem states that the Fourier transform of the convolution of $f(u)$ and $g(u)$ is equal to the product of their Fourier transforms, $F(k)$ and $G(k)$, respectively. Then the convolution theorem states that

$$F(k)G(k) = 1/\sqrt{2\pi} \int_{-\infty}^{\infty} C(x)e^{ikx}dx \tag{1.41}$$

1.5 Diffraction

Diffraction is the spreading of waves perpendicular to the direction of propagation. When the waves encounter an obstacle, this spreading causes the waves to bend into the geometric shadow of the obstacle.

The classic explanation of diffraction starts with the realization, first proposed by Huygens, that any wavefront can be broken up into a series of point sources, each of which radiates new circular or spherical waves, so-called Huygens wavelets, into the forward direction. Precise theories, such as Kirkoff's, see e.g. (Born and Wolf 1999), analytically specify the functionality of the angular dependence of the wavelets from the forward direction, but we will not need such precision here. The Huygens concept is illustrated in figure 1.8, which shows how a plane wave of finite width propagates through space. We envision that the original plane wave at time t breaks into a series of point sources for circular Huygens wavelets. These propagate forward and their common envelope is the new plane wave at time $t + \Delta t$. Note, however, that the circular Huygens wavelets at the edges cause spreading of the wave beyond the edges of the plane wave. This spreading is diffraction.

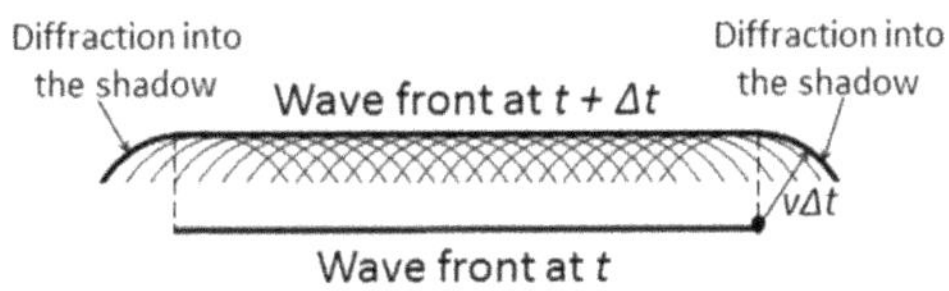

Figure 1.8. Illustration of the Huygens wavelet construction to describe the propagation of a plane wave.

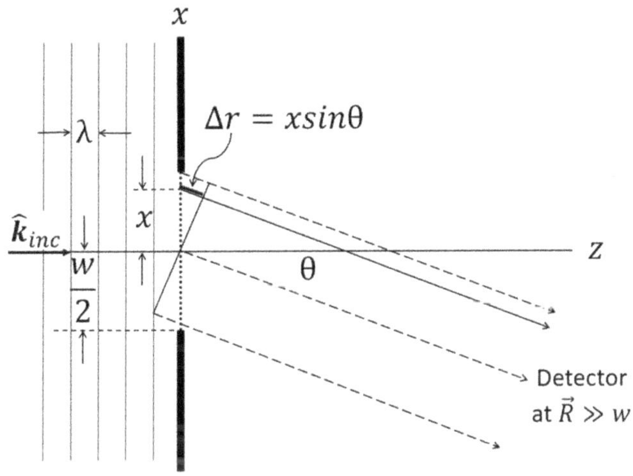

Figure 1.9. Diagram of a single slit of width w. Waves of wavelength λ are incident from the left of the slit along the z-axis. The width of the slit lies along the x-axis. The wave in the slit has been broken up into Huygens wavelet point sources, and some of the rays from those sources at an angle of θ from the z-axis have been drawn. The nearly parallel rays meet at a detector far away. A wavelet source at x above the z-axis is shown to have a path difference of $\Delta r = x \sin \theta$ relative to the ray at the center of the slit.

1.5.1 Single slit diffraction

Perhaps the most common textbook example of diffraction is that due to waves passing through a long, narrow slit (Hecht 2002, Born and Wolf 1999). The detection of the diffraction is assumed to be at a very large distance from the slit relative to both the slit width and the wavelength, in the so-called *far field.* Diffraction in the far field is termed *Fraunhofer*, hence the name *single slit, Fraunhofer diffraction.* The slit is long so that that dimension can be ignored, and we then have an essentially two-dimensional problem, the propagation and slit width directions (one dimension apiece). Fortunately, this simple example is of prime importance for our development of an understanding of scattering, so we proceed with a derivation of the diffraction properties here.

Figure 1.9 shows a diagram of the profile of a single slit of width w with plane waves incident from the left. We imagine a wave exactly in the slit opening and break it up into Huygens wavelet point sources. These sources re-radiate wavelets that shine uniformly over a range of angles θ relative to the forward direction. The task now is to add these waves as they come together a great distance $R \gg w$ from the source in the far field.

To add the wavelets in the far field the complex notation for waves, equation (1.18), can be used to keep track of the phases. Each wavelet has the same amplitude, and we will use scalar waves to keep things simple without affecting the result. Furthermore, we make another reasonable assumption that the wavelet sources are temporally coherent, i.e. coordinated in time, oscillating in phase across the width of the slit. This is usually true because in most experiments to observe Fraunhofer diffraction, the slit is quite narrow, on the order of the wavelength. This allows us to drop the time dependence in the phase of the propagating wave because every wave is in sync.

Equation (1.18) is now used to describe an arbitrary wavelet emitted a distance x from the center of the slit. Without loss of generality, the phase of the incident wave can be set to equal zero at the center of the slit. Then, because the incident wave vector is perpendicular to the slit, the phase is zero across the entire slit. The incident wave within the slit now continues to propagate by breaking up into Huygens wavelet point sources.

Consider propagation at an angle θ relative to the z-axis toward the detector far away. The detector is at $\boldsymbol{R}$ relative to the origin at the center of the slit, hence any given wavelet source is at position vector $\boldsymbol{r} = \boldsymbol{R} + \Delta\boldsymbol{r}$, see figure 1.9. The part of the wavelet propagating at θ is propagating with wave vector $\boldsymbol{k}$ parallel to the position vector so $\boldsymbol{k}\cdot\boldsymbol{r} = kr$. With all this, the wave at the detector due to a single wavelet in the slit is

$$E = E_0 e^{ikr} \tag{1.42}$$

For now we will not concern ourselves with the value of the wave amplitude, E_0. From the geometry of the system, $\Delta r = x \sin\theta$. Then equation (1.42) yields

$$E = E_0 e^{ikR} e^{ixk\sin\theta} \tag{1.43}$$

The total field at the detector is the sum of the fields due to each wavelet source. The sources are specified uniquely by their continuous positions x; thus the addition is achieved by integrating equation (1.43) over the width of the slit, i.e. over x as

$$E = E_0 e^{ikR} \int_{-w/2}^{w/2} e^{ixk\sin\theta}\, dx \tag{1.44}$$

Now recall that we have done this integral before, see equation (1.31) above. There, we were calculating the Fourier transform of the square pulse. Except for relatively unimportant multiplicative factors of $h/\sqrt{2\pi}$ and $E_0 e^{ikR}$ and that k is replaced by $k \sin\theta$, the formulae are the same. Thus, the process is the same, and we propose that diffraction and Fourier transformation are closely related. The proposal refines with the realization that the transmission profile of the slit, i.e. a plot of the incident wave amplitude transmitted as a function of x, call it $T(x)$, is a square pulse, viz.

$$\begin{aligned} T(x) &= E_0, \ -w/2 \leqslant x \leqslant w/2 \\ &= 0, \ x < -w/2 \text{ or } x > w/2 \end{aligned} \tag{1.45}$$

This allows us to claim that the diffracted wave amplitude for the slit is the Fourier transform of the amplitude transmission profile for the slit! Thus, we write

$$E(q') \sim \int_{-\infty}^{\infty} T(x)e^{iq'x}dx \tag{1.46}$$

where, for convenience, we define $q' = k \sin\theta$ as the *Fourier wave vector.*

But why stop here? Make a bold generalization and leave out 'for the slit'. Indeed, it can be shown most generally that:

> **The diffracted wave amplitude in the far field for an arbitrarily shaped aperture is, to within a constant factor, the Fourier transform of the aperture's amplitude transmission profile.**

Note that this is a statement about *amplitudes*, the intensity is proportional to the amplitude squared.

Now rejoin the problem at hand, single slit diffraction. Equation (1.44) becomes via the logic of equations (1.31)–(1.34)

$$E(q') = E_0 e^{ikR} \sin u/u = E_0 e^{ikR}\,\mathrm{sinc}(u) \tag{1.47}$$

with the dimensionless variable

$$u = q'w/2 \tag{1.48}$$

Detectors, be they photomultiplier tubes, photodiodes or our own eyes in the case of light, don't detect the wave amplitude directly. Instead, they detect the energy flux, which, as we have seen above, equation (1.14), is proportional to the square of the field amplitude. Then equation (1.47) leads to

$$I(u) - \mathrm{I}(0)(\sin u/u)^2 = \mathrm{I}(0)(\mathrm{sinc}(u))^2 \tag{1.49}$$

where $\mathrm{I}(0) = E_0^* E_0$ is the intensity at $u = 0$, hence $\theta = 0$, the exact forward direction (see below). The observed intensity is called the *diffraction pattern*[5].

These results are illustrated in figure 1.10. The zeros of the sinc function are at $u = m\pi$ where $m = 1,2,3, \ldots$. Thus, using equations (1.46) and (1.48), the minima of the diffraction pattern are at

$$\sin\theta = m\lambda/w \tag{1.50}$$

or for small θ

$$\theta \simeq m\lambda/w \tag{1.51}$$

This means that the so-called central or primary maximum is $2\lambda/w$ radians wide and the minima are separated by λ/w. The maxima of the secondary maxima are not exactly half way between the minima, as careful application of equation (1.49) will show.

---

[5] Interchanging the independent variable between q' and u depends on our point of view.

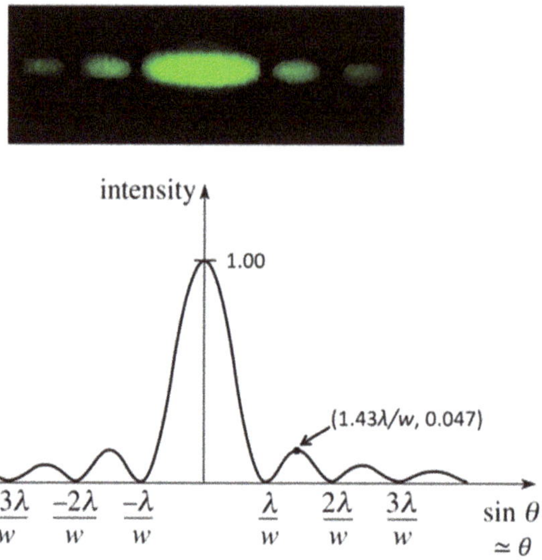

Figure 1.10. (Top) A photograph of an optical diffraction pattern. (Bottom) The single slit diffraction pattern for light with a wavelength of λ passing through a slit of width w.

In the derivation above, I have been rather flippant with the amplitude E_0. A precise treatment has detail that I wish to avoid and is secondary to my purpose here. However, the resulting forward intensity, $I(0)$, has important functionalities that need to be understood from a physical point of view, so we address those here. We expect $I(0)$ to be proportional to the light intensity incident I_{inc} on the slit; hence $I(0) \sim I_{\text{inc}}$. The diffracted intensity observed in the far field should decrease with increasing distance from the slit, R. For a semi-infinite slit of finite width w but infinitely long (a one-dimensional source), the distance functionality is $1/R$, (not $1/R^2$ as for a point source, which is zero dimensional), hence $I(0) \sim I_{\text{inc}}/R$. Finally, at $\theta = 0$, all the waves from the slit add constructively, in phase, at the point of observation. Thus, the total amplitude is proportional to the width of the slit and, since intensity is proportional to amplitude squared, the intensity is proportional to the width, squared. Hence, we obtain

$$I(0) \sim I_0 w^2 / R \tag{1.52}$$

Note also that the slit, being one-dimensional, only has width. One might say that the width is the volume of the slit. This square of the volume size, forward diffraction proportionality will continue to hold for other diffraction situations. Finally, remember that we are working in the Fraunhofer, far field limit where the slit width is limited by $w \ll R$.

Regarding diffraction, one could argue that equation (1.51) is the only reasonable possibility. Consider that we are looking for a characteristic angle of diffraction (bending). Angles are dimensionless. The slit is completely specified by its width, which is a length and nothing else. To get a dimensionless angle from this length another, unrelated, length scale is needed, something associated with the wave. The only such scale is the wave's wavelength. A ratio of these two length scales can yield a dimensionless angle: which ratio? We expect diffraction to be minimal when the wavelength is small. Thus, the characteristic angle of diffraction should be proportional to λ/w, the same result as the mathematical theory. This general argument is good for all waves, be they light, sound, electrons or waves on the ocean. Moreover, the comparison of length scales will become an important theme of our development of scattering theory.

1.5.2 Convolutions and diffraction

It was shown above that the diffracted wave amplitude $E(q')$ is directly proportional to the Fourier transform of the aperture transmission profile, $T(x)$. We also know that the diffracted intensity is the square of the diffracted wave amplitude. Now recall the convolution theorem. In equation (1.41) set $k = q'$ and then $E(q') = F(q') = G(q')$ to write

$$I(q') = |\, E(q')|^2 = 1/\sqrt{2\pi} \int_{-\infty}^{\infty} [T(u) \otimes T(u)] e^{iq'x} dx \tag{1.53}$$

Recall that the convolution in the brackets of the integrand of equation (1.53) is a function of x as described in equation (1.40). Equation (1.53) shows that the diffracted intensity profile is the Fourier transform of the aperture transmission profile convoluted with itself. Thus, there are two ways to calculate intensity diffraction patterns in the far field: the Fourier transform of the aperture transmission profile can be calculated and then squared, or the aperture profile can be convoluted with itself and then Fourier transformed to get the intensity diffraction pattern directly. Each method has its attributes, and it is good to have both at one's disposal.

To build a physical intuition for these two paths, it is worth the effort to illustrate them for the single slit diffraction. The first method, squaring the Fourier transform of the aperture profile, has been accomplished above. Here, the second method is used.

Often the convolution is best done with the aid of a graph. Figure 1.11 shows the square pulse amplitude transmission profile, equation (1.45), for a single slit aperture

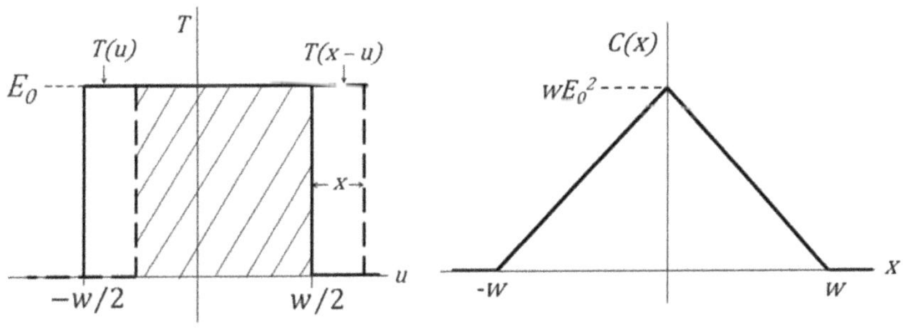

Figure 1.11. (Left) Diagram of the square pulse convoluting with itself for positive x. The cross hatched area is the convolution. (Right) The triangle function which is the convolution of the square pulse.

plotted as both $T(u)$ and $T(x-u)$, where u is the integration variable in the convolution integral, equation (1.40) above. Obviously, the integrand is non-zero only where the two functions overlap. Moreover, we know that an integral is an area under a curve, which for this product integrand is simply the rectangular overlap region with area $E_0^2(w-x)$ for $x>0$. For $x<0$ the area is $E_0^2(w+x)$. Thus, the convolution of the square pulse of width w and height E_0 with itself is the triangle function

$$\begin{aligned} T(u) \otimes T(u) &= 1/\sqrt{2\pi}E_0^2(w-x), \quad w > x > 0 \\ &= 1/\sqrt{2\pi}E_0^2(w+x), \quad 0 > x > -w \end{aligned} \tag{1.54}$$

The Fourier transform of the triangle function is

$$F(k) = E_0^2/2\pi \int e^{ikx}(w \mp x)dx \tag{1.55}$$

$$= E_0^2/2\pi\left[-i\frac{w}{k} \mp \left(-i\frac{x}{k} + \frac{1}{k^2}\right)\right]e^{ikx} \tag{1.56}$$

Application of the limits 0 to w for the minus sign and $-w$ to 0 for the plus sign, the Euler equation (equation (1.15)), some algebra and the trig identity $2\sin^2\theta/2 = 1 - \cos\theta$ yields

$$F(k) = (1/2\pi)E_0^2(w\sin(u)/u)^2 = (1/2\pi)E_0^2w^2(\mathrm{sinc}(u))^2 \tag{1.57}$$

where

$$u = wk/2 \tag{1.58}$$

The results in equations (1.57) and (1.58) are functionally equivalent to the relations (1.48), (1.49) and (1.52) for single slit diffraction. Both sets have the square of the *sinc* function, equivalent dimensionless variables u and overall magnitudes proportional to the incident intensity and the width of the slit squared.

Figure 1.12 gives a summary of these machinations. The process starts in the upper left with the square pulse that could be the transmission profile of a slit. It is part of a Fourier transform pair with the sinc function to its right. The square of the sinc function would be the intensity profile for diffraction through the slit, or scattering from a same shape obstacle; it is the diffraction pattern. The other route to this diffraction pattern is to convolute the square pulse with itself to get the triangle function immediately below. This convolution is part of another Fourier transform pair with the diffraction pattern; the square of the sinc function. In chapter 5, we will find that this convolution, indeed any convolution, is the autocorrelation function of the item with itself.

The point of all this is that diffraction can be calculated in two ways:

1. Square the Fourier transformation of the aperture profile, or
2. Fourier transformation of the convolution of the aperture profile convoluted with itself.

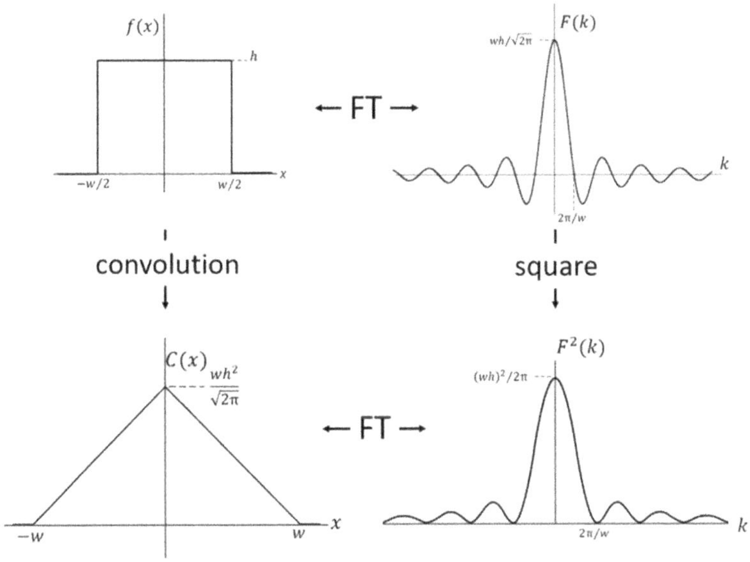

Figure 1.12. Clockwise from upper right: square pulse; Fourier transform (FT) of the square pulse to yield the sinc function; the sinc function squared; the Fourier transform of the sinc function squared, also the convolution of the square pulse with itself, the triangle function. The square pulse and the sinc function are Fourier transform (FT) pairs as are the triangle function and the sinc function squared.

1.5.3 Phasor description of single slit diffraction

The complex notation was introduced for two reasons: the integrals that sum up waves, such as the Fourier integrals, are easier to perform with exponentials than with the trig functions, and complex numbers can be represented by phasors that add together like vectors and thus give a graphical description and understanding to wave addition. Here the graphical picture is developed.

Once again consider diffraction from a single slit of width w. Equation (1.44) shows that the field at the detector is the integral sum of the phasors exp $(ixk\sin\theta) = \exp(ixq')$ from across the slit aperture. Note that the relative phase φ of the wavelets from across the aperture varies with the position x via $\varphi = xq'$. To get a picture for what this integral sum is doing, let the angle of diffraction be at the first minimum, which by (1.51) is at $\theta \simeq \lambda/w$. The phase of the phasor representing the wavelet emanating from the center of the slit, where $x = 0$, is zero, but the phase of the wavelets from the edges of the slit where $x = \pm w/2$ are $\varphi = xq' = xk \sin\theta \simeq xk\lambda/w = (\pm w/2)(2\pi/\lambda) = \pm\,\pi$, respectively. For wavelets emanating from other positions x, the phase changes continuously with x. This situation is depicted in figure 1.13 where both the slit transmission profile and, below it, the phasors of a number of wavelets from across the slit are shown. Examination of these phasors shows that they point uniformly in all directions, and hence sum to zero. This leads to a minimum in the diffraction pattern. It is important to remember that the phasors are all in phase at the slit. The phasors drawn here represent the phase at the point of detection in the far field.

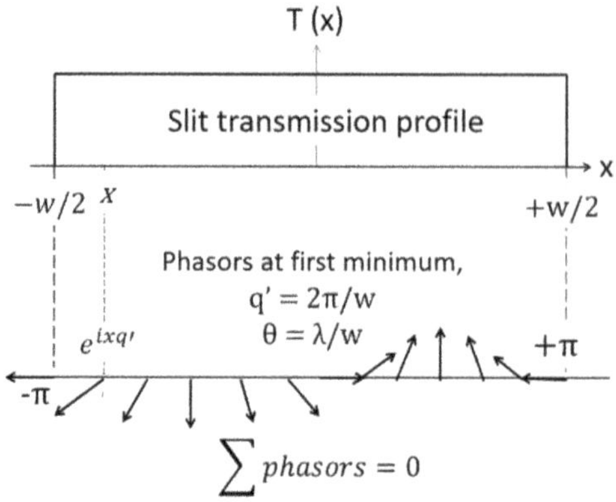

Figure 1.13. Diagram of a single slit transmission profile, $T(x)$ versus x, and below it the phasors, $\exp(ixq')$, of the wavelets at the observation point in the far field that have emanated from the various positions x across the slit. The sum of these phasors is zero; hence the diffraction pattern is at the first minimum corresponding to $q' = 2\pi/w$ and $\theta = \lambda/w$.

Now return to the scattering wave vector magnitude q' and recall that inverse q' is a length scale. For example, at the diffraction first minimum, $2\pi/q' = w$, the width of the slit. It is easy to generalize this to see that minima occur at

$$m(2\pi/q') = w \tag{1.59}$$

for $m = 1,2,3 \ldots$.

We can make use of the length scale $2\pi/q'$ and define a '$2\pi/q'$-region' with diameter equal to $2\pi/q'$. Within any $2\pi/q'$-region, the sum of the phasors is zero; figure 1.13 is an example of a single $2\pi/q'$-region. With this, consider the situation drawn in figure 1.14 where again we see a single slit aperture transmission profile. For a large diffraction angle θ, q' is large so that the $2\pi/q'$-regions are small. Figure 1.14 shows a number of $2\pi/q'$-regions centered on the center of the slit. The contribution to the diffracted intensity from each $2\pi/q'$-region is zero. The situation drawn in figure 1.14, however, is one not at a minimum. Thus, the $2\pi/q'$-regions don't perfectly fill the slit and there is 'leftover' slit at each end. It is the phasors in these leftover regions at the edges (surfaces) of the slit that contribute to a non-zero intensity in the complete diffraction. By centering these $2\pi/q'$-regions on the center of the profile, we see that *the only non-zero contribution to the diffracted field is from the edges; the interior is effectively dark!* It is the edges of the slit that yield the secondary maxima between the successive minima at $m\lambda/w$.

This picture suggests that the decline in diffracted intensity with increasing diffraction angle, hence increasing q', is due to the uniform interior of the slit no longer contributing to the intensity. What intensity does make it to the detector is due to the edges, i.e. the surface of the slit. This picture is attractive because it is sympathetic with concepts we will develop later, which show that scattering does not occur unless there are fluctuations in density. Note that the density of wavelets is uniform in the interior of the slit, and only at the edges does their density change; hence a fluctuation in density. No doubt by now the reader has noticed that our

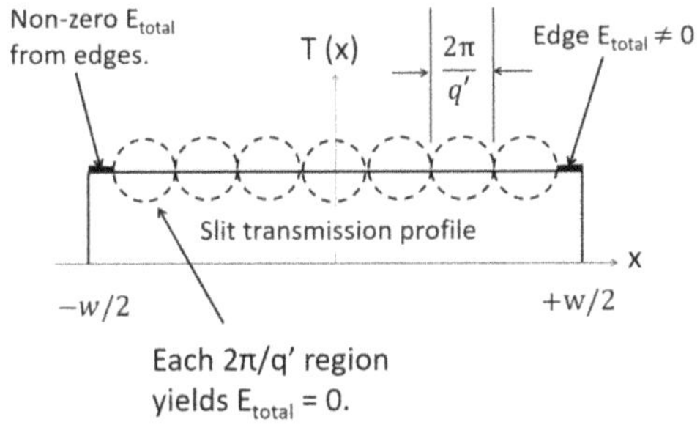

Figure 1.14. Diagram of a single slit transmission profile with $2\pi/q'$-regions drawn on it. Each $2\pi/q'$-region contributes zero to the total diffracted intensity in the far field. In this case, however, the $2\pi/q'$-regions don't perfectly fill the slit leaving leftover regions at the edges. In these edge regions the phasors don't completely cancel, hence there is some diffracted intensity.

assignment of $2\pi/q'$-regions is not unique; we could have started at the edge and worked into the middle to get leftover regions. Why do we choose to leave the edges leftover and not the center? Because it works.

1.5.4 Babinet's principle

With the single slit, we have studied diffraction through a finite sized, transparent hole in an otherwise infinite, opaque barrier. What about diffraction around[6] a finite sized, opaque barrier surrounded by an infinite, transparent region? My intended reciprocity in the previous sentence sets the stage for Babinet's principle (Hecht 2002), which is:

> **The diffraction pattern in the far field limit for a wave passing through a finite sized, transparent hole in an otherwise infinite, opaque barrier is identical to the diffraction pattern for the same waves passing a finite sized, opaque barrier of the same shape as the hole surrounded by an infinite, transparent region.**

Thus, for example, if you illuminate a long, narrow fiber of width w with a parallel beam of fairly monochromatic light and view the light after it has gone a few meters, you will see exactly the same pattern as you would have had you illuminated a long slit of the same width. The situation for the fiber is best seen using a narrow incident light beam so that the diffraction in the far field can spread well beyond the width of the incident beam.

The proof of Babinet's principle is conceptually pleasing and will be useful later in our study of the extinction paradox. Consider two complimentary screens, i.e. screens made so that the openings in one correspond to obstacles in the other and vice versa. Let E_1 and E_2 be the scalar fields at an observation point when either screen is in place, and E_0 be the light field observed when no screens are in place. Then one must have

[6] Material waves, such as sound and water waves, diffract 'around' a material object, while light waves diffract but not 'around'. See section 8.4.

$$E_1 + E_2 = E_0 \tag{1.60}$$

This is a statement of Babinet's principle (Born & Wolf, 1999).

To understand the significance of equation (1.60) return to the slit and the fiber both with width w, and recognize that they are complimentary screens. If either is illuminated by the same source, equation (1.60) holds. Let the source be a beam with wavelength λ and finite width $w_{\text{beam}} \gg w$. Let the observation point be far away relative to λ, w_{beam} and w. If the observation point is not straight ahead but at an angle $\theta \gg \lambda/w_{\text{beam}}$, then for the unobstructed beam $E_0 \simeq 0$ at that point. Then it follows from Babinet's principle, equation (1.60), that

$$E_1 = -E_2 \tag{1.61}$$

The negative sign implies that the two fields will be 180° out of phase with each other. However, their magnitudes are equal, hence their intensities, which are the squares of the magnitudes, are equal, $I_1 = I_2$, and hence the diffraction pattern will be the same for both.

Notice that with Babinet's principle, we are, for the first time in this discourse, shining a wave, a beam of light, onto an object. That's called 'light scattering'!

1.6 Foundations of scattering

At the end of this chapter, we can reflect back and realize we have come a long way from the simple statement that waves are described by sinusoidal functions seen in figure 1.1. The purpose of this chapter has been to establish the fundamental connection between waves and diffraction. Indeed, things that diffract are waves, and when waves propagate, they diffract. However, all the waves studied in this chapter do not interact with any materials other than being blocked by them to produce shadows into which the waves can diffract. The electromagnetic waves of light interact with matter in remarkable and sometimes subtle ways that can be described on both macroscopic and atomic levels. This interaction is the source of light scattering and absorption with manifold consequences. Nevertheless, the wave character of light will always shine through to govern the propagation of the waves both within the scatterer and out to the detector. The subsequent chapters will show that the foundation of scattering is diffraction.

References

Born M and Wolf E 1999 *Principles of Optics: Electromagnetic Theory of Propagation, Interference and Diffraction of Light* (Cambridge: Cambridge University Press)

Griffiths D J 1999 *Introduction to Electrodynamics* (Upper Saddler River, NJ: Prentice-Hall)

Hecht E 2002 *Optics* (Reading, MA: Addison-Wesley)

Chapter 2

Introduction to scattering and absorption

Cross sections describe the probability for scattering or absorption of radiation such as light. As the name implies, the dimension of cross section is area, i.e. length squared. The name is very appropriate too, for a scattering cross section is analogous to a *geometric cross section*, which is the projected area of a three-dimensional object onto a plane perpendicular to the viewing direction. One could say that the cross section is the effective area exposed to or seen from a given direction. If we 'shine' raindrops, baseballs or bullets onto an object, the scattering cross section is the geometric cross section, but we will see that waves are a different matter and the cross section is much more interesting[1].

2.1 The total cross section

Imagine a beam of light spread uniformly over an area A_{beam}. Now place a single, non-absorbing particle in the beam, and ask: what fraction of the incident radiation in the beam is scattered by that particle? A reasonable answer would be that there is an 'effective area' for the scattering ability of the particle, and the ratio of that area divided by the area of the beam is the fraction scattered. This effective area is the *total scattering cross section* of the particle, C_{sca}[2].

If the incident beam is described by a power P_{inc} and a cross sectional area A_{beam}, hence an incident intensity of $I_{\text{inc}} = P_{\text{inc}}/A_{\text{beam}}$, then the total scattering cross section can yield the total scattered power via the ratio of this effective scattering area to the area of the beam. Thus,

$$P_{\text{sca}} = P_{\text{inc}}\frac{C_{\text{sca}}}{A_{\text{beam}}} = I_{\text{inc}}C_{\text{sca}} \tag{2.1}$$

[1] I have assumed 'classical' particles. Precisely speaking, raindrops, baseballs and bullets have a wave nature too and should be treated accordingly when necessary.

[2] It is not uncommon to find σ used to denote cross section, especially in the physics literature.

doi:10.1088/978-0-7503-5157-7ch2

Equation (2.1) describes the total power scattered, not the scattered intensity, because the total scattering cross section says nothing about how the scattered light is distributed through space.

The same 'effective area' argument can be made for absorption, so the absorption cross section, C_{abs}, has the dimensions of the area too. Unlike scattering which redirects the radiation, absorption changes the radiant energy into another form of energy, most commonly heat.

With a total scattering cross section, no attention is paid to where the light goes once it is scattered. This means that the intensity at a given point cannot be specified. However, we often want to know this, and then we can use the spherical polar angles (r, θ, ϕ) to specify the direction. Given the direction, we also want to know how much light scatters to a detector in that direction. Any detector will have a finite detecting area, and this area will subtend a solid angle relative to the scattering source. Thus, it is worthwhile to review the notions of angles and solid angles.

2.2 Angles and solid angles

In two dimensions, an arc of length S on a circle of radius r subtends an angle of θ, as drawn in figure 2.1, given by

$$\theta = S/r \tag{2.2}$$

The unit of the angle so defined is the *radian*. A complete circle has a circumference of $2\pi r$ so the total radians in a circle is 2π. Hence, 2π radians = 360° and one radian = 57.3° ≃ 60°.

In an analogous manner, in three dimensions an area of size A on a sphere of radius r subtends a solid angle of Ω given by

$$\Omega = A/r^2 \tag{2.3}$$

The unit of the solid angle so defined is the *steradian*. A complete sphere has a surface area of $4\pi r^2$ so the total steradians in a sphere is 4π. In fact, any closed shape in three-dimensional space subtends a total of 4π steradians for any interior position.

When the angles are small, the arcs and pieces of the sphere can be taken as flat. Figure 2.2 shows an area of dimensions S_1 by S_2. This area is r away from a point of observation O. The radian angles subtended by the length and width are $\theta_1 = S_1/r$ and $\theta_2 = S_2/r$. The area of the patch is $A = S_1S_2$. Thus, the solid angle subtended by the area is

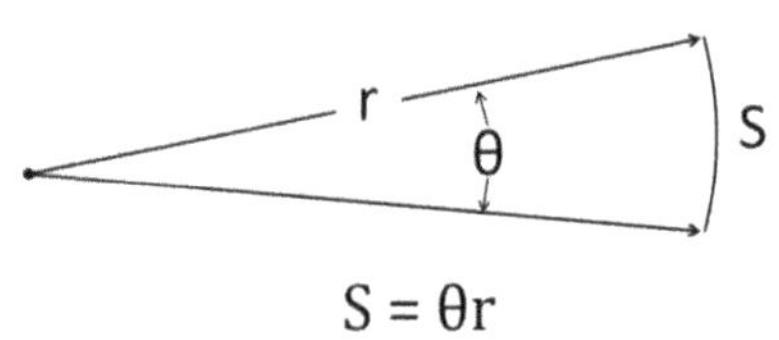

Figure 2.1. Definition of the radian angle θ.

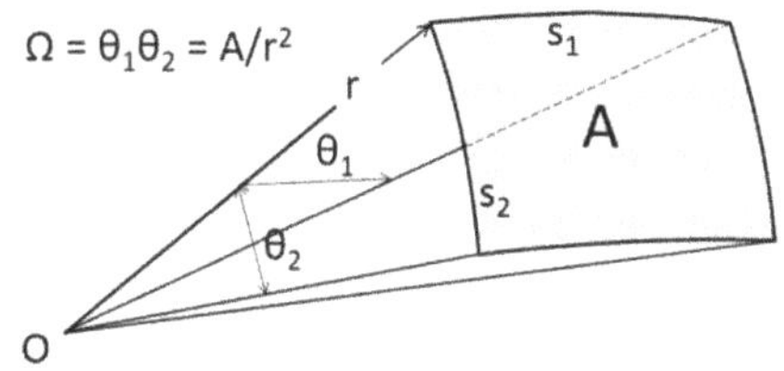

Figure 2.2. The solid angle Ω subtended by the area A at a distance r from a reference point O.

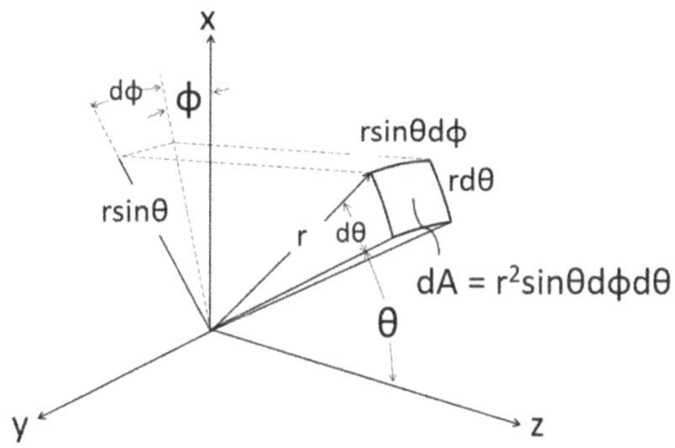

Figure 2.3. The differential of area, dA, in spherical polar coordinates. The azimuthal angle φ is in the xy-plane; the polar angle θ is in the volume directly between the z-axis and the position vector $\boldsymbol{r}$ such that $\cos\theta = \hat{z}\bullet\hat{r}$.

$$\Omega = A/r^2 = \frac{S_1 S_2}{r^2} = \theta_1\theta_2 \tag{2.4}$$

This shows that the unit steradian can be thought of as radians squared, however, both are dimensionless with regard to physical quantities like length and mass.

Now consider the spherical polar coordinates as drawn in figure 2.3. The differential radian angles $d\varphi$ and $d\theta$ conspire to create a differential area dA at a distance r from the origin. The dimensions of this area follow from application of equation (2.2) solving for S. With these dimensions, the area is $dA = r^2 \sin\theta d\varphi d\theta = r^2 d\varphi d(\cos\theta)$, where a negative sign has been dropped from the second equality. With this, the solid angle in the spherical polar coordinates is

$$d\Omega = d(\cos\theta)d\phi \tag{2.5}$$

2.3 The differential scattering cross section

The total cross section gives the scattering into the entire 4π steradians. Now what is needed is a refinement of this to a specific direction within a solid angle range $d\Omega$. This need leads to the *differential scattering cross section*, $dC(\theta, \phi)_{\text{sca}}/d\Omega$, with dimensions length squared per steradian. It follows from our development that

$$C_{\text{sca}} = \int_{4\pi} \frac{dC_{\text{sca}}}{d\Omega} d\Omega \tag{2.6}$$

Following the argument that led to equation (2.1), the power scattered to a detector in some arbitrary direction is

$$P_{sca} = P_{inc}\frac{dC_{sca}/d\Omega}{A_{beam}}d\Omega_{det} = I_{inc}\frac{dC_{sca}}{d\Omega}d\Omega_{det} \tag{2.7}$$

where $d\Omega_{det} = A_{det}/r^2$ is the solid angle subtended by the detector of area A_{det} a distance r away from the scatterer. Since the intensity at the detector is P_{sca}/A_{det}, one can write

$$I_{sca} = \frac{I_{inc}}{r^2}\frac{dC_{sca}}{d\Omega} \tag{2.8}$$

This has an inverse distance squared functionality, as the detector recedes from the source. Also, remember that this applies to a single particle.

2.3.1 The phase function

A non-dimensional representation of the differential scattering function can be found with the *phase function* defined as

$$p(\theta, \varphi) = \frac{4\pi}{C_{sca}}\frac{dC(\theta, \varphi)_{sca}}{d\Omega} \tag{2.9}$$

Notice that the phase function is normalized with both the total cross section and the total solid angle. With these, the integral of $p(\theta,\varphi)$ over all 4π steradians of space is unity.

2.3.2 The asymmetry parameter

A useful parameter to quantify the possible asymmetry of light scattering to different scattering angles is the aptly named the *asymmetry parameter*, defined as the scattered intensity weighted average of the cosine of the scattering angle. The asymmetry parameter is

$$g = \langle\cos\theta\rangle = \int_{4\pi} p(\theta, \varphi)\cos\theta d\Omega \tag{2.10}$$

This parameter is bounded as $-1 \leqslant g \leqslant 1$. If $g = 0$, the scattering is symmetric in the scattering plane containing θ. If $g > 0$, forward scattering dominates, which is typical for particles larger than the wavelength of light.

2.4 Extinction, albedo and the efficiencies

Radiation can be either scattered, absorbed or a combination of both, but nothing else. Both scattering and absorption remove radiant energy from the incident beam, and their total effect depletes or extinguishes the beam. To quantify this total effect the *extinction cross section* is defined as

$$C_{ext} = C_{sca} + C_{abs} \tag{2.11}$$

The *albedo*, ω, is the ratio of scattering to extinction

$$\omega = C_{\text{sca}}/C_{\text{ext}} \tag{2.12}$$

It is bounded as $0 \leqslant \omega \leqslant 1$. It is a measure of brightness (or darkness). When ω is near unity, the particle or ensemble of particles is very bright, white; when ω is close to zero, the situation is very dark, black. One common example of albedo is the specified brightness given on a ream of paper. A brightness of 95 means an albedo of 0.95, very white.

An *efficiency*, or *efficiency factor*, Q, is defined as a cross section, C, normalized by (i.e. divided by) the geometric projected area of the particle, G, onto a plane perpendicular to the incident direction. Thus, we have the dimensionless ratios

$$Q = C/G \tag{2.13}$$

There are scattering, absorption and extinction efficiencies. The efficiency is aptly named because it compares the scattering or absorption ability to the geometric size exposed to the radiation. This comparison is a measure of how 'efficient' the scattering or absorption is. For example, were it not for the subtle effects due to the wave and electromagnetic natures of light, one might expect that any of these cross sections would be directly proportional to the projected geometric cross section and, in fact, equal to it. However, this is often very far from the case. A scattering $Q < 1$ implies that for some reason the particle is less active at scattering than one would have expected from its actual size. The converse is true if $Q > 1$.

2.5 Attenuation by an ensemble of particles

Attenuation of an incident beam of light by an ensemble of particles is an important physical process. We expect this to depend on particle properties including the scattering and absorption cross sections and the properties of the ensemble, such as the particle number density.

Physical intuition or experience can lead us to realize that the total light transmitted through a system of particles approaches zero in an asymptotic manner. Moreover, that same intuition or experience leads us to realize that the fractional loss of light is the same for any given length as long as the system is homogeneous. The only function that has these two properties is an exponential decaying with distance x. Thus, we write

$$I(x) = I(0)e^{-\tau x} \tag{2.14}$$

where $I(x)$ is the beam intensity after traveling a distance x, $I(0)$ is the incident intensity (at $x = 0$), and τ is a constant that must have dimensions of inverse length. Equation (2.14) is the *Beer–Lambert law*, or sometimes just Beer's law, etc (Monk 1937).

The inverse of the constant τ has dimensions of length and as such represents the '$1/e$ distance' for the attenuation, the attenuation length. This is the distance over which the light attenuates to $1/e \simeq 37\%$ of its initial value. τ itself is called the *turbidity*. Again taking a physical perspective, the attenuation can only depend on the scattering and absorption of the particles, which sum to the extinction cross

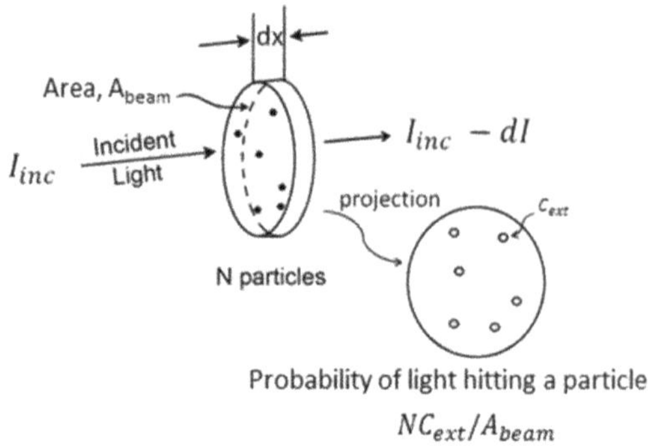

Figure 2.4. Diagram of light incident on a small section of a system with N particles each with an extinction cross section C_{ext}.

section C_{ext}, and the number density of the particulate system, n. The dimensions of C_{ext} are length squared; the dimensions of n are inverse length cubed. To make τ with inverse length dimensions from these, we conclude

$$\tau = nC_{\text{ext}} = n(C_{\text{sca}} + C_{\text{abs}}) \tag{2.15}$$

An analytical derivation of the Beer–Lambert law can be obtained by starting with figure 2.4. There, a beam of light with cross sectional area A_{beam} is losing a small amount of intensity dI after traveling through a small section of the extincting system with length dx. N particles with an extinction cross section of C_{ext} will stop a fraction of the light, $NC_{\text{ext}}/A_{\text{beam}}$. This fraction is equal to the fractional intensity loss (hence, the negative sign)

$$\frac{dI}{I} = -\frac{NC_{\text{ext}}}{A_{\text{beam}}} \tag{2.16}$$

Multiplying the numerator and denominator of the right hand side of equation (2.16) by dx (hence a multiplication by $dx/dx = 1$) and recognizing that the volume of the small section is $V = A_{\text{beam}}\, dx$ yields

$$\frac{dI}{I} = -nC_{\text{ext}}dx \tag{2.17}$$

where $n = N/V$, the number density of the particles. The solution of equation (2.17) is equation (2.14) with $\tau = nC_{\text{ext}}$ as in equation (2.15).

Often with commercial instruments such as UV–Vis spectrometers, one encounters the term *absorbance*, A, defined as

$$A = \log_{10}[I(0)/I(x)] \tag{2.18}$$

Furthermore, there is often the convention that $x = 1.0$ cm, which is a standard optical cuvette width.

Finally, if one sets $C_{\text{abs}} = 0$ in equation (2.15), then $\tau = nC_{\text{sca}}$. This is a Rayleigh ratio, which will be described in section 6.8.

2.6 Multiple scattering

The Beer–Lambert law can be derived from a statistical argument if the light is viewed as a beam of photon particles. This could be somewhat construed as contrary to the wave-particle conceptualization of light, which uses the wave picture to describe light when it is propagating and the particle (photon) concept when the light is either being emitted or detected (absorbed). However, one could claim that at the moment of scattering from, or absorption by, a scatterer, the light is being detected and possibly re-emitted, so light is a photon particle while it is scattering or getting absorbed. Regardless of these (important) metaphysical considerations, we find it very useful in the following to view light as a photon acting like a particle when it is encounters a material particle (it can be a wave while propagating between encounters with material particles). The description below follows that given in Mokhtari *et al* (2005).

Consider a beam of light entering a volume in which particles that can scatter and absorb light reside. Let these particles be randomly distributed. The incident beam intensity is $I(0)$, which is proportional to the rate of photons in the beam. After the light has propagated a distance x through the volume, what is the rate of photons, proportional to $I(x)$, in the beam? To answer this, assume that the encounters of the photons with scattering/absorbing particles in the volume are a Gaussian random process (Sorensen *et al* 1978). For a Gaussian random process, the probability $P(s)$ that a given photon has s encounters with the ensemble of particles during its passage along the entire length x is given by the Poisson distribution

$$P(s) = e^{-\langle s\rangle}\frac{\langle s\rangle^s}{s!} \tag{2.19}$$

In equation (2.19), $\langle s\rangle$ is the average number of photon–particle encounters for the photons in the beam as they propagate a distance x.

When light passes through the volume with particles, the medium, only photons that have no encounters, i.e. $s = 0$, pass through the distance x out of the far end of the volume. Thus, by equation (2.19), the probability of no encounters is

$$P(0) = e^{-\langle s\rangle} \tag{2.20}$$

The light intensity passing through the length x is equal to this probability times the incident intensity, $I(0)$, to yield

$$I(x) = I(0)e^{-\langle s\rangle} \tag{2.21}$$

Equation (2.21) affords us the opportunity to measure $\langle s\rangle$. The photon mean free path between encounters should depend on the number density of the particles and the combined scattering and absorption cross section of a particle, the extinction cross section, C_{ext}. A dimensional analysis leads to the photon mean free path as

$$l = (nC_{ext})^{-1} = \tau^{-1} \tag{2.22}$$

The second equality follows from equation (2.15) and we find that the turbidity is the inverse of the photon mean free path (even though photons do not propagate!). The average number of photon–particle encounters $\langle s \rangle$ is related to the photon mean free path and the total path length in the medium by

$$\langle s \rangle = x/l \tag{2.23}$$

Equations (2.22) and (2.23) yield

$$\langle s \rangle = nC_{\text{ext}}x = \tau x \tag{2.24}$$

Substitution of this into equation (2.21) yields

$$I(x) = I(0)e^{-nC_{\text{ext}}x} = I(0)e^{-\tau x} \tag{2.25}$$

which is the Beer–Lambert law, equation (2.14). Later, in section 6.8, we shall see that the photon mean free path is also the Rayleigh ratio.

We have established the veracity of this formulation with experiments (Mokhtari *et al* 2005). The experiments used polystyrene particles with a diameter of 9.6 μm suspended in water. The refractive index of the particles relative to the water medium was 1.196 at the laser wavelength of $\lambda = 488$ nm. The containing cell had a 2 mm optical path-length. Different volume fractions f_v of particles hence different number densities were studied. Figure 2.5 shows the experimental scattered intensities $I(q)$ versus q, where $q = (4\pi/\lambda)\sin(\theta/2)$ and θ is the scattering angle. The data are compared to Mie scattering calculations for these spherical particles using the BHMIE code (Bohren and Huffman 1983) integrated over a small size distribution. Figure 2.5 shows that when the polystyrene microsphere colloidal

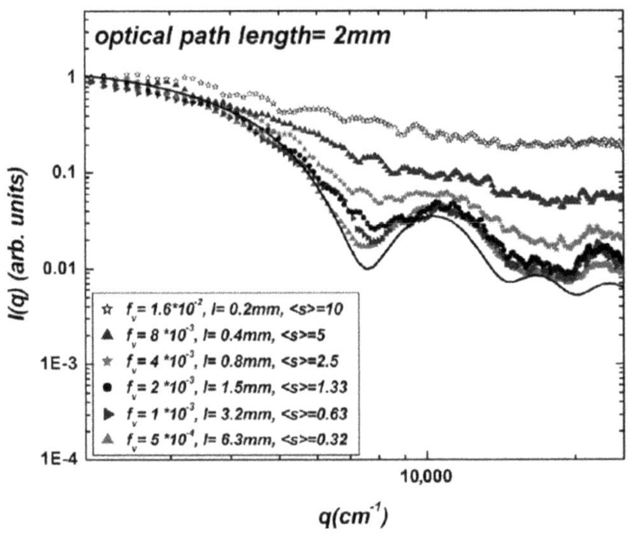

Figure 2.5. Normalized light scattering intensities plotted versus $q = (4\pi/\lambda)\sin\left(\frac{\theta}{2}\right)$ at different volume fractions f_V for 9.6 μm diameter polystyrene spheres in water with a geometric size dispersion of 7.4%. The optical path length of the cell is 2 mm. The photon mean free path l and the average number of scattering events $\langle s \rangle$ are calculated for each concentration. The solid curve is the Mie scattering theory for these particles averaged over the size distribution. Reproduced from Mokhtari *et al* (2005).

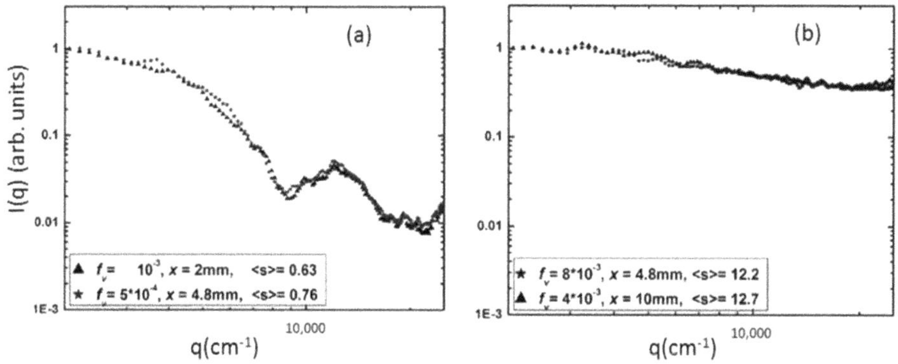

Figure 2.6. Normalized light scattering intensity plotted versus $q = (4\pi/\lambda)\sin\left(\frac{\theta}{2}\right)$ for 9.6 μm diameter polystyrene particles in water. The two panels have different average numbers of photon scattering events, $\langle s \rangle$: (a) 0.7 ± 0.06, (b) 12.4 ± 0.3. Note that for each $\langle s \rangle$ there are two different combinations of the optical path length x of the cells and the volume fraction of the particles f_V. Adapted from Mokhtari *et al* (2005).

suspension is dilute, the scattered intensity has a ripple structure and is well represented by the theoretical calculation. However, as the volume fraction of the colloidal solution increases, the ripple structure washes out and the scattered intensity becomes more isotropic. This is an indication of multiple scattering.

The washing out of the ripples and increasing isotropy of the scattering correlate well with the decrease in the photon mean free path l and the concomitant increase in the average number of scattering events $\langle s \rangle$ for different volume fractions as indicated in figure 2.5. These were calculated from the scattering cross section, which is very close to $2\pi R^2$, and the volume fraction f_V of the particles from which the number density can be determined ($n = 6f_V/\pi D^3$). The implication is that to have essentially only single scattering in an experiment, the photon mean free path should be greater than the optical path length. Via equation (2.20), this is equivalent to the average number of scattering events $\langle s \rangle$ being much less than one. Figures 2.6(a) and (b) demonstrate this conclusion. They also show the universal nature of $\langle s \rangle$ because graphs with similar $\langle s \rangle$ yet ranging volume fractions and path lengths behave the same.

These results can be very useful in the lab. Typically, lab work involves particles with no imaginary part to the refractive index; hence no absorption, only scattering. Then a simple measurement or even an estimate of the transmittance through the sample, $I(x)/I(0)$, yields the average number of scattering events $\langle s \rangle$ in the sample via equation (2.21). From this the relative amount of double scattering, $P(2)/P(1)$, or multiple scattering, $P(s \geqslant 2)/P(1)$, can be determined from the Poisson equation, equation (2.19), or estimated by comparing the photon mean free path to the path length. If these are relatively large, the experimenter should beware that the experiment is not in the single scattering limit as usually assumed and necessary for a straightforward interpretation of the scattering data. Put most simply, if only about one-third of the light is transmitted, $\langle s \rangle \approx 1$ and $l \approx x$, and the experiment is on the edge of multiple scattering.

Another useful way to detect multiple scattering is by looking at the depolarization of the scattered light. While the subject of light polarization is yet to be discussed in this book, see chapter 3, it is well-known that light is a vector wave and hence can have various states of polarization. Single scattering of an incident light beam that is linearly polarized from particles with spherical symmetry, or scattering from fluctuations in continuous media like gases and liquids, especially near a critical point, will remain linearly polarized. However, if the light, the photons, scatter more than once, the polarization will randomize to some degree. In a typical laboratory situation, the incident light is linearly polarized in the vertical direction and the incident and scattered directions are in the horizontal plane. Then, one can define a depolarization ratio as

$$\rho_V = I_{VH}/I_{VV} \tag{2.26}$$

where I_{VH} is the intensity of the scattered light when the incident light is vertically polarized and the scattered light is detected with a linear horizontal polarizer in front of the detector, and I_{VV} is when the incident light is vertically polarized and the scattered light is detected with a linear vertical polarizer in front of the detector. If there is no multiple scattering, $I_{VH} = 0$ so $\rho_V = 0$. However, multiple scattering will lead to a finite depolarization, $\rho_V > 0$. A relation between the average number of photon scattering events, $\langle s \rangle$, and the depolarization ratio has been derived and a graph of this is given in figure 2.7 (Sorensen *et al* 1978). Relative I_{VH} and I_{VV} intensities can be easily measured with a linear polarizer and then ratioed to get ρ_V. Then, figure 2.7 can be used to determine the average number of scattering events and hence the extent of multiple scattering.

These results can also lead the experimenter to a means to eliminate the multiple scattering problem. Equation (2.23) implies that the average number of scattering events can be made smaller by reducing either the number concentration or the scattering cross section of the scatterers, the latter perhaps by index matching to the surrounding fluid. Very often, however, neither of these can be done without

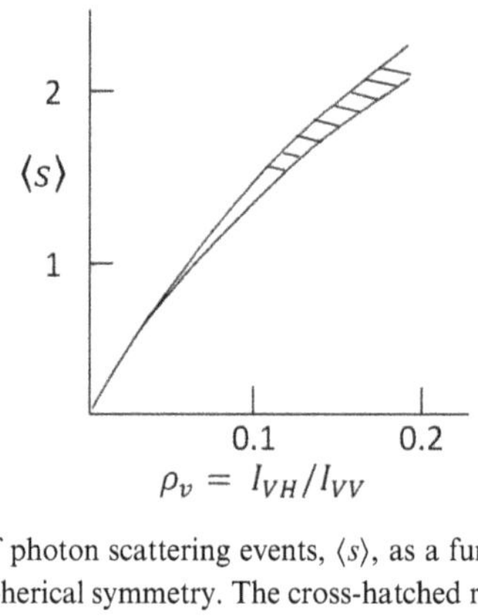

Figure 2.7. The average number of photon scattering events, $\langle s \rangle$, as a function of the depolarization ratio ρ_V for scattering from particles with spherical symmetry. The cross-hatched region indicates uncertainty based on the assumptions that went into the calculation. Figure adapted with permission from Sorensen *et al* (1976). Copyrighted (1978) by the American Physical Society.

affecting the system adversely. But the path length x remains so that by reducing the path length in the scattering container, $\langle s \rangle$ can be reduced and multiple scattering effectively eliminated.

The statistical derivation given here of the Beer–Lambert law treats light as photon particles and does not account for light's wave nature. On the other hand, the Beer–Lambert law can be derived from the Maxwell equations by assuming independent scattering for dispersed particles that are spherically symmetric and optically isotropic (Mishchenko 2002, Mishchenko *et al* 2004). Independent scattering means what it says; the consequence is that that the scattered light from each particle is incoherent and only intensities scattered by particles need to be considered, not the wave amplitudes with their associated phases. This is inherent in the approach that uses the photon description of the light.

References

Bohren C F and Donald R H 1983 *Absorption and Scattering of Light by Small Particles* (New York: Wiley)

Mishchenko M I 2002 Vector radiative transfer equation for arbitrarily shaped and arbitrarily oriented particles: a microphysical derivation from statistical electromagnetics *Appl. Opt.* **41** 7114–34

Mishchenko M I, Hovenier J W and Mackowski D W 2004 Single scattering by a small volume element *J. Opt. Soc. Am. A: Opt. Image Sci. Vis.* **21** 71–87

Mokhtari T, Sorensen C M and Chakrabarti A 2005 Multiple-scattering effects on static light-scattering optical structure factor measurements *Appl. Opt.* **44** 7858–61

Monk G S 1937 *Light: Principles and Experiments* (New York: McGraw-Hill)

Sorensen C M, Mockler R C and O'Sullivan W J 1978 Multiple-scattering from a system of Brownian particles *Phys. Rev.* A **17** 2030–35

Sorensen C M, Mockler R C and O'Sullivan W J 1976 Depolarized correlation function of light double scattered from a system of Brownian particles *Phys. Rev.* A **14** 1520–32

Chapter 3

Polarization

Electromagnetic radiation is a vector wave. This means that the thing that is waving (oscillating) has both magnitude and direction. The direction of the oscillating quantity is called the *polarization*. Both the vector fields $\boldsymbol{E}$ and $\boldsymbol{B}$ oscillate in an electromagnetic wave. As seen in the discussion of the Poynting vector illustrated in figure 1.2, their perpendicular directions are coupled together to yield the propagation direction via the right hand rule. Thus, given the propagation direction, the specification of one field direction determines the other. This coupling, as well as the fact that nearly all interactions of light with matter[1] are due to the electric field, leads to the convention that defines the direction of the electric field as the polarization direction. Then equation (1.18) can be generalized as

$$\boldsymbol{E} = \hat{\boldsymbol{p}} E_0 e^{i(\boldsymbol{k}\boldsymbol{\cdot}\boldsymbol{r} \mp \omega t)} \tag{3.1}$$

where $\hat{p}$ is a unit vector that designates the polarization direction.

Scattering can change the polarization state of the light. Thus, if we are to understand scattering, we must be able to quantitatively describe polarization and understand how these changes occur. With understanding comes opportunity to use the measured changes to infer the nature of the scatterer.

3.1 Polarized light

Figure 3.1 shows the electric fields of light propagating along the positive z-axis. Two polarizations are shown, one each, in the x- and y-directions. The phases of the waves are such that the fields are at their maximum values where they are drawn. A number of polarization situations can be presented with the help of figure 3.1.

a. If $E_x \neq 0$ and $E_y = 0$, the light is linearly polarized in the x-direction.
b. If $E_x = 0$ and $E_y \neq 0$, the light is linearly polarized in the y-direction.

[1] Matter with a great many lightweight, electric monopoles called electrons.

doi:10.1088/978-0-7503-5157-7ch3

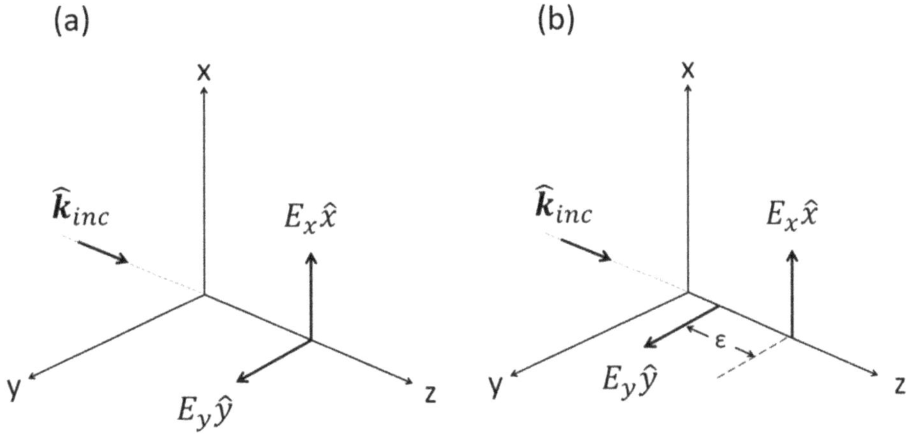

Figure 3.1. The electric fields of two light waves propagating along the positive z-axis. Two polarizations are shown, one each, in the x- and y-directions. The fields are at their maximum values where they are drawn.

c. In figure 3.1(a), if $E_x = E_y \neq 0$ and the two fields are coherent with each other, their addition yields light that is linearly polarized 45° from the x-direction and y-directions. The term 'coherent' means that the fields oscillate together for a significant period of time, and that they are 'in step' with each other.
d. If $E_x = E_y \neq 0$ and the two fields are not coherent with each other, the light is said to be 'unpolarized'.
e. If $E_x \neq 0$, $E_y \neq 0$, $E_x \neq E_y$ and the two fields are not coherent with each other, the light is said to be partially polarized.
f. If $E_x = E_y \neq 0$ and the two fields are coherent with each other but out of phase with each other, see figure 3.1(b), the light is elliptically polarized. If the phase difference is $\varepsilon = 90°$ ($\pi/2$ radians), the light is circularly polarized.

These situations can be specified quantitatively by the Stokes vectors, which will be described below.

Thermal sources of light such as incandescent or fluorescent light bulbs and sources in nature such as the Sun and stars emit 'unpolarized light' (situation (d) above). Such sources are, in fact, emitting a great number of waves that have polarization directions random to each other. Since polarization is a vector, such randomly polarized light can be resolved into equal amounts of two perpendicular components. For example, a light beam from an unpolarized source traveling in a horizontal direction to the north has equal amounts of vertical and east/west horizontal polarizations.

Sources of linear polarizations are less common. Many laser sources are linearly polarized. Synchrotron radiation from circularly moving charges in synchrotrons, auroras or accretion disks of black holes is linearly polarized. Reflection can polarize light completely or to some lesser degree depending on the angle of reflection, hence the market for Polaroid sunglasses. And certainly, scattering can cause unpolarized light to gain some degree of polarization or change the state of the polarization.

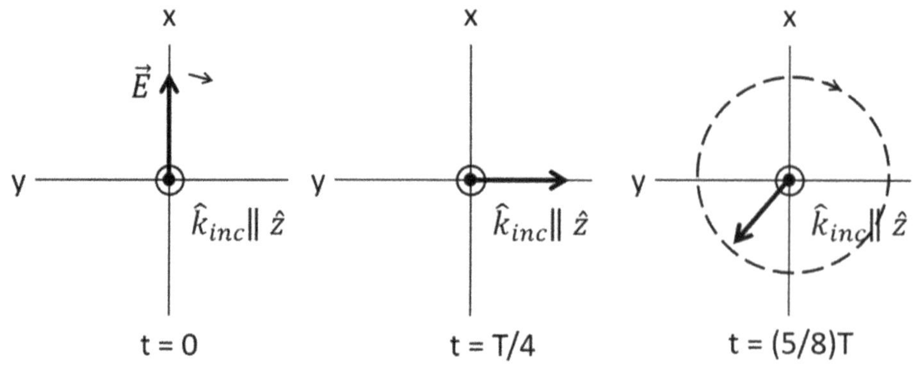

Figure 3.2. Diagram of the electric field at $z = 0$ as a function of time of right circular polarized light propagating along the positive z-axis out of the plane of the page toward the observer. The electric field vector is drawn at three times where $T = 2\pi/\omega$ is the period.

Elliptical polarization, for which circular polarization is a special case, warrants a little more discussion. One way to represent elliptical polarization is with the expression

$$\boldsymbol{E} = \hat{x}E_x\cos(kz - \omega t) + \hat{y}E_y\sin(kz - \omega t) \tag{3.2}$$

This equation represents two linearly polarized light waves propagating in the $+z$-direction. However, these linear polarization waves are coherent as indicated by the fact that their time dependencies do not randomly change with respect to each other. Thus, they will add coherently. Moreover, they are 90° ($\pi/2$) out of phase. This means that at $z = 0$, the total field starts at $t = 0$ directed along the positive x-axis, but one quarter of a period, $T = 2\pi/\omega$, later, it's along the negative y-axis, and so on. If one views the combined wave approaching while looking down the negative z-direction, as shown in figure 3.2, the superposition at one point in space is a vector that rotates with time in the clockwise direction. If $E_x \neq E_y$, the light is right elliptically polarized; if $E_x = E_y$, the light is right circularly polarized. A negative sign in equation (3.2) between the cosine and sine terms would yield left-handed polarization[2]. Other combinations of phase shifts and field magnitudes can yield various elliptical and circular polarizations of either right or left handedness.

3.2 Polarization by an oscillating electric dipole

An electric field applied to an electron causes a force on the electron. This force accelerates the electron. An accelerated electron radiates an electromagnetic wave of its own. Scattering involves the excitation of electrons to oscillation by an incident electromagnetic wave, and the concomitant re-radiation of a new wave due to the oscillation. Thus, it is important to understand the polarization of an electromagnetic wave due to an oscillating charge. This is a classic problem in

[2] This is the 'traditional' convention, but the opposite convention is sometimes found. See Bohren and Huffman (1983).

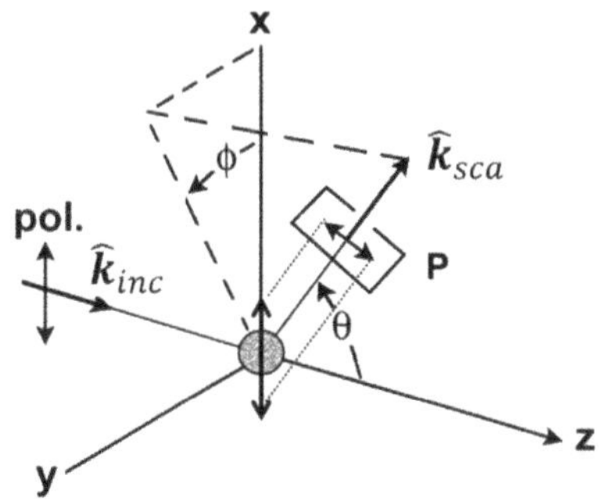

Figure 3.3. Diagram of a wave with incident direction $\hat{k}_{\text{inc}} = \hat{z}$ and polarization along the x-axis causing a charged particle at the origin to oscillate along the same direction. The oscillating charge radiates another (scattered) wave in the scattered direction, $\hat{k}_{\text{sca}}$. The scattered wave has a polarization direction that is the projection of the oscillation direction, $\hat{x}$, onto a plane, P, perpendicular to the scattered direction, $\hat{k}_{\text{sca}}$.

electromagnetic theory known as the oscillating electric dipole with the resultant electric dipole radiation (Griffiths 1999, Jackson 1999).

Consider a charged particle, such as an electron, at the origin of a coordinate system induced to oscillate by an incident beam of light propagating in the $+z$-direction with a polarization along the x-direction, as drawn in figure 3.3. The electric field of the incident radiation will cause a force hence an acceleration on the charge anti-parallel to the field (anti-parallel because the electron's charge is negative). The result of this will be the charge oscillating along the x-direction to radiate another electromagnetic wave. Both the relative magnitude E and direction, hence polarization, $\hat{p}$, of the radiated electromagnetic wave amplitude when it propagates in the scattered direction, $\hat{k}_{\text{s}}$, is determined by the double vector cross product

$$E\hat{p} \sim \hat{k}_{\text{sca}} \times \hat{x} \times \hat{k}_{\text{sca}} \tag{3.3}$$

The polarization direction given by this double cross product works out to be the direction of the projection of the oscillation direction, in this case the x-direction, onto a plane perpendicular to the propagation direction. This projection gives the relative magnitude as well. This geometric picture is often more convenient than the complex double cross product in equation (3.3).

Because the scattered direction is the radial direction, it can be expressed in terms of the spherical-polar angles θ and ϕ. Then the double cross product of equation (3.3) can be calculated, the square of which is the relative intensity

$$I \propto 1 - \cos^2\phi \sin^2\theta \tag{3.4}$$

For re-radiation in the plane perpendicular to the oscillation direction, $\phi = 90°$, this yields a constant radiation intensity independent of the angle θ. This isotropy is an important result. On the other hand, for $\phi = 0$ the intensity is proportional to $\cos^2\theta$ in the (x, y) plane. One can show that this same functionality holds for all planes passing through the x-axis. Then, one can envision the x-axis as an axis of rotation

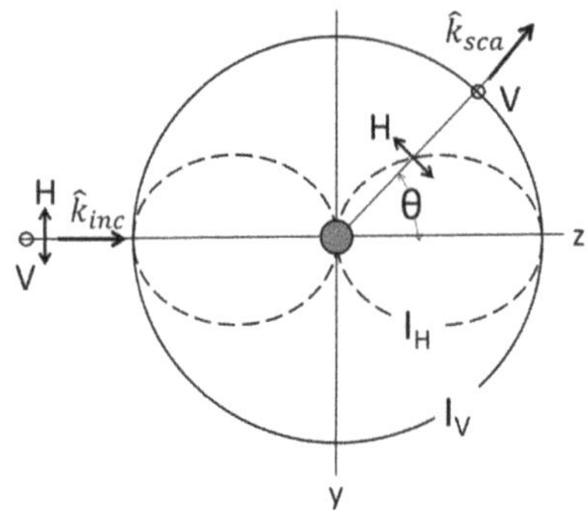

Figure 3.4. Dipole radiation pattern for an oscillating charge or small particle. The view is looking down onto the horizontal plane from above along the vertical x-axis. Light incident from the left has both horizontal H (dashed) and vertical V (solid) polarizations. A polar plot of the scattered intensity is isotropic for I_V and has a $\cos^2\theta$ dependence for I_H. Both scattered polarizations are perpendicular to $\hat{k}_{sca}$.

for $\cos^2\theta$ as drawn in figure 3.4. This shape, a polar plot of the scattered intensity, is the classic *dipole radiation pattern*. Figure 3.4 uses a notation common to the lab or the field in which the yz scattering plane is horizontal and x-axis is vertical. Then one can have horizontal H and vertical V polarizations

We shall see that these results apply to scattering by small particles (relative to the wavelength) of any shape, i.e. they apply to Rayleigh scattering.

3.3 The Stokes vector and the Mueller matrix

The polarization state for scattering by large particles is more complex than for single charges and Rayleigh scattering and thus requires a more detailed formalism, which we now present.

The state of the polarization of light can be completely described by its *Stokes parameters*, which are a convenient, precise (but non-unique) representation for mathematical description and manipulation of the polarization state. For light propagating along the z-axis in a coordinate system as in figure 3.1 with the vertical x-axis, the Stokes parameters are defined as (Born and Wolf 1975, Hecht 2002)

$$I = \langle E_{0x}^2\rangle + \left\langle E_{0y}^2\right\rangle \tag{3.5a}$$

$$Q = \left\langle E_{0y}^2\right\rangle - \langle E_{0x}^2\rangle \tag{3.5b}$$

$$U = \left\langle 2E_{0x}E_{0y}\cos\varepsilon\right\rangle \tag{3.5c}$$

$$V = \left\langle 2E_{0x}E_{0y}\sin\varepsilon\right\rangle \tag{3.5d}$$

In equations (3.5) E_0 is the amplitude of either the x- or y-components of the electromagnetic wave, ε is the difference in phase between these components, and the brackets $\langle\cdots\rangle$ indicate an average over a time comparable to the observation time. The phase difference ε is defined by

$$\vec{E} = \hat{x}E_x\cos(kz - \omega t) + \hat{y}E_y\cos(kz - \omega t - \varepsilon) \tag{3.6}$$

Note that $E_{0x}^2 = I_x$ the x-polarized intensity, etc, and so the first Stokes parameter is the total intensity of the light wave. It is common usage to normalize the Stokes parameters by the intensity so that the first parameter is 1.

For mathematical manipulations the Stokes parameters are cast into a column vector called the *Stokes vector* as

$$\begin{pmatrix} I \\ Q \\ U \\ V \end{pmatrix} \tag{3.7}$$

Now apply equations (3.5) to various polarization situations.

1. Unpolarized light or 'natural' light has equal amounts of x- and y-polarizations so $Q = 0$. There is also no coherence in the phase between the x- and y-components. Thus, the phase difference between the two components ε varies randomly over intervals of time less than the time scale of the average implied by the brackets in equations (3.5), which is the observation time. Thus, $\langle\cos\varepsilon\rangle = \langle\sin\varepsilon\rangle = 0$ and so $U = V = 0$. Then, the normalized Stokes vector is $(I, Q, U, V)^T = (1, 0, 0, 0)^T$ for unpolarized light[3].
2. Light linearly polarized in the x-direction (vertical in figure 3.4) is represented by $(1, -1, 0, 0)^T$ and in the y-direction (horizontal) by $(1, 1, 0, 0)^T$.
3. Light linearly polarized in the +45° direction has equal x- and y-components of $E_0/\sqrt{2}$ that are in phase so $\varepsilon = 0$, thus $\langle\cos\varepsilon\rangle = 1$, $\langle\sin\varepsilon\rangle = 0$, to yield $U = 1$, $V = 0$ for a Stokes vector of $(1, 0, 1, 0)^T$.
4. Right circular polarized light has equal x- and y-components with a constant phase difference of $\varepsilon = 90°$ ($\pi/2$); thus, it is represented by $(1, 0, 0\ 1)^T$; left circularly light is $(1, 0, 0, -1)^T$.

These common polarization Stokes vectors are summarized in figure 3.5.

In general, for linearly polarized light at an angle of γ from the $+y$-axis (the horizontal axis), the Stokes vector is as given in figure 3.6.

For the common laboratory situation in which the incident and scattered directions are in the horizontal plane, we often speak of horizontal and vertical polarizations. Then it is convenient to describe the Stokes vectors in a second, complementary, but equivalent, description. In this description, I is once again the total intensity of the light, Q is the difference between the horizontally and vertically polarized intensities, U is the difference between the +45° and −45° polarized intensities, and V is the difference between the right-handed and left-handed circularly polarized intensities.

It is also not uncommon to be in situations where the scattering plane isn't necessarily horizontal (or vertical). Then it is useful to describe the directions as in-plane and perpendicular to the scattering plane (defined by the incident and

[3] The superscript T means 'transpose', which turns the row vector into a column vector.

Unpolarized	Vertical Linear Polarized	Horizontal Linear Polarized	+ 45° Linear Polarized	− 45° Linear Polarized	Right Circular Polarized	Left Circular Polarized
$\begin{pmatrix}1\\0\\0\\0\end{pmatrix}$	$\begin{pmatrix}1\\-1\\0\\0\end{pmatrix}$	$\begin{pmatrix}1\\1\\0\\0\end{pmatrix}$	$\begin{pmatrix}1\\0\\1\\0\end{pmatrix}$	$\begin{pmatrix}1\\0\\-1\\0\end{pmatrix}$	$\begin{pmatrix}1\\0\\0\\1\end{pmatrix}$	$\begin{pmatrix}1\\0\\0\\-1\end{pmatrix}$

Figure 3.5. Stokes vectors for some common polarizations.

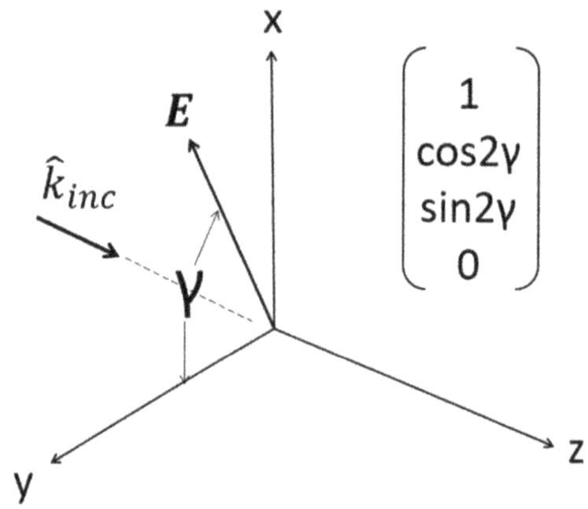

Figure 3.6. Polarized vector field ***E*** at an angle γ from the horizontal axis and its Stokes vector.

scattering directions) or parallel and perpendicular, or co-polarized and cross-polarized, respectively. Thus, $(I, Q, U, V)^T$ is equation (3.8)

$$\begin{pmatrix} I_{\|} + I_{\perp} \\ I_{\|} - I_{\perp} \\ I_{+45} - I_{-45} \\ I_R - I_L \end{pmatrix} \quad \text{or} \quad \begin{pmatrix} I_H + I_V \\ I_H - I_V \\ I_{+45} - I_{-45} \\ I_R - I_L \end{pmatrix} \tag{3.8}$$

The Stokes parameters are related through the following inequality,

$$I^2 \geqslant Q^2 + U^2 + V^2 \tag{3.9}$$

The equality holds when the light is, by definition, completely polarized. Given this, one can define the *degree of polarization* as

$$P = \sqrt{Q^2 + U^2 + V^2}/I \leqslant 1, \tag{3.10}$$

and the degree of linear polarization as

$$P_l = \sqrt{Q^2 + U^2}/I \leqslant 1 \tag{3.11}$$

This is readily measured as (Bohren and Huffman 1983)

$$P_l = \frac{I_{\max} - I_{\min}}{I_{\max} + I_{\min}} \tag{3.12}$$

where $I_{\max}$ and $I_{\min}$ are the maximum and minimum intensities, respectively, measured after the scattered light passes through a linear polarizer that is rotated.

The *linear depolarization ratio* (LDR) is the ratio of the cross-polarized scattered intensity to the co-polarized scattered intensity, sometimes represented by μ_L

$$\mathrm{LDR} \equiv \mu_L = I_{\text{cross-pol}}/I_{\text{co-pol}} \tag{3.13}$$

'Co-polarized' means the incident and detected polarizations are parallel to each other, and 'cross-polarized' means the incident and detected polarizations are perpendicular to each other. The LDR is often used in the back direction, where 'co' and 'cross' are obvious and straightforward to apply.

Mathematical manipulations of the polarization begin with the simple fact that when two light beams that are incoherent with respect to each other are combined, the sum of their Stokes vectors describes the resulting polarization. When light interacts with an optical material or element, such as a polarizer or wave plate, or a scattering particle, the manner in which its polarization state, as described by its Stokes vector, changes can be described by multiplication of the Stokes vector by a 4 × 4 matrix called the *Mueller matrix*. The matrix multiplication yields a new Stokes vector, which represents the new polarization state produced after the interaction with the optical material. Examples of two useful Mueller matrices are those for horizontal and vertical polarizers, respectively, given in equation (3.14) below

$$\frac{1}{2}\begin{pmatrix} 1 & 1 & 0 & 0 \\ 1 & 1 & 0 & 0 \\ 0 & 0 & 0 & 0 \\ 0 & 0 & 0 & 0 \end{pmatrix} \qquad \frac{1}{2}\begin{pmatrix} 1 & -1 & 0 & 0 \\ -1 & 1 & 0 & 0 \\ 0 & 0 & 0 & 0 \\ 0 & 0 & 0 & 0 \end{pmatrix} \tag{3.14}$$

3.4 The scattering matrix

The Mueller matrix for scattering is usually called the *scattering matrix*. It transforms the incident Stokes vector into the scattered Stokes vector via matrix multiplication. The elements of the scattering matrix have common functionalities on size, refractive index, morphology, scattering angle, etc, the same functionalities that lead, as we will see, to the differential scattering cross section. The matrix multiplication is shown in equation (3.15), where the subscripts 'inc' and 'sca' mean 'incident' and 'scattered', respectively.

$$\begin{pmatrix} I_{\text{sca}} \\ Q_{\text{sca}} \\ U_{\text{sca}} \\ V_{\text{sca}} \end{pmatrix} = \frac{1}{(kr)^2}\begin{pmatrix} S_{11} & S_{12} & S_{13} & S_{14} \\ S_{21} & S_{22} & S_{23} & S_{24} \\ S_{31} & S_{32} & S_{33} & S_{34} \\ S_{41} & S_{42} & S_{43} & S_{44} \end{pmatrix}\begin{pmatrix} I_{\text{inc}} \\ Q_{\text{inc}} \\ U_{\text{inc}} \\ V_{\text{inc}} \end{pmatrix} \tag{3.15}$$

In equation (3.15), $k = 2\pi/\lambda$, the wave number of the light, and r is the distance between the scatterer and the detector. As with the Mueller matrix, the elements of the scattering matrix are dimensionless. The coefficient $1/k^2r^2$ accounts for the fact that the scattered intensity must fall off as the inverse square of the distance, r. The other part of this coefficient involving k maintains the dimensionless nature of the Mueller matrix. Its origin stems from the electromagnetic derivation of the scattering and an approximation that has been made that the scattering is in the far field, such that $r \gg \lambda$ (Bohren and Huffman 1983).

Application of equation (3.15) to incident light plane polarized vertically with Stokes vector $(1, -1, 0, 0)^T$ yields a Stokes vector whose first element is the scattered intensity of

$$I_{sca} = I_{inc}(kr)^{-2}(S_{11} - S_{12}) \tag{3.16}$$

Now recall equation (2.8), which defined the differential cross section $dC_{sca}/d\Omega$

$$I_{sca} = \frac{I_{inc}}{r^2}\frac{dC_{sca}}{d\Omega} \tag{2.8}$$

Comparison of equations (3.16) and (2.8) yields, for vertically polarized incident light,

$$\frac{dC^{v}_{sca}}{d\Omega} = \frac{(S_{11} - S_{12})}{k^2} \tag{3.17}$$

Application of equation (3.15) to unpolarized incident light with Stokes vector $(1, 0, 0, 0)^T$ yields a scattered intensity of

$$I_{sca} = I_{inc}(kr)^{-2}S_{11} \tag{3.18}$$

Comparison of equations (3.18) and (2.8) yields, for unpolarized incident light,

$$\frac{dC^{u}_{sca}}{d\Omega} = \frac{S_{11}}{k^2} \tag{3.19}$$

In the rather unusual notation for the differential cross section in equations (3.17) and (3.19) the superscripts v and u mean that the cross section is for vertically polarized and unpolarized incident light, respectively. Note that there are other differential cross sections dependent on other incident polarizations.

Perhaps the simplest scattering situation is Rayleigh scattering with a scattering matrix of (Mishchenko *et al* 2002)

$$\frac{1}{2}k^6R^6F(m)\begin{pmatrix} \cos^2\theta + 1 & -\sin^2\theta & 0 & 0 \\ -\sin^2\theta & \cos^2\theta + 1 & 0 & 0 \\ 0 & 0 & 2\cos\theta & 0 \\ 0 & 0 & 0 & 2\cos\theta \end{pmatrix} \tag{3.20}$$

In (3.20), θ is the scattering angle and $\varphi = 90°$, see figure 3.3. Also in (3.20) R is the radius of the spherical particle and $(m) = |(m^2 - 1)/(m^2 + 2)|^2$, the Lorentz–Lorenz factor with m being the complex refractive index, see section 6.2. This matrix is to be used with equations (3.17) and (3.18) where the factor of $(kr)^{-2}$ in (3.15) has been accounted for. The θ functionalities are a consequence of the projection of the incident polarization onto a plane perpendicular to the scattered direction as embodied in equation (3.4), which yields for $\varphi = 0$, $I_{sca} \sim 1$ and for $\varphi = 90°$, $I_{sca} \sim \cos^2\theta$. Application of equation (3.17) using the matrix elements of (3.20) yields $I_{sca} \sim 1$, a constant, and to (3.18) $I_{sca} \sim 1 + \cos^2\theta$, both of which are consistent with figure 3.4. Another useful exercise is to multiply (3.20) with the Stokes vectors for unpolarized and polarized incident light and compare the results with figure 3.4.

Increasing the complexity takes us to the Rayleigh–Debye–Gans (RDG) scattering situation, discussed extensively below, for particles of arbitrary shape and size but with an index of refraction well matched to the surrounding medium, i.e. $m \to 1$[4]. This RDG scattering matrix when orientationally averaged has the same form as (3.20) but multiplied by the normalized structure factor, $S(q)$, where $q = \left(4\pi/\lambda\right)\sin\left(\frac{\theta}{2}\right)$, of the shape (see chapter 4). Note that this implies angular dependencies of the matrix elements can originate from both polarization and structural sources. Fractal aggregates, which are usually well described by RDG, behave in a similar manner (Singham and Bohren 1993).

For spherical particles of arbitrary size and refractive index (Mie scattering), there are four independent matrix elements and the scattering matrix takes the form

$$\begin{pmatrix} S_{11} & S_{12} & 0 & 0 \\ S_{12} & S_{11} & 0 & 0 \\ 0 & 0 & S_{33} & S_{34} \\ 0 & 0 & -S_{34} & S_{33} \end{pmatrix} \tag{3.21}$$

Note that for spherical particles $S_{22} = S_{11}$ and $S_{33} = S_{44}$. Any deviation from these equalities can be used as an indication of non-sphericity. Also, once again, equation (3.17) connects these matrix elements to the differential cross section.

A very general scattering matrix occurs for an ensemble of arbitrary particles with random orientations, average separations much larger than their linear sizes (hence 'dilute') and with no interparticle multiple scattering. Then there are 10 independent matrix elements and the sole angular dependence is with the scattering angle θ, which describes the angle between the incident and scattered light. This scattering matrix is given in expression (3.22).

$$\begin{pmatrix} S_{11} & S_{12} & S_{13} & S_{14} \\ S_{12} & S_{22} & S_{23} & S_{24} \\ -S_{13} & -S_{23} & S_{33} & S_{34} \\ S_{14} & S_{24} & -S_{34} & S_{44} \end{pmatrix} \tag{3.22}$$

If, as well, particles and their mirror image particles are present in equal numbers and randomly oriented, or each particle has a plane of symmetry, the scattering matrix has six independent elements and takes the form

[4] The exact condition is when the internal coupling parameter $\rho' < 1$, see section 7.1.2.2.

$$\begin{pmatrix} S_{11} & S_{12} & 0 & 0 \\ S_{12} & S_{22} & 0 & 0 \\ 0 & 0 & S_{33} & S_{34} \\ 0 & 0 & -S_{34} & S_{44} \end{pmatrix} \tag{3.23}$$

In their exposition of this subject, Mishchenko *et al* (2002) point out that having an ensemble of particles with mirrored counterparts or mirror symmetry might be 'a rather special case'. Nevertheless the matrix in (3.23) provides a good numerical description of scattering for many systems encountered in practice. Such an ensemble will form a macroscopically isotropic scattering medium, hence the matrix elements will only be a function of the angle between the incident and scattered directions, with the scattering angle θ.

Scattering in the exact forward and backward directions eliminates the scattering plane and more symmetries ensue. For the exact forward direction the matrix (3.23) becomes diagonal with $S_{22}(0) = S_{33}(0)$. In the exact backward direction, the matrix (3.23) is also diagonal with S_{33} (180°) = $-S_{22}(180°)$ and $S_{44}(180°) = S_{11}(180°) - 2S_{22}(180°)$.

Figure 3.7 gives an example of the normalized scattering matrix element data for an ensemble of water droplets with a mean radius of $R = 0.8$ micron distributed in a

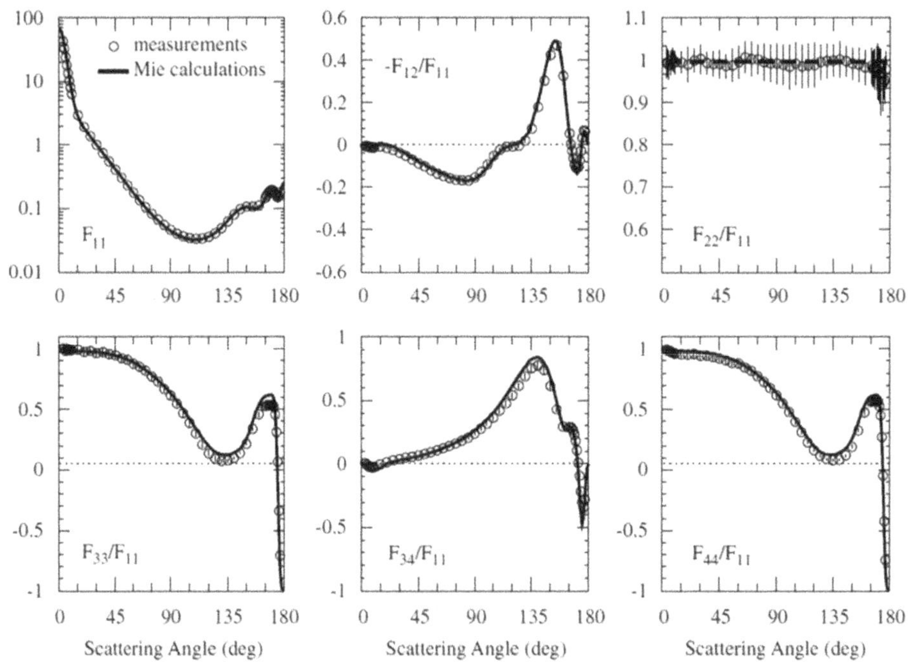

Figure 3.7. Scattering matrix elements $F_{ij} = S_{ij}$ for water droplets scattering $\lambda = 488$ nm light. F_{11} is normalized to 1° at 30°. Other elements are normalized by F_{11}. Solid lines correspond to fits to Mie calculations with a log-normal size distribution with a most probable radius of R = 0.8 μm and a geometric width of σ = 1.5. Reprinted from (Munoz *et al* 2010), Copyright (2010), with permission from Elsevier.

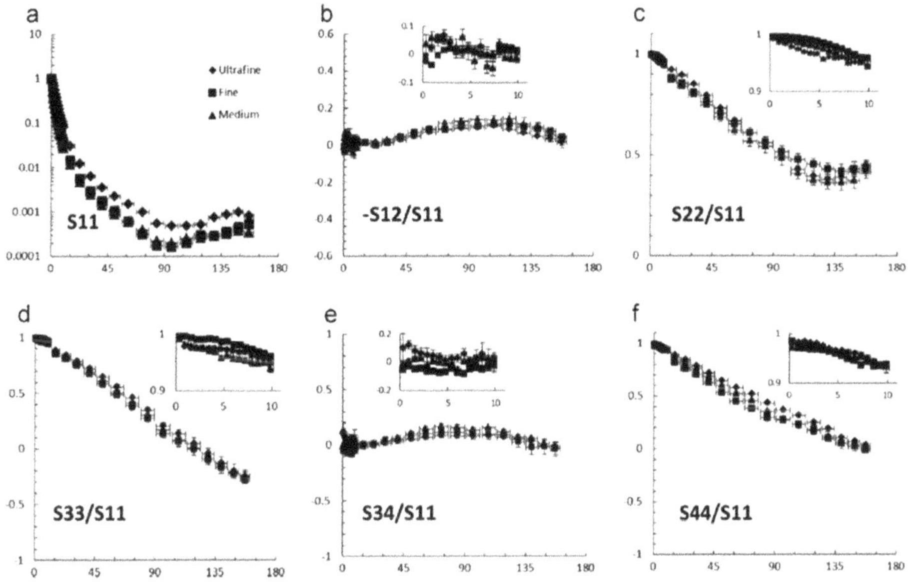

Figure 3.8. The scattering matrix elements for three grades of Arizona Road Dust with mean sizes on the order of 10 microns versus the scattering angle. The incident light had a wavelength of λ = 532 nm. S_{11} of three ARZD samples are normalized at the smallest angle of 0.32°. Reprinted figure with permission from (Wang *et al* 2015). Copyright (1973) by the American Physical Society.

log-normal size distribution with geometric width $\sigma = 1.5$ (Munoz *et al* 2010). The fit to Mie calculations is quite good as expected for spherical particles. Also, note the ratio F_{22}/F_{11} is very close to unity, indicative of spherical particles.

Figure 3.8 gives an example of scattering matrix element data for irregularly shaped particles, in this case Arizona Road Dusts with mean sizes on the order of 10 microns (Wang *et al* 2015). The scattered intensity matrix element S_{11} is qualitatively similar to that for spheres, figure 3.7, whereas the other elements are quite different. Now note that the ratio S_{22}/S_{11} falls significantly from unity at small angles with increasing angle, unlike spheres.

3.5 Polarization upon scattering

Consider unpolarized light with Stokes vector $(1, 0, 0, 0)^T$ incident upon a scattering medium as illustrated in figure 3.9. If the scattering matrix is in its most general form, expression (3.21), the resulting Stokes vector of the scattered light will be $(S_{11}, S_{12}, -S_{13}, S_{14})^T$. The possible non-zero values of the off diagonal elements of the scattering matrix in this vector uncovers the general result that *scattering in some manner polarizes the light*.

We now consider the variety of possible experimental situations of incident light polarization and detected polarizations.

For the scattering situation immediately above, if the incident light were unpolarized, the scattered intensity would be proportional to S_{11} if the scattered light fell directly upon the detector. If instead a vertical polarizer was placed in front

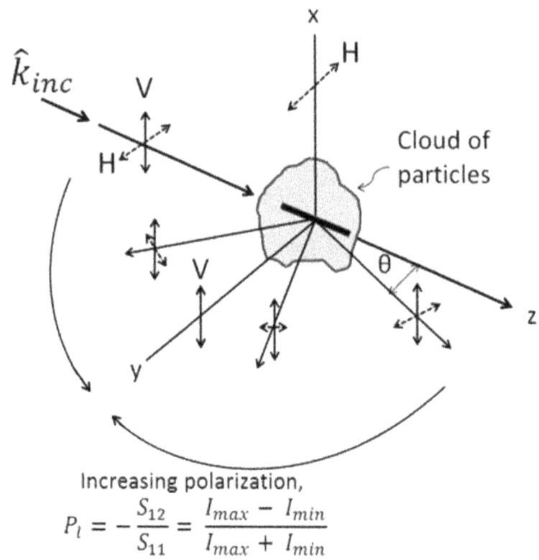

Figure 3.9. Diagram of unpolarized light (with equal amounts of linear polarizations in the vertical, V solid double arrow, and horizontal, H dashed double arrow, directions) incident on an aerosol of particles that subsequently scatter light at angles θ in the horizontal (yz) scattering plane. The degree of linear polarization is maximum with $P_l = 1$ at scattering angles of 90°.

of the detector, application of the Mueller matrix for a vertical polarizer, expression (3.14), to the scattered light Stokes vector yields $S_{11} - S_{12}$ for the detected intensity. Similarly, a horizontal polarizer would pass light with a Stokes vector with $S_{11} + S_{12}$ for the detected intensity. We now have

$$I_{UU} \propto S_{11} \tag{3.24a}$$

$$I_{UV} \propto \tfrac{1}{2}(S_{11} - S_{12}) \tag{3.24b}$$

$$I_{UH} \propto \tfrac{1}{2}(S_{11} + S_{12}) \tag{3.24c}$$

In the proportionalities (3.24), the first subscript describes the incident polarization, the second the detected polarization, U designates unpolarized incident light or no polarizer in front of the detector, V designates vertically polarized incident light or a vertical polarizer in front of the detector, and H designates horizontally polarized incident or a horizontal polarizer in front of the detector. These proportionalities have the same proportionality constant so ratios of these would yield equalities. Note the consistency of equations (3.24a) and (3.18).

Next consider vertically polarized light with the Stokes vector $(1, -1, 0, 0)^T$ incident upon a scattering medium with the less general scattering matrix, but still widely applicable, expression (3.23). The resulting Stokes vector of the scattered light would have a first element, proportional to the scattered intensity, of $S_{11} - S_{12}$. This would be detected as it fell directly on the detector. If instead a vertical polarizer was placed in front of the detector, application of the Mueller matrix for a vertical polarizer, expression (3.14), to the scattered light Stokes vector would yield

$S_{11} - 2S_{12} + S_{22}$ for the detected intensity. Similarly, a horizontal polarizer would pass light with a Stokes vector with $S_{11} - S_{22}$ for the detected intensity. We now have

$$I_{VU} \propto S_{11} - S_{12} \tag{3.25a}$$

$$I_{VV} \propto \tfrac{1}{2}(S_{11} - 2S_{12} + S_{22}) \tag{3.25b}$$

$$I_{VH} \propto \tfrac{1}{2}(S_{11} - S_{22}) \tag{3.25c}$$

Lastly, consider horizontally polarized light with the Stokes vector $(1, 1, 0, 0)^T$ incident upon a scattering medium with the less general scattering matrix, but still widely applicable, expression (3.23). The resulting Stokes vector of the scattered light would have a first element, proportional to the scattered intensity, of $S_{11} + S_{12}$. This would be detected as it fell directly on the detector. If instead a vertical polarizer was placed in front of the detector, application of the Mueller matrix for a vertical polarizer, expression (3.14), to the scattered light Stokes vector would yield $S_{11} - S_{22}$ for the detected intensity. Similarly, a horizontal polarizer would pass light with a Stokes vector with $S_{11} + 2S_{12} + S_{22}$ for the detected intensity. We now have

$$I_{HU} \propto S_{11} + S_{12} \tag{3.26a}$$

$$I_{HV} \propto \tfrac{1}{2}(S_{11} - S_{22}) \tag{3.26b}$$

$$I_{HH} \propto \tfrac{1}{2}(S_{11} + 2S_{12} + S_{22}) \tag{3.26c}$$

As above, these proportionalities have the same proportionality constant so ratios of these would yield equalities.

The degree of linear polarization, P_l, defined in equation (3.11) is measured via equation (3.12). However, the combination of matrix elements that comprise P_l depends on the incident polarization. For example, application of the less general but common scattering described by the Mueller matrix in equation (3.23) to unpolarized light yields $(S_{11}, S_{12}, 0, 0)^T$ for the scattered Stokes vector. Application of equation (3.10) to this yields the degree of linear polarization $P_l = -Q/I = -S_{12}/S_{11}$[5], which is measured via equation (3.12). For Rayleigh scatterers expression (3.19) leads to

$$P_l = \frac{1 - \cos^2\theta}{1 + \cos^2\theta} \tag{3.27}$$

where θ is the scattering angle. This is consistent with equation (6.12) as discussed in chapter 6 on Rayleigh scattering and illustrated in figure 6.3.

For incident light linearly polarized in the vertical, i.e. x-direction, as is often the case in a laser light scattering lab, the Stokes vector is $(1, -1, 0, 0)^T$. Multiplication of this by the fairly general scattering matrix in expression (3.23) yields a scattered Stokes vector of $(S_{11} - S_{12}, S_{12} - S_{22}, 0, 0)^T$. Then the degree of linear polarization is

[5] The negative sign is a convention that is included so that usually $P_l > 0$.

$P_l = (S_{22} - S_{12})/(S_{11} - S_{12})$. If the particles are spherical, expression (3.21) applies with $S_{11} = S_{22}$, hence $P_l = 1$ at all scattering angles. The scattered light is polarized in the vertical direction, the same as for Rayleigh scatterers and arbitrarily shaped objects in the RDG limit.

One can also define depolarization ratios to indicate how scattering changes the polarization. If the incident light is completely polarized vertically, one can define the *depolarization ratio* as

$$\rho_v = I_{VH}/I_{VV} \tag{3.28}$$

This is also a form of the *linear depolarization ratio*, μ_L, equation (3.13), above (Mishchenko *et al* 2006) and equation (2.26). In equation (3.28), I_{VH} is the intensity of the scattered light when the incident light is vertically polarized and the scattered light is detected with a linear horizontal polarizer, and I_{VV} is when the incident light is vertically polarized and the scattered light is detected with a linear vertical polarizer. Notice that unlike the situations above, a polarizer is now in front of the detector. For Rayleigh, RDG and particles with spherical symmetry, we expect $\rho_v = 0$, i.e. the scattered light is completely polarized in the vertical direction. Then with the appearance of a finite amount of horizontal polarization the incident polarization has been changed and the scattering is no longer perfectly polarized but is depolarized as indicated by $\rho_v > 0$. One can show by application of the general scattering matrix in (3.22) and the polarizer Mueller matrices of (3.14) or application of relations (3.25(b) and (c)) to equation (3.28) that

$$\rho_v = \frac{S_{11} - S_{22}}{(S_{11} - 2S_{12} + S_{22})} \tag{3.29}$$

The numerator of (3.29) is zero for perfect spheres but becomes finite when the spherical symmetry breaks and hence presents a sensitive indicator of non-sphericity. We also note that the ratio I_{HV}/I_{HH} is the same as equation (3.29) with the exception that the term $-2S_{12}$ becomes $+2S_{12}$.

These results apply to the particle properties so long as there is no multiple scattering from an ensemble of particles. In chapter 2, depolarization due to multiple scattering was described.

3.6 Microphysical description for the scattered light polarization for spheres

Here, we will gain an intuition for the scattered light polarization that comes about for a single, non-point-like particle in terms of the microphysical picture of scattering. We will follow the approach presented by Berg *et al* closely (Berg *et al* 2008). As promoted later in this book, the foundation of the microphysical model is to represent any scattering particle as a system of point scatterers. These point scatterers see both the incident light wave and the light scattered to them from all the other scatterers in the particle, a process we have called internal coupling. Then each point scatterer acts as a source that radiates its contribution to the detector to create the scattered wave in the far field.

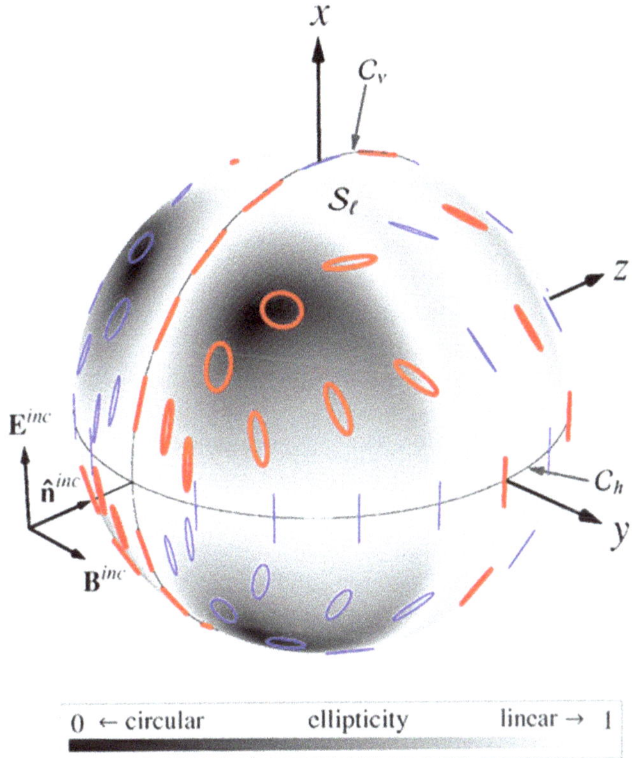

Figure 3.10. The polarization state of the scattered wave for a sphere with $kR = 4$, $m = 1.25 + 0i$ and $\rho = 2$ and $\rho' = 1.26$. The plot shows the vibration ellipses that the head of the polarization vector scribes out over time at various points on a large imaginary spherical surface S_l in the far field. The polarization ellipticity is shown in gray shades on S_l. Darker shades indicate more elliptical, i.e. closer to circular, lighter shades mean closer to linear polarization. The bold red ellipses correspond to left-handed rotation of the field, whereas thin blue ellipses correspond to right-handed rotation. The sense of rotation is right-handed if the scattered electric field vector rotates in the counterclockwise direction when looking into the oncoming wave. From Berg *et al* (2008).

Figure 3.10 shows an incident light wave propagating along the positive z-axis with linear polarization in the x-direction (vertical). It is incident on a sphere with the size parameter $kR = 4$ and real refractive index $m = 1.25$, hence the phase shift parameter $\rho = 2$ and an internal coupling parameter $\rho' = 1.26$, see section 7.1.2.2. The scattering was calculated from the Mie equations. The scattered polarizations are cast upon a large imaginary spherical surface S_l in the far field. The ellipticity of the polarization is indicated by gray shades, and lighter shades correspond to more linear polarization, whereas darker shades correspond to more circular polarization. Polarization ellipses are also shown for various points on S_l. The left-handed rotation of an ellipse is indicated by a thick red line, and the right-handed rotation is indicated by a thin blue line. This figure also shows that the wave is linearly polarized along the horizontal and vertical circumferences of S_l, C_h and C_v, respectively.

Figure 3.10 shows that the scattered light is linearly polarized only in the xz (vertical) and yz (horizontal) planes. Reflection symmetry is seen across the yz plane as the degree of ellipticity is equal for the same distance above and below the plane but the handedness inverts. The geometry is not symmetric about the xy-plane, however, because as the internal field propagates along the z-axis, it will change when the parameters ρ' and $\rho > 1$. The three plane symmetry invoked by ρ' and $\rho < 1$ returns the situation to RDG scattering.

The reflection symmetries of a sphere's internal field can be connected to the overall structure of the polarization state of its scattered wave in the far field. For example, consider an observation point that lies on the C_h contour. Figure 3.11 shows two wavelets located at points $\mathbf{r}_j$ and $\mathbf{r}_k$ inside a sphere. These points are mirror points about the yz-plane, and the interior electric fields shown at the points, $\mathbf{E}_\text{int}(\mathbf{r}_j)$ and $\mathbf{E}_\text{int}(\mathbf{r}_k)$, obey the symmetry properties derived by Berg *et al* with equal x-components and equal but opposite y-components. Shown at $\mathbf{r}$ on C_h are the contributions to the scattered field from these interior points $\mathbf{E}_j$ and $\mathbf{E}_k$, respectively. Figure 3.11 illustrates that the projections of $\mathbf{E}_j$ and $\mathbf{E}_k$ onto the $\hat{\theta}$ direction cancel each other out, and hence, it is only the φ-components that survive which in turn depend only on the x-component of the internal electric field. The result is linearly polarized light along the x-direction, the same as the incident polarization, on the contour C_h. A similar cancellation process can be used to explain the linear polarization of the scattered wave for points on the C_v contour. Moreover, from this viewpoint one can now understand that the ellipticity of the polarization state is due to the break in symmetry that results from taking the observation point off the C_h or C_v contours. This break in symmetry prevents the unique cancellation of components that results in the linear polarization state and introduces a relative phase shift between the components.

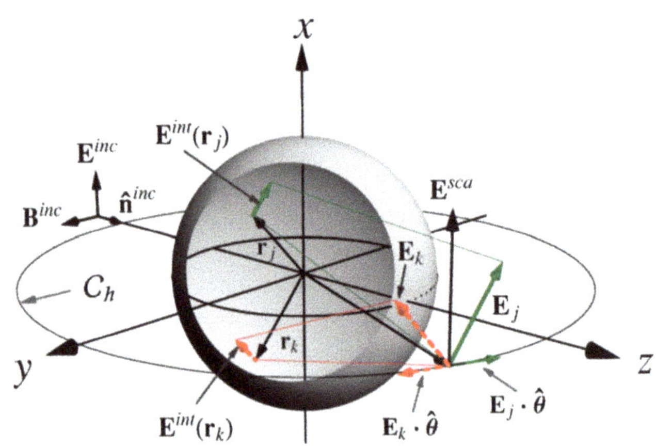

Figure 3.11. Illustration of the wavelet component cancellation that is responsible for the linear polarization state of the scattered wave on the C_h contour. Internal mirror points $\mathbf{r}_j$ and $\mathbf{r}_k$ about the yz-plane are shown along with the internal electric field vectors $\mathbf{E}_\text{int}(\mathbf{r}_j)$ and $\mathbf{E}_\text{int}(\mathbf{r}_k)$ at these points. Each of these internal wavelets contribute $\mathbf{E}_j$ and $\mathbf{E}_k$ to the scattered field, where the cancellation of their θ-components is evident leaving only a linear polarization in the x-direction. From Berg *et al* (2008).

References

Berg M J, Sorensen C M and Chakrabarti A 2008 Reflection symmetry of a sphere's internal field and its consequences on scattering: a microphysical approach *J. Opt. Soc. Am.* A **25** 98–107

Bohren C F and Donald R H 1983 *Absorption and Scattering of Light by Small Particles* (New York: Wiley)

Born M and Wolf E 1975 *Principles of Optics: Electromagnetic Theory of Propagation, Interference, and Diffraction of Light* (Oxford: Pergamon)

Griffiths D J 1999 *Introduction to Electrodynamics* (Upper Saddler River, NJ: Prentice-Hall)

Hecht E 2002 *Optics* (Reading, MA: Addison-Wesley)

Jackson J D 1999 *Classical Electrodynamics* (New York: Wiley)

Mishchenko M I, Travis L D and Lacis A A 2006 *Multiple Scattering of Light by Particles* (Cambridge: Cambridge University Press)

Mishchenko M I, Larry D T and Andrew A L 2002 *Scattering, Absorption, and Emission of Light by Small Particles* (Cambridge: Cambridge University Press)

Munoz O, Moreno F, Guirado D, Ramos J L, Lopez A, Girela F, Jeronimo J M, Costillo L P and Bustamante I 2010 Experimental determination of scattering matrices of dust particles at visible wavelengths: the IAA light scattering apparatus *J. Quant. Spectrosc. Radiat. Transf.* **111** 187–96

Singham S B and Bohren C F 1993 Scattering of unpolarized and polarized-light by particle aggregates of different size and fractal dimension *Langmuir* **9** 1431–35

Wang Y, Chakrabarti A and Sorensen C M 2015 A light-scattering study of the scattering matrix elements of Arizona Road Dust *J. Quant. Spectrosc. Radiat. Transf.* **163** 72–9

Chapter 4

The structure factor

The structure factor is the square of the Fourier transform of the real space structure of the object that is scattering the wave radiation. Since no information is lost in the Fourier transformation, it contains all the information there is about the structure of the scattering object. This information is written in reciprocal space with dimensions of inverse length. However, squaring the Fourier transform to obtain the structure factor unfortunately causes a loss of phase information. This squaring often occurs during detection of the wave[1]. Nevertheless, much information remains.

Recall from chapter 1 that the square of the Fourier transform of the real space structure is the intensity diffraction pattern of the diffracting object. Also, recall Babinet's principle so that diffraction patterns created by holes are identical to diffraction patterns created by same-shape obstacles. Common usage calls diffraction by objects 'scattering'. Thus, if we are only dealing with waves, for instance if the waves are light and the electric[2] polarizabilities of the scattering objects are small, as designated by their refractive indices, then scattering is the structure factor of the object that is scattering, and that scattering is simply diffraction. And we don't need Maxwell's equations.

Of course, real objects for which we want to describe and understand light scattering properties are usually quite polarizable, as indicated by substantial refractive indices. Nevertheless, the imprint of the structure factor, the imprint of diffraction, remains as a foundation as their nascent scattering properties.

4.1 A system of scatterers

For most of us, our initial conception of optics is likely a vision of light rays reflecting and refracting at an interface between the vacuum and a transparent medium. The angles of reflection and refraction and the intensities of the reflected

[1] Square law detectors and homodyne detection.

[2] And magnetic.

doi:10.1088/978-0-7503-5157-7ch4

and refracted beams are solely functions of the angle of incidence, the polarization and the refractive index of the medium and are given by the law of reflection, Snell's law and the Fresnel equations. This vision is one of continuum electrodynamics in which the medium is represented as infinitely divisible and characterized electrodynamically by a single variable, the refractive index.

However, we know that matter is composed of atoms that contain relatively lightweight and charged electrons that readily respond to electric fields, such as those found in a beam of light. Moreover, when electrons are driven into oscillation by the applied oscillating optical field, they reradiate an electromagnetic field of their own. So despite the great accuracy and utility of the continuum approach, we are compelled to ask, what really happens when a beam of light in vacuum encounters a material object?

The answer has been given by Ewald and Oseen in their famous Ewald–Oseen extinction theorem (Born and Wolf 1975). This subject is discussed in detail in section 7.4.1. This theorem is based on a microphysical approach in which a material is represented as a vacuum filled with point scatterers responsive to an applied electric field. In this approach, an incident wave penetrates unhindered the void space between the scatterers within the material object without any change in either its direction or wavelength. It does, however, set the point scatterers into oscillation at the frequency of the incident wave. These scatterers subsequently reradiate electromagnetic waves of their own. Now the fun begins! These secondary waves propagate to other scatterers in the medium so that the scatterers sense not only the incident wave but also the waves of all the other scatterers in the system. In an iterative, self-consistent manner, the result of all this interaction, this internal coupling, is three new waves: (1) an extinction wave propagating in the incident direction with the same field magnitude as the incident wave but exactly $\pi(180^{\circ})$ out of phase. This wave cancels (extinction) the incident wave in the medium. (2) A reflected wave obeying the law of reflection. (3) A refracted wave obeying Snell's law. The macroscopic index of refraction is shown to be a consequence of the polarizability of the scatterers and their number density.

Cool! But what has this got to do with scattering? Everything. Scattering, on the one hand, and reflection and refraction on the other, are different limits of the same physical process, the interaction of light with matter[3]. Consider an example. A light beam from a laser is about 1 mm in diameter. Shine this laser at a glass ball 100 mm in diameter. What will happen? It will reflect and refract according to the laws of continuum electrodynamics. Now imagine the glass ball gets smaller. When it is 1 micron in diameter, what will happen? It will scatter the light! Did anything fundamentally change as the glass sphere grew smaller? No, the physics is the same, but the analysis methods could be different. The continuum electrodynamics for a small beam and a big sphere uses the Maxwell equations with boundary conditions applied in an approximate planar geometry and achieves the Fresnel equations. The continuum electrodynamics for a big beam and a small sphere uses the same Maxwell equations only now in with boundary conditions suitable for the spherical

[3] Perhaps better said, the *elastic* interaction of light with matter, which means the light energy is conserved, hence the frequency remains the same during the interaction.

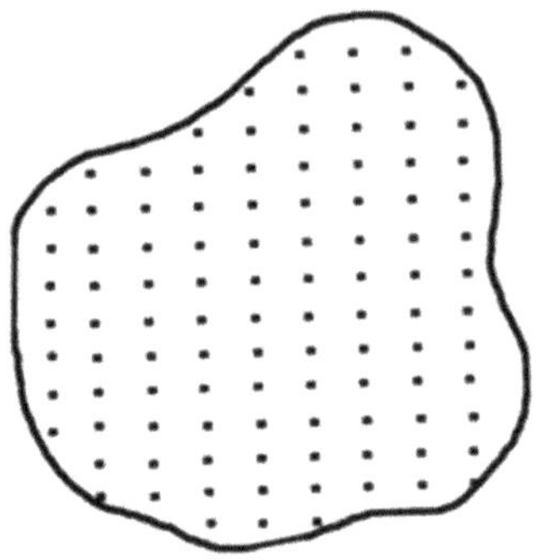

Figure 4.1. A system of scatterers.

shape of the glass ball and achieves the Mie equations. But the physics, embodied in the continuum description of Maxwell's equations, is the same.

This physics is the same at the atomic level too. Ewald–Oseen continues to hold regardless of the size of the particle. This fact leads us to a very useful concept: a *system of scatterers*. We will represent a particle as its geometric shape filled with the vacuum and *point scatterers*. An example for an arbitrarily shaped object is given in figure 4.1. I remark that such a division of a continuous body into infinitesimals is the method of calculus. Then, according to Ewald–Oseen, an incident wave will penetrate unaffected into the system of scatterers. The scatterers will respond by reradiating waves of their own. The calculations to follow will involve these reradiated waves. Very often in the developments to follow the waves will be treated as scalar waves with no polarization that scatter from the scatterers isotropically. Polarization effects can be added later. Note the similarity to the Huygens wavelet method (see section 1.5).

4.2 The scattering wave vector

The foundation of wave scattering, and light scattering in particular, is the phase relations of the waves scattered from the different volume elements of the scattering object. The system of scatterers of figure 4.1, in which the point scatterers represent infinitesimal volume elements, facilitates the conceptualization of the foundation. Here, we derive a most important relationship: the relationship between the positions of the scatterers in the scattering system and the incident and scattered wave vectors, on the one hand, and the phase of the scattered waves at the detector due to the scatterers on the other.

Consider an electromagnetic wave traveling from left to right in the positive z-direction with wave vector $\boldsymbol{k}_{\text{inc}}$ as drawn in figure 4.2. We use the complex notation, section 1.3, but neglect the time dependence of the phase as we are only concerned with phase changes upon propagation through space. The wave encounters a point scatterer at $\boldsymbol{r}$. Then the field at the scatterer has the following proportionality

$$E(\boldsymbol{r}) \sim e^{i\boldsymbol{k}_{\text{inc}}\cdot\boldsymbol{r}} \tag{4.1}$$

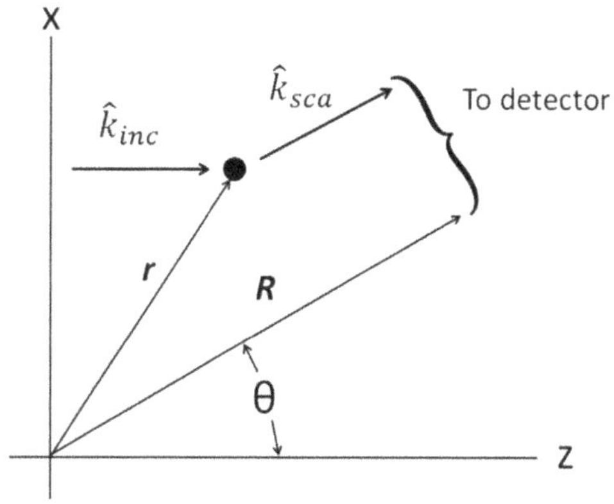

Figure 4.2. Light incident upon a single point scatterer at $\boldsymbol{r}$ and then scattered towards a detector on the far field at $\boldsymbol{R}$. This single point scatterer could be one of many in the system of scatterers of figure 4.1.

The phase of this wave, $\varphi = \boldsymbol{k}_{\text{inc}} \cdot \boldsymbol{r}$, is measured relative to a phase of zero at the origin.

Let the wave scatter at an angle θ with wave vector $\boldsymbol{k}_{\text{sca}}$. Assume that the scattering is elastic, which means the wave neither losses nor gains energy during the scattering. Then the frequency and wavelength do not change, hence the magnitudes of the scattering and incident wave vectors are equal, viz.

$$|\boldsymbol{k}_{\text{sca}}| = |\boldsymbol{k}_{\text{inc}}| = k = 2\pi/\lambda \tag{4.2}$$

Next, place a detector at position vector $\boldsymbol{R}$ with $R \gg r$ so that $\boldsymbol{R}$ is essentially parallel to $\mathbf{k}_{\text{sca}}$. As the wave propagates from the scatterer to the detector, the phase changes by $\mathbf{k}_{\text{sca}} \cdot (\mathbf{R} - \mathbf{r})$. Then the field at the detector is

$$E\,(\boldsymbol{R}) \sim e^{i\mathbf{k}_{\text{inc}} \cdot \boldsymbol{r}} e^{i\boldsymbol{k}_{\text{sca}} \cdot (\boldsymbol{R} - \boldsymbol{r})} \tag{4.3}$$

The position of the detector is arbitrary, but the positions $\boldsymbol{r}$ of all the scatterers defines the shape of the scattering object as depicted in figure 4.1. Thus, we rearrange equation (4.3) to bring these $\boldsymbol{r}$ terms together as

$$E(\boldsymbol{R}) \sim e^{i(\boldsymbol{k}_{\text{inc}} - \boldsymbol{k}_{\text{sca}}) \cdot \boldsymbol{r}} e^{i\boldsymbol{k}_{\text{sca}} \cdot \boldsymbol{R}} \tag{4.4}$$

Now we read equation (4.4) to recognize that the position of the detector $\boldsymbol{R}$ and the direction of the scattering $\mathbf{k}_{\text{sca}}$ are set by the experimenter. On the other hand, as said above, the set of $\boldsymbol{r}$ values defines the scattering object. To determine the scattering by that object it is necessary to sum over all $\boldsymbol{r}$. This implies that the term multiplying $\boldsymbol{r}$ in the exponential is an important quantity. With this realization, we define the *scattering wave vector* as

$$\boldsymbol{q} = \boldsymbol{k}_{\text{inc}} - \boldsymbol{k}_{\text{sca}} \tag{4.5}$$

The scattering wave vector is the single most important quantity in this book. Its dimensions are inverse length, hence q^{-1} is a length scale in the direction of $\hat{\boldsymbol{q}}$. The length q^{-1} is the *length scale of the scattering experiment*. As such, the scattering

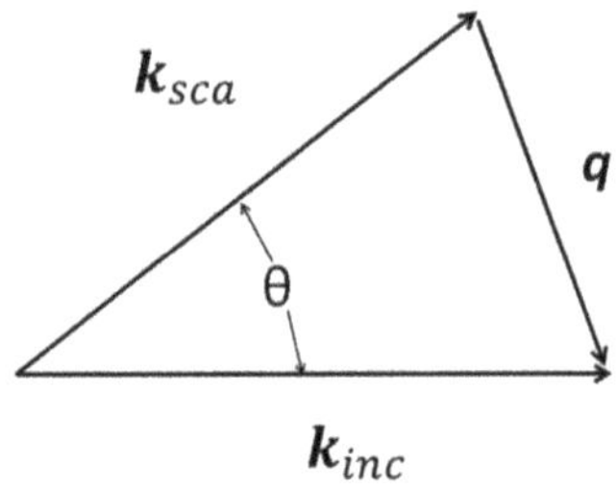

Figure 4.3. Diagram of the scattering wave vector $\boldsymbol{q}$ and its relationship to the incident and scattering wave vectors and the scattering angle θ.

experiment cannot resolve anything much less than q^{-1}. We simplify equation (4.4) to the single relevant term

$$E(\boldsymbol{q}) \sim e^{i\boldsymbol{q}\cdot\boldsymbol{r}} \tag{4.6}$$

This is the wave amplitude at the detector due to a point scatterer, or an infinitesimal volume element dV, at $\boldsymbol{r}$. A diagram of $\boldsymbol{q}$ is given in figure 4.3. From this we find the magnitude $q = |\boldsymbol{q}|$ to be

$$q = 2k\sin(\theta/2) = (4\pi/\lambda)\sin(\theta/2) \tag{4.7}$$

With this, we see that in the laboratory we can vary the magnitude of q between 0 and $2k$ simply by changing the scattering angle.

4.3 The structure factor

Above, it was shown that the phase behavior of the scattered field due to a point-like element of an object, a 'scatterer', scattering isotropically is given by equation (4.6). In equation (4.6), we only keep track of the phase since that is all we need for the development that follows.

The total scattered field for a system of N such scatterers in an object is given by

$$E_{\text{sca}}(\boldsymbol{q}) \sim \sum_{i}^{N} e^{i\boldsymbol{q}\cdot\boldsymbol{r}_i} \tag{4.8}$$

Then the scattered intensity $I = E^{*}E$ is

$$I_{\text{sca}}(\boldsymbol{q}) \sim \sum_{i}^{N}\sum_{j}^{N} e^{i\boldsymbol{q}\cdot(\boldsymbol{r}_i-\boldsymbol{r}_j)} \tag{4.9}$$

Note that $I(0) \sim N^2$. We now define the *structure factor* as[4]

[4] Some authors use a factor of N^{-1}, as we will below in section 4.5, or no factor preceding the sums in equation (4.10), so stay on your toes when reading the literature.

$$S(\boldsymbol{q}) = N^{-2}\sum_{i}^{N}\sum_{j}^{N} e^{i\boldsymbol{q}\cdot(\boldsymbol{r}_i-\boldsymbol{r}_j)} = N^{-2}\left|\sum_{i}^{N} e^{i\boldsymbol{q}\cdot\boldsymbol{r}_i}\right|^2 \tag{4.10}$$

Note that the structure factor is dimensionless. It follows that

$$S(0) = 1 \tag{4.11}$$

and from (4.9) and (4.10) the scattered intensity has the proportionality

$$I_{\text{sca}}(\boldsymbol{q}) \sim N^2 S(\boldsymbol{q}) \tag{4.12}$$

4.4 The structure factor as a Fourier transform squared of the density distribution function

It is very desirable to pass the sums over N scatterers into integrals over the volume that contains the N scatterers. This is done largely because integrals are easier to do than sums. To do this, notice that the sum operation is dimensionless, but the integral over a volume, $\int dV \equiv \int d\boldsymbol{r}$, has dimensions of volume. The difference between these is a number density, $n = N/V$. Thus, the sum over N is replaced by an integral of the position dependent number density of scatterers, i.e. we take advantage of the simple fact

$$N = \sum_{i}^{N} = \int n\,(\boldsymbol{r})d\boldsymbol{r} \tag{4.13}$$

A general description of the number density $n(\boldsymbol{r})$ is given by

$$n(\boldsymbol{r}) = \sum_{i}^{N} \delta\,(\boldsymbol{r} - \boldsymbol{r}_i) \tag{4.14}$$

where $\delta(\boldsymbol{r})$ is the three-dimensional Dirac delta function, see appendix A. This pretty little formula yields zero when the position vector $\boldsymbol{r}$ does not coincide with a position of a scatterer $\boldsymbol{r}_i$, but when it does, the delta function lights up with an area equal one singularity. Then equation (4.13) is satisfied. Moreover, sums can be replaced by integrals such as

$$\int n\,(\boldsymbol{r})e^{i\mathbf{q}\cdot\mathbf{r}}d\mathbf{r} = \sum_{i}^{N} e^{i\mathbf{q}\cdot\mathbf{r}i} \tag{4.15}$$

From this and equation (4.8) we find

$$E_{\text{sca}}(\boldsymbol{q}) \sim \int n(\boldsymbol{r})e^{i\mathbf{q}\cdot\mathbf{r}}d\mathbf{r} \tag{4.16}$$

and from equation (4.10)

$$S(\boldsymbol{q}) = N^{-2}\left|\int n(\mathbf{r})e^{i\mathbf{q}\cdot\mathbf{r}}d\mathbf{r}\right|^2 \tag{4.17}$$

Given the utility of integration, equation (4.17) augments equation (4.10) as our defining relation for the structure factor[5].

The astute reader will notice that we have used a position vector $\boldsymbol{r}$ without defining the position relative to what: we need a coordinate system with an origin for $\boldsymbol{r}$. Changing origins will not affect the physical density, but will cause a phase shift of $e^{i\boldsymbol{q}\cdot\boldsymbol{R}}$, where $\boldsymbol{R}$ is the vector displacement from one origin to another, in equation (4.16). However, this phase will not appear in the observable structure factor of equation (4.17). This arbitrary origin gives us license to pick the most convenient origin, and we shall choose the center of mass of the system of scatterers. Restating: the position vector $\boldsymbol{r}$ is relative to the center of mass of the system of scatterers.

Equations (4.14)–(4.17) above contain a great deal of useful insight upon application of further thought. The number density, equation (4.14), specifies the exact position of every point-like scatterer that comprises the object or system. Thus, it completely describes the structure of the object or system. The integrals in both equations (4.16) and (4.17) have the form of a weighted average; the weight function is the number density $n(\boldsymbol{r})$ so that the average of the exponential $e^{i\boldsymbol{q}\cdot\boldsymbol{r}}$ weighted by the structure is being found. We will write this weighted average as

$$\langle e^{i\mathbf{q}\cdot\mathbf{r}}\rangle = N^{-1}\int n(\boldsymbol{r})e^{i\mathbf{q}\cdot\mathbf{r}}d\mathbf{r} \tag{4.18}$$

where $\langle\cdots\rangle$ means 'average'. Equations (4.17) and (4.18) combine to yield

$$S(q) = |\langle e^{i\mathbf{q}\cdot\mathbf{r}}\rangle|^2 \tag{4.19}$$

Equation (4.19) shows that the structure factor is the square of the average of the scattering phase term over the structure of the system at the detector. As such, it contains a great deal of information about the structure, i.e. size, shape, configuration, etc, of the system. Hence the name 'structure factor'. But it's more! Recall our discussion of Fourier transforms. Compare the Fourier transform definition of equation (1.29) to the structure factor of equation (4.17). We see that the structure factor is also the square of the Fourier transform of the system structure, described by $n(\boldsymbol{r})$[6].

Further compare these two perspectives for the structure factor. From the structure perspective, the number density $n(\boldsymbol{r})$ acts on the phase term $e^{i\boldsymbol{q}\cdot\boldsymbol{r}}$ to create a structure weighted average phase term; whereas from the Fourier perspective, the phase term Fourier transforms the structure embodied in the position dependent number density.

4.5 The structure factor as a Fourier transform of the density autocorrelation function

The atomic structure of condensed matter systems, such as solids and liquids, can be determined by scattering waves, such as x-rays or neutrons, from the system. Inversion of the scattering as a function the scattering wave vector $\boldsymbol{q}$, which is a function of the scattering angle, yields a description of the atomic structure in terms

[5] Normalized by N^{-2}.

[6] Give or take a factor of 2π.

of a conditional probability known as the density correlation function, to be described in detail below. This method, which essentially dates back to the work of the Braggs (Sidebottom 2012), has proven to be a very powerful method which has opened our view of the atomic and molecular realm. Although the topic of this book is light scattering from particles, it is very useful to connect with the perspective, formalism and physical intuition that this previous method affords.

To begin, recall equation (4.10) and redefine the structure factor as

$$S(\boldsymbol{q}) = N^{-1}\sum_{i}^{N}\sum_{j}^{N} e^{i\boldsymbol{q}\cdot(\boldsymbol{r}_i - \boldsymbol{r}_j)} \tag{4.20}$$

This definition of $S(\boldsymbol{q})$ differs by a simple factor of N from that in (4.10), and we shall not re-designate it, but instead let the context indicate which factor is being used. Note that $S(0) = N$.

The double sum of equation (4.20) can be reconfigured into another double sum: a sum over *self-terms*, $i = j$, and distinct or *cross-term* sum, $i \neq j$. When $i = j$, then $\vec{r}_i = \vec{r}_j$ and the exponent of the self-term is zero, the exponential of which is one, so the self-term when summed over N pairs and normalized by N, is equal to one. Hence,

$$S(\boldsymbol{q}) = 1 + N^{-1}\sum_{i \neq j}^{N} e^{i\boldsymbol{q}\cdot(\boldsymbol{r}_i - \boldsymbol{r}_j)} \tag{4.21}$$

Note that there are $N^2 - N$ terms in the distinct term, so it typically dominates the self-term.

Recall from (4.10) that the sum in equation (4.21) is a double sum and then use the density operator of equation (4.14) to pass the sums to integrals to obtain

$$S(\boldsymbol{q}) = 1 + N^{-1}\left|\int n(\boldsymbol{r})e^{i\boldsymbol{q}\cdot\boldsymbol{r}}d\boldsymbol{r}\right|^2, \tag{4.22}$$

where the integral is over all space. In fact, this integral includes the self-term. However, from my reading of the literature, it appears that this fact is ignored because the second term dominates. Nevertheless, by separating the self-term out, one can keep an eye on it. The integral is, as noted above, a Fourier transform. Let this Fourier transform, FT, be written as

$$\hat{n}(\boldsymbol{q}) = FT\{n(\boldsymbol{r})\} = \int n(\boldsymbol{r})e^{i\boldsymbol{q}\cdot\boldsymbol{r}}d\boldsymbol{r} \tag{4.23}$$

Then using the convolution theorem, see section 1.4.5, we have

$$\hat{n}(\boldsymbol{q})^2 = FT\{n(\boldsymbol{r}) \otimes n(\boldsymbol{r})\} \tag{4.24}$$

With this, the structure factor of equation (4.22) becomes the Fourier transform of the convolution of the density operator with itself, viz.

$$S(\boldsymbol{q}) = 1 + N^{-1}FT\{n(\boldsymbol{r}) \otimes n(\boldsymbol{r})\} \tag{4.25a}$$

The question then arises; does this convolution have physical meaning?

Before we proceed to answer this question, let us return to the remark below equation (4.21) that 'there are $N^2 - N$ terms in the distinct term, so it typically dominates the self-term'. Indeed, it is convention to resolve into self and cross terms and so, another common convention is not to resolve these two terms. This alters equation (4.25) to

$$S(\boldsymbol{q}) = N^{-1}FT\{n(\boldsymbol{r}) \otimes n(\boldsymbol{r})\} \tag{4.25b}$$

In general, equation (4.25a) occurs in the x-ray structure factor of condensed matter literature, while the form (4.25b) occurs in the particulate light scattering literature.

4.6 The density autocorrelation function

The arrival of the convolution above forecasts the invention of a very important function for condensed matter physics, the *density autocorrelation function* (Egelstaff 1967, McQuarrie 1973). The density autocorrelation function describes a conditional probability: given a particle at one point in space what is the probability of another a given distance away? To answer this question, recognize that the density operator is an unnormalized probability of finding a particle at position $\boldsymbol{r}$. To normalize divide equation (4.14) by N. Then the probability of finding a single particle at position $\boldsymbol{r}$ is

$$P(\boldsymbol{r}) = n(\boldsymbol{r})/N = N^{-1}\sum_{i}^{N}\delta(\boldsymbol{r} - \boldsymbol{r}_i) \tag{4.26}$$

To understand physically why (4.26) is a probability, consider that the Dirac delta function is infinitely high and infinitely narrow such that its area is unity. Thus, as the position vector $\boldsymbol{r}$ sweeps through space and happens to encounter a particle at $\boldsymbol{r}_i$, equation (4.26) 'turns on' with magnitude $1/N$, which is the probability of finding one of N particles.

The joint probability of given a particle at $\boldsymbol{u}$ there is another a position $\boldsymbol{r}$ away at $\boldsymbol{u} - \boldsymbol{r}$ is the product of single particle probabilities, $P(\boldsymbol{u})P(\boldsymbol{u} - \boldsymbol{r})$. This joint probability can be averaged over the entire system by integrating over the system volume. Thus, the average joint probability (un-normalized) is

$$G(\boldsymbol{r}) = \int P(\boldsymbol{u})P(\boldsymbol{u} - \boldsymbol{r})d\boldsymbol{u} \tag{4.27}$$

Note that this is a convolution. Under the assumption of isotropy, $P(\boldsymbol{u} - \boldsymbol{r}) = P(\boldsymbol{r} - \boldsymbol{u})$. Then use of the first equality of equation (4.26) yields

$$G(\boldsymbol{r}) = N^{-2}\int n(\boldsymbol{u})n(\boldsymbol{r} - \boldsymbol{u})d\boldsymbol{u}, \tag{4.28}$$

which contains the convolution of the density operator with itself[7]. This joint probability $G(\boldsymbol{r})$ is the *density autocorrelation function* (McQuarrie 1973) with

[7] Note that equation (4.28) combined with the delta functions of equation (4.26) is problematic in that delta functions are infinitely narrow, so it would be rare for both sums of delta functions to be non-zero while integrating. A pragmatic fix for this is to coarse grain the volume element $d\boldsymbol{u}$ to be finite with size comparable to the separation between the points represented by the delta functions. Then the integral would see a continuous system much like atomic matter appears continuous on human length scales.

dimensions of inverse volume. Solving (4.28) for the convolution and then substituting into (4.25a) leads to

$$S(\boldsymbol{q}) = 1 + N \int G(\boldsymbol{r}) e^{i\boldsymbol{q}\cdot\boldsymbol{r}} d\boldsymbol{r} \tag{4.29}$$

Now recall that in obtaining equation (4.21) the self-term was extracted, so the joint probability $G(\boldsymbol{r})$ is for distinct particles, a fact not usually indicated but acknowledged as given.

Equation (4.29) shows that the structure factor is directly related to the Fourier transform of the density autocorrelation function. Thus, we have two useful ways to calculate the structure factor as either from the square of the Fourier transform of the density or from the Fourier transform of the density autocorrelation function (not squared). We have also learned the density autocorrelation function is related to the convolution of the density operator with itself. Finally, we remark that $G(\boldsymbol{r})$ could just as well be called the *pair correlation function*, since it looks at the correlation of pairs of particles.

Now is a good time to revisit figure 1.12 to see the elements of the structure factor as either the Fourier transform squared of the density function or equivalently the Fourier transform of the convolution of the density function with itself, equivalent to the density autocorrelation function.

Our development differs somewhat compared to others (Egelstaff 1967, McQuarrie 1973) that start with a statistical mechanical approach to the density autocorrelation function. Notice that the dimensions of $G(\boldsymbol{r})$ are inverse volume. The volume involved is the scattering volume, V_{sca}, the volume illuminated by the waves whose scattered intensities are described by the structure factor. To connect to other treatments define another, dimensionless density autocorrelation function as

$$g(\boldsymbol{r}) = V_{\text{sca}} G(\boldsymbol{r}) \tag{4.30}$$

Then equation (4.29) becomes

$$S(q) = 1 + n \int g(\boldsymbol{r}) e^{i\boldsymbol{q}\cdot\boldsymbol{r}} d\boldsymbol{r} \tag{4.31}$$

where n is the average density over the entire scattering volume. Note that had equation (4.25b) been used, equation (4.31) would not include the additive 1.

To continue our effort to conform, consider that in a uniform system $g(r)$ is constant and equal to 1. To see this recall $S(0) = N$ and at $\boldsymbol{q} = 0$, equation (4.31) yields $1 + N \simeq N$ for $N \gg 1$ and $g(r) = 1$. What is interesting is how deviations in the system uniformity affect the scattering. This is reflected in how $g(r)$ deviates from unity. With this point of view, we subtract and add unity from $g(\boldsymbol{r})$ in the integrand of equation (4.31) to find

$$S(\boldsymbol{q}) = 1 + n \int [g(\boldsymbol{r}) - 1 + 1] e^{i\boldsymbol{q}\cdot\boldsymbol{r}} d\boldsymbol{r} \tag{4.32}$$

The third term in the integrand is a three-dimensional Dirac delta function. Thus

$$S(\boldsymbol{q}) = 1 + n(2\pi)^3\delta(\boldsymbol{q}) + n\int[g(\boldsymbol{r}) - 1]e^{i\boldsymbol{q}\cdot\boldsymbol{r}}d\boldsymbol{r} \tag{4.33}$$

The three terms in equation (4.33) from left to right represent the small self-term, the structure factor of the entire scattering volume, which is relatively very large in real space so very narrow in q-space (see section 5.5.2 for a description of a finite scattering volume), and the Fourier transform of the non-uniform part of the system density autocorrelation function.

It is good now to recall that equation (4.33) was derived under the assumption of weak scattering. This means the scattering volume sees only the incident wave, i.e. there is no internal coupling or multiple scattering between one part of the scattering volume and another. Weak, non-multiple scattering often occurs with x-rays and neutrons. The scattering of x-rays is relatively weak because, unlike visible light, the refractive index for condensed matter at x-ray wavelengths is very close to 1.0. Moreover, because these radiations have wavelengths on the order of a few angstroms, they are ideal for probing the structure of condensed atomic and molecular systems. The form of equation (4.33) is especially useful for amorphous systems, such as liquids, glasses and gels. Crystalline materials are more conveniently described by a different formulation often referred to as x-ray or neutron diffraction. Of course, it is all diffraction, just a different formalism.

An example of the application of equation (4.33) is shown in figure 4.4. The upper part of this figure shows the structure factor for scattering of x-rays from liquid argon (Yarnell et al 1973). The lower figure shows the Fourier inversion of the structure factor to obtain the density autocorrelation function. The density autocorrelation function shows a strong anti-correlation at small r. This is due to the hard-sphere-like repulsion between the argon atoms when they try to approach to center-to-center distances less than their diameter. The ensuing peak at larger r just beyond this anti-correlation indicates a positive correlation interpreted as the nearest neighbor shell of atoms. It can be shown that the area under this peak is related to the number of atoms within this shell. Subsequently, at yet larger r, minor ripples occur with yet more neighboring shells, but quickly the ripples, which indicate order, die out to indicate that there is no long-range order in the liquid.

4.7 And another form for the structure factor

We return to equation (4.17) with normalization by N^{-2}. As shown immediately above, one can use the convolution theorem to convert the square of the Fourier transform in equation (4.17) to a Fourier transform of a convolution, which is a density autocorrelation function. Moreover, if we do not extract any self-terms, we can achieve a formula very similar to equation (4.29) to be

$$S(\boldsymbol{q}) = \int G(\boldsymbol{r})e^{i\boldsymbol{q}\cdot\boldsymbol{r}}d\boldsymbol{r} \tag{4.34}$$

Recall that $d\boldsymbol{r} = r^2 dr d\varphi d(\cos\theta)$. We now assume isotropic symmetry, then $G(\boldsymbol{r})$ becomes $G(r)$, direction independent. With no φ functionality in $G(r)$ the integration

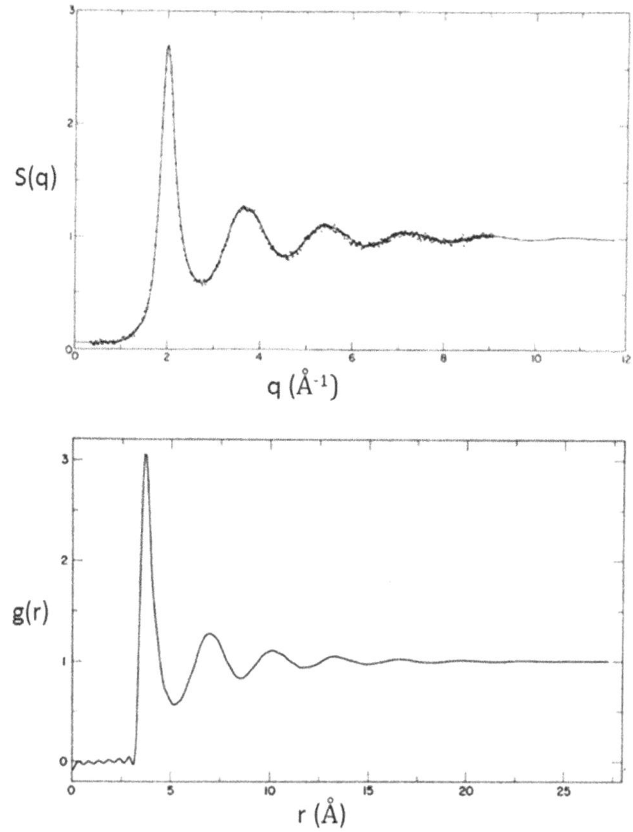

Figure 4.4. (Top) The structure factor for scattering of x-rays from liquid argon. The line is a fit to the data. (Bottom) The Fourier inversion of the structure factor to obtain the density autocorrelation function. Reprinted figure with permission from Yarnell et al (1973). Copyright (1973) by the American Physical Society.

over φ yields 2π. For the $d(\cos\theta)$ integral we pick for convenience $\boldsymbol{q}$ along the z-axis. Then $\boldsymbol{q}\cdot\boldsymbol{r} = qr\cos\theta$, and the integral over $\cos\theta$ leads to

$$S(q) = 4\pi\int G(r)\left[\frac{\sin qr}{qr}\right]r^2dr \tag{4.35}$$

Notice that the sinc(u) function, equation (1.34), has appeared in the integrand. Equation (4.35) is a very useful result used throughout the literature when spherical, isotropic symmetry applies.

4.8 The Guinier regime

The structure factor has some useful limits when the scattering length scale q^{-1} is either much less than, comparable to, or much greater than the size or a size scale of the scattering particle or system. The limit of small q hence q^{-1} being much greater than any particle length scale is trivial and already given in equation (4.11), $S(0) = 1$, a constant independent of q, hence θ. The situation at slightly larger q when q^{-1} is comparable to the particle size is very useful and discussed here.

If the particle has spherical symmetry, or is rotationally randomly averaged over the time of the measurement, or the scattering takes place from an ensemble of similar particles at random orientations, then there is no angular functionality in the density distribution. Then the integral in equation (4.16) becomes

$$E_{\text{sca}}(\boldsymbol{q}) \sim \int n(\mathbf{r})e^{i\mathbf{q}\cdot\mathbf{r}}d\mathbf{r} = \int n(r)e^{i\mathbf{q}\cdot\mathbf{r}}r^2drd(\cos\theta)d\varphi \tag{4.36}$$

There is no φ functionality in $n(r)$ and any φ functionality due to polarization, or the lack of it (scalar waves) leads to a constant, $8\pi/3$ or 2π, respectively, which for us here is inconsequential. The $d(\cos\theta)$ integral yields the sinc(u) function as it did in the derivation of equation (4.35), above, to obtain

$$E_{\text{sca}}(q) \sim 4\pi\int n(r)\left[\frac{\sin qr}{qr}\right]r^2dr \tag{4.37}$$

Now expand the sine function for small qr as

$$\sin qr \simeq qr - \frac{1}{3!}(qr)^3 \tag{4.38}$$

Substitution into equation (4.37) use of $N = \int n(\mathbf{r})d\mathbf{r}$ and not concerned with constants yields

$$E_{\text{sca}}(q) \sim N(1 - q^2\langle r^2\rangle/6) \tag{4.39}$$

where

$$\langle r^2\rangle = \frac{4\pi}{N}\int[n(r)r^2]r^2dr \tag{4.40}$$

and where $\langle\cdots\rangle$ means 'average'. Equation (4.40) may not look like an average, but take it 'backwards' with $4\pi r^2dr = d\mathbf{r}$ and $N = \int n(\mathbf{r})d\mathbf{r}$ to rewrite it as[8]

$$\langle r^2\rangle = \frac{\int n(\mathbf{r})r^2d\mathbf{r}}{\int n(\mathbf{r})d\mathbf{r}} \tag{4.41}$$

Equation (4.41) shows that $\langle r^2\rangle$ is indeed an average, weighted by the number density distribution, of the square of the position vectors relative to the center of mass of the entire system of scatterers. Note that integral in equation (4.41) is over the entire volume of the system or particle. Thus, it is a volume average; it is less than the perimeter size. Since it is quite common and useful, it has a name; it is the *Radius of Gyration*, R_g, of the system of scatterers and hence of the rotationally averaged system or particle itself

[8] Had we used linearly polarized waves and the factor of $8\pi/3$, the result would have been the same.

$$R_g = \sqrt{\langle r^2 \rangle} \tag{4.42}$$

With equation (4.42), we see that R_g is a root-mean-square, 'rms', size. For example, a sphere of radius R has a radius of gyration of $R_g = \sqrt{3/5}R$.

Combining equations (4.39) and (4.42) with the formulae for the structure factor, equations (4.16) and (4.17), with $qR_g < 1$, yields

$$S(q) = (1 - q^2R_g^2/3) \text{ for } qR_g < 1 \tag{4.43}$$

which is known as the *Guinier equation* (Guinier and Fournet 1955). This result is very useful for determining the size of whatever is scattering the radiation.

4.9 The structure factor of the sphere

We use equation (4.17) to find the structure factor for a sphere. The density distribution for a uniform sphere is

$$n(r) = n,\ r \leqslant R \tag{4.44a}$$

$$= 0,\ r > R \tag{4.44b}$$

where R is the radius of the sphere. We choose, without loss of generality, a coordinate system such that the polar angle θ is the angle between $\vec{r}$ and $\vec{q}$ so that $\vec{q} \cdot \vec{r} = qr\cos\theta$. Then the integral in the structure factor equation (4.17), which we will call E', becomes

$$E' = n\int_0^R \int_{-1}^1 \int_0^{2\pi} e^{i\mathrm{q}\cdot\mathrm{r}} r^2\, dr\, d(\cos)d\varnothing \tag{4.45}$$

Performing the angle integrals and use of Euler's formula, equation (1.15), yields

$$E' = \frac{4\pi n}{q}\int_0^R r \sin qr\, dr \tag{4.46}$$

Integrate by parts to obtain

$$E' = 4\pi n\left[\frac{\sin qR}{q^3} - \frac{R\cos qR}{q^2}\right] \tag{4.47}$$

Now recognize that the number density of scatterers is $n = N/V$ and $V = (4\pi/3)R^3$ is the volume of the sphere. Substitution of these yields

$$E' = \frac{3N}{(qR^3)}[\sin qR - qR\cos qR] \tag{4.48}$$

Then by equation (4.17)

$$S(q) = \left[\frac{3}{u^3}(\sin \mathrm{u} - u\cos u)\right]^2 \tag{4.49}$$

where

$$u = qR \tag{4.50}$$

a dimensionless variable.

We will call equation (4.49) the *Rayleigh–Debye–Gans* formula for the structure factor for a sphere. It also has the exact functional form for wave diffraction from a sphere. In section 7.1.2, we will describe how this applies to actual scattering from a sphere.

We remark that this derivation contains both self and cross terms and hence is consistent with equations (4.25b) and (4.35).

4.10 The structure factor as diffraction: generalization to arbitrary dimension

As scientists, we gain an aesthetic pleasure when a wide variety of phenomena are seen to be connected. Here we connect single slit diffraction of our undergraduate training with the spherical particle structure factors recently learned (perhaps), above. Beyond the aesthetic, connections have practical importance because they broaden our physical intuition.

The structure factor of an object is simply the far-field, Fraunhofer diffraction pattern for that object. Single slit diffraction is the structure factor for a homogeneous, one-dimensional, square pulse. Recall that pulses (obstacles) and apertures have identical far-field diffraction via Babinet's principle. In a similar manner, the structure factor for a sphere is diffraction from a homogeneous, three-dimensional, spherical pulse. We see that our usage of the term 'pulse' connotes an abrupt edge. In two dimensions, a circular obstacle (the Babinet brother of a circular aperture) is a homogeneous pulse, and its diffraction pattern is well known in optics textbooks (circular aperture diffraction and the Airy pattern (Hecht 2002)).Thus, from a general point of view, we have structure factors, or equivalently, Fraunhofer diffraction patterns, for one-, two-, and three-dimensional pulses as

$$S(q) = \left[\frac{\sin u}{u}\right]^2 = [j_0(u)]^2 \text{ 1d} \tag{4.58a}$$

$$S(q) = \left[\frac{2J_1(u)}{u}\right]^2 \text{ 2d} \tag{4.58b}$$

$$S(q) = \left[\frac{3}{u^3}(\sin u - u\cos u)\right]^2 = \left[\frac{3j_1(u)}{u}\right]^2, \text{ 3d} \tag{4.58c}$$

respectively. In the above equations, $j_0(u)$ and $j_1(u)$ are the first and second spherical Bessel functions, respectively, and $J_1(u)$ is a Bessel function of the first kind of order one (Abramowitz and Stegun 1972). The dimensionless variable u in one and two dimensions is $u = q'R = kR\sin\theta$ and in three dimensions is $u = qR = 2kR\sin(\theta/2)$ (these are essentially equal for small θ), where R is now, from a general perspective, the half width of the pulse or aperture.

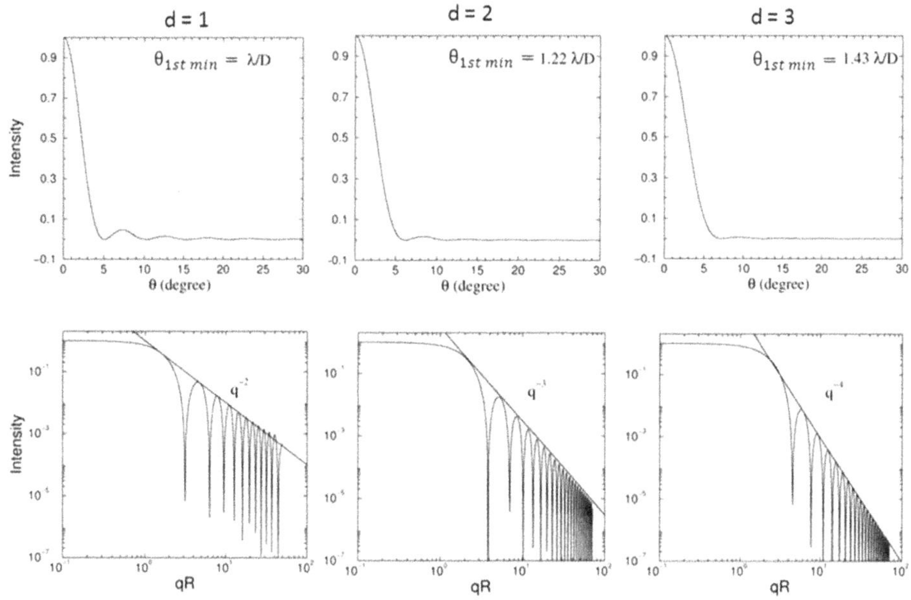

Figure 4.5. Plots of one-, two- and three-dimensional structure factors, i.e. diffraction patterns, for a *d*-dimensional 'spherical' pulse. Both linear, as a function of scattering angle θ, and double logarithmic, as a function of $u = qR$ where $R = D/2$ is the pulse half width (*d*-dimensional sphere radius) are shown. By Babinet's principle, section 1.5.4 and Hecht (2002), these diffraction patterns apply to both obstacles and apertures of the same shape and size.

Each structure factor has a Guinier regime, which can be written in a general form for d dimensions as

$$S(q) \simeq 1 - u^2/(d + 2) = 1 - q^2 R_g^2/d \tag{4.59}$$

where q is either q' or q depending on the dimension[9]. The first minima are at $\theta = \lambda/2R$, $1.22\lambda/2R$ and $1.43\lambda/2R$, for dimensions of $d = 1$, 2 or 3, respectively. Furthermore, each structure factor has a Porod regime (Guinier and Fournet 1955) with an envelope that obeys the power law

$$S(q) \simeq u^{-(d+1)} \tag{4.60}$$

These structure factors, (diffraction patterns) are plotted both as a function of scattering angle θ (diffraction angle) on linear graphs, and as a function of $u = qR$ on double log graphs in figure 4.5. There we see the common characteristics that unify these different diffraction phenomena.

Figure 4.5 is an example of elegance in science. It is also our first example of Q-space analysis.

[9] Note that the radius of gyration for a d-dimensional sphere of radius R is $R_g = [d/(d + 2)]^{1/2} R$.

References

Abramowitz M and Irene A S 1972 *Handbook of Mathematical Functions, with Formulas, Graphs, and Mathematical Tables* (New York: Dover)
Born M and Wolf E 1975 *Principles of Optics: Electromagnetic Theory of Propagation, Interference, and Diffraction of Light* (Oxford: Pergamon)
Egelstaff P A 1967 *An Introduction to the Liquid State* (London: Academic)
Guinier A and Fournet G é 1955 *Small-angle Scattering of X-rays* (New York: Wiley)
Hecht E 2002 *Optics* (Reading, MA: Addison-Wesley)
McQuarrie D A 1973 *Statistical Thermodynamics* (New York: Harper & Row)
Sidebottom D L 2012 *Fundamentals of Condensed Matter and Crystalline Physics: An Introduction for Students of Physics and Materials Science* (Cambridge: Cambridge University Press)
Yarnell J L, Katz M J, Wenzel R G and Koenig S H 1973 Structure factor and radial-distribution function for liquid argon at 85 K *Phys. Rev.* A **7** 2130–44

Chapter 5

The scaling approach to the structure factor

We have seen that the diffractive scattering of waves into the far-field from an object is equivalent to Fraunhofer diffraction by the object, which itself is the Fourier transform, squared, of the structure of the scattering object. Each part of the scattering object sends a wave to the far field described by $e^{i\boldsymbol{q}\cdot\boldsymbol{r}}$, a form that can be represented by a phasor of unity magnitude in the complex plane at an angle of $\boldsymbol{q}\cdot\boldsymbol{r}$ from the real axis. Scattering, diffraction and Fourier transformation are all represented by vector summing these phasors as the vector $\boldsymbol{r}$ sweeps across the object.

The previous chapter showed that the structure factor could be calculated by summing the phasors $e^{i\boldsymbol{q}\cdot\boldsymbol{r}}$, usually with an integral. Useful results were obtained. Thus, we are well armed to calculate any structure factor we please within our abilities to solve integrals. However, one neat thing about diffraction is that it teaches us, if we are willing, to compare length scales. For example, where else could a diffraction minimum be other than the simplest ratio of the only length scales of the system, λ/w or λ/R, to within a factor of unity? Moreover, length scales are subtlety couched in figures 1.12 and 1.13. There, advantage was taken of the fact that the dimension of the scattering wave vector q is inverse length; hence inverse q, q^{-1}, has dimensions of length. Indeed, the phasor $e^{i\boldsymbol{q}\cdot\boldsymbol{r}}$ shows us that it will not change much unless $\boldsymbol{r}$ changes in a direction parallel to $\boldsymbol{q}$ a distance significant relative to q^{-1}. Figures 1.12 and 1.13 make use of this fact and demonstrate that the relevant comparison of length scales is q^{-1} compared to the size of the slit.

This comparison of scales, the scale of the diffracting object and the scale of the wave diffraction, q^{-1}, leads to a method that we call the *scaling approach* (Oh and Sorensen 1999), the subject of this chapter. It is strictly for waves; we remain free of the complications of electric polarizability. However, as should be becoming apparent, understanding waves is the foundation for our tack towards understanding light scattering.

The q^{-1} length scale is set by the experimenter via application of a source with wavelength λ, and then, with much more facility, variation of the scattering angle. The term 'scaling approach' is used because the method essentially compares the length scale of the diffraction, i.e. the scattering, to the length scales of the scattering object. The scaling approach is applicable to diffraction, Fourier transformation and structure factor determination. For light scattering the results apply under the assumption of no internal coupling between the different parts of the scattering object (one could call this internal multiple scattering), which is to say that scattering from one part of the scattering object does not affect the scattering by another part. Furthermore, realize here that a scattering object could be a single particle, an aggregate of particles, an aerosol or a colloid of many particles, or a condensed matter material composed of many atoms and molecules. The conceptual simplicity of the scaling approach leads to physical understanding and insight for the scattering process and the interpretation of the structure factor.

5.1 The scaling approach concepts

To gain an intuition for the summing of phasors involved in diffractive scattering (or for that matter, Fourier transformation), return to the canonical problem of diffraction from a single slit discussed in chapter 1. Figure 1.12 of that chapter showed the profile of a slit and the phasors of the waves within that slit when the diffraction angle was $\theta = \lambda/w$ to achieve the first minimum for diffraction. The minimum occurs because, as shown in the figure, the sum of all the phasors is zero. Figure 1.13 continues the analysis of phasors for diffraction by a slit by showing that, at an arbitrary diffraction angle such that $\theta \gg \lambda/w$, there are many $2\pi/q$ regions each with a total phasor addition of zero. This implies that *the uniform interior of the slit contributes nothing to the diffracted wave; only the edges of the slit make a non-zero contribution to the diffracted wave.*

Figure 5.1 takes the argument of figure 1.13 further. Figure 5.1 again considers a situation when q is large such that $2\pi/q$ is smaller than the width of the slit w. Thus, many $2\pi/q$ regions fill the transmission profile, each one contributing exactly zero to the diffracted field. Also shown in figure 5.1 is what we will call a '*q-region*' of extent q^{-1}. Within this q-region, figure 5.1 shows that the phasors are approximately in phase.

The lessons learned from the analysis of figures 1.12, 1.13 and 5.1 above lead to what we (Oh and Sorensen 1999, Sorensen 2001) have called *the scaling approach* to determine the structure factor (the diffraction, the Fourier transform) of an object. This approach starts with the concept:

The inverse of the scattering wave vector magnitude q^{-1} is the length scale of the scattering experiment.

Notice that I use the term 'scattering' in this important statement because this is a book on scattering, and we are headed in that direction. However, the various near synonyms apply such as 'diffraction', 'diffractive scattering', 'structure factor' and 'Fourier transform'.

5.2 The scaling approach rules

Consider a system of N identical point scatterers, such as drawn in figure 4.1, each of which scatter waves isotropically. The scatterers are distributed uniformly throughout the system volume. The total scattered amplitude in the far field is proportional

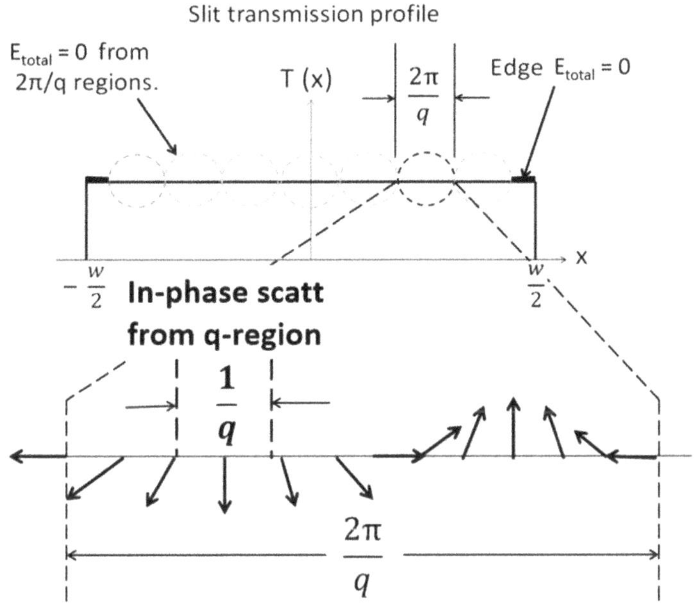

Figure 5.1. Single slit transmission profile with regions of size $2\pi/q$ superimposed upon it. The phasor diagram below shows that the total wave from each $2\pi/q$ region is zero. Consequently, only the edges of the slit make a non-zero contribution to the total diffracted wave. Note, however, that in smaller regions of size q^{-1} the phasors add approximately in phase.

to the sum of the phasors $e^{i\boldsymbol{q}\cdot\boldsymbol{r}}$; hence the scattered intensity, which is proportional to the structure factor, is given by

$$I(\boldsymbol{q}) \sim N^2 S(\boldsymbol{q}) = \left| \sum_i^N e^{i\boldsymbol{q}\cdot\boldsymbol{r}_i} \right|^2 \tag{5.1}$$

In what follows, either the intensity $I(\boldsymbol{q})$ or the structure factor $S(\boldsymbol{q})$ will be described depending upon the application. Note that the intensity possibly depends on a number of other things such as incident intensity, refractive index, number of scattering entities, distance to the detector, etc, all of which are not relevant to the discussion that follows. Here our purpose is to display the close connection of the scattered intensity to the structure factor and will often refer to $I(q) = N^2 S(q)$ as the intensity structure factor[1].

The scaling approach determines the proportionalities[2] of this intensity by codifying the lessons learned above into the following *scaling approach rules*:

1. If N scatterers are within q^{-1} of each other, the phase of the N scattered waves will be essentially the same; hence the phasors will add constructively.

[1] Also recall from sections 4.4 and 4.5 that the N^2-normalization in equation (5.1) to define the structure factor $S(q)$ is not universal across the literature.

[2] Relax! We are seeking proportionalities and functionalities. Factors of two, etc, are ignored in the scaling approach.

This leads to a total scattered amplitude proportional to N, thus the total scattered intensity proportional to N^2. We now define a *q-region* as that region for which the scatterers are within q^{-1} of each other. A q-region is q^{-1} in size.

2. If N scatterers have all pairs separated by distances greater than q^{-1}, the phases of the N scattered waves will be random, hence the phasors will add randomly. This random walk of phasors leads to a total scattered amplitude proportional to $\sqrt{N}$, thus the total scattered intensity will be proportional to N. Perhaps said more simply: when waves scatter with random phases, one need only add their intensities to get the total intensity.
3. Uniform, continuous regions contribute nothing to the scattered intensity. This can be augmented by stressing that only fluctuations in density of the scatterers lead to a total non-zero scattering, and it is very useful to know that fluctuations include edges.

5.3 The scaling approach applied to various shapes

As a first application of the scaling approach, consider a system of scatterers in a d-dimensional, spherical region of radius R. I quickly remark that the sphericity of the system can be a result of the system actually having a spherical geometry or a system that is orientationally averaged. Given the spherical symmetry, only the magnitude of the scattering wave vector is needed. Let the density of the scatterers be uniform by placing them on a regular lattice with $2a$ the lattice spacing, as drawn in figure 5.2. Then the only non-uniformity in density of the scatterers is on the surface of the system in the form of a discontinuity. To apply the scaling approach, cover the system with imaginary q-regions of size q^{-1}. All the scatterers in the same q-region yield scattered waves at the detector that are in phase with each other. Then the intensity of the waves at the detector from a q-region is (rule 1, above)

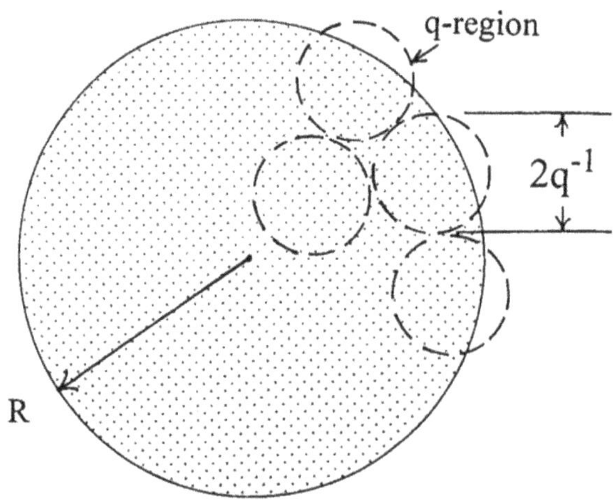

Figure 5.2. Drawing of a system of point scatterers in a 'spherical' system of scatterers of size R. Also drawn are a few q-regions of size $2q^{-1}$. From Sorensen (2001).

$$I(q) \sim N_q^2 \tag{5.2}$$

where N_q is the number of scatterers in the q-region, proportional to the volume of the q-region. There are many such q-regions in the system of radius R, but only those on the surface of the system contribute to the scattering, because it is only on the surface of this uniform system where there is a fluctuation in density (rule 3, above). Let n_q be the number of q-regions of size q^{-1} needed to cover the surface of the system. Because of their size, they must all be separated by distances greater than q^{-1}, thus their scattered waves add randomly (rule 2, above) to yield

$$I(q) \sim n_q \tag{5.3}$$

The combination of equations (5.2) and (5.3) yields

$$I(q) \sim n_q N_q^2 \tag{5.4}$$

Now we apply these general formulae to the case of the d-dimensional sphere. For this example there are two length scales for the system of scatterers, so there are three relevant regimes for q.

(a) $q < R^{-1}$. At very small q such that $q < R^{-1}$, or equivalently $q^{-1} > R$, the length scale of the scattering experiment q^{-1} is larger than any length scale of the system. Then the entire system is enclosed within a single q-region. Thus, $n_q = 1$ and $N_q = N$ and equation (5.4) yields

$$I(q) \sim N^2 \tag{5.5}$$

For this very small q the entire system scatters waves in phase to the detector to yield a total field proportional to the total number N of scatterers in the system, and thus the scattered intensity is proportional to the N^2. Note that N is proportional to the volume of the system, so a volume squared is predicted as well.

It is also very important to notice how q^{-1} represents the resolution of the scattering; that is, the scattering can't see features smaller than q^{-1}! From this perspective, all N scatterers in the volume of the system of scatterers are on the surface of the system. Furthermore, the scattered intensity does not change as q varies so long as q^{-1} remains larger than R.

(b) $R^{-1} < q < a^{-1}$. Now q^{-1} is less than the overall size of the system. Thus, one can envision the system covered throughout its volume by smaller imaginary q-regions of size q^{-1}, as shown in figure 5.2. The q-regions in the interior of the system contribute nothing to the scattered field because the scatterers are uniform in the interior. This is the application of rule 3, above, and consistent with the lessons we learned from the single slit, recall figures 5.1, 1.12 and 1.13. On the other hand, q-regions on the surface have discontinuities, hence they contribute to the scattered field.

The calculation here will consider the most general scaling of material, whether compact, like a solid object, or ramified, like an aggregate, by considering how the number of point scatterers that comprise the scattering object scale with linear dimension. Thus, we use the following scaling relations

$$N = k_m(R/a)^{D_m} \tag{5.6}$$

$$N_s = k_s(R/a)^{D_s} \tag{5.7}$$

These define both the coefficients k_m and k_s, of order unity dependent upon the point scatterer lattice, and the *mass and surface scaling dimensions, D_m and D_s, respectively*. For example, a sphere in $d = 3$ dimensional space (i.e. a normal sphere) has $D_m = d = 3$ and $D_s = d - 1 = 2$.

Given the scaling relations, the number of scatterers in a q-region, N_q, can be determined from the total N in the system of scatterers and the ratio of the length scale of the q-region, q^{-1}, to the length scale of the system, R, raised to the mass scaling dimension power, i.e.

$$N_q \sim N(q^{-1}/R)^{D_m} \tag{5.8}$$

A similar argument finds the number of q-regions on the surface of the system to be proportional to the ratio of R to q^{-1} raised to the surface scaling dimension power, i.e.

$$n_q \sim (R/q^{-1})^{D_s} \tag{5.9}$$

When applied to equation (5.4), these yield

$$I(q) \sim N^2(qR)^{-2D_m + D_s} \tag{5.10}$$

Note the exponent $-2D_m + D_s$. The size of the system is represented by both N and R. To eliminate R for N in equation (5.10) use the scaling relations (5.6) and (5.7) to find

$$I(q) \sim k_0 N_s(qa)^{-2D_m + D_s} \tag{5.11}$$

where $k_0 = k_m^2/k_s$. Note the surface dependence with N_s.

(c) $R^{-1} < a^{-1} < q$. In this regime, all the point scatterers are farther apart than the size of the q-regions, q^{-1}, so all the scattered waves have random phases at the detector, and the intensities add, but only those on the surface contribute to the scattering. Thus, $I(q) \sim N_s$. Continuity with equation (5.11) at $q = a^{-1}$ implies

$$I(q) \sim k_0 N_s \tag{5.12}$$

In summary

$$I(q) \sim N^2 \text{ for } q < R^{-1} \tag{5.13a}$$

$$\sim k_0 N_s(qa)^{-2D_m + D_s} \text{ for } R^{-1} < q < a^{-1} \tag{5.13b}$$

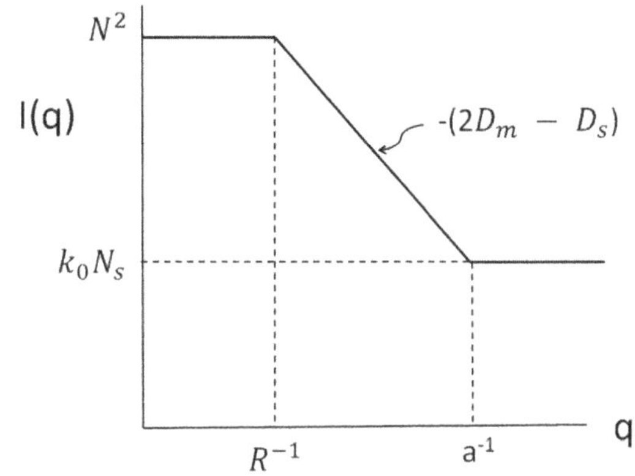

Figure 5.3. Log–log plot of the scattered intensity, normalized to be equal to N^2 at small q versus q, equations (5.13), for a system of N point scatterers separated by $2a$ in a 'spherical' region of size R with mass and surface scaling dimensions D_m and D_s, respectively. There are N_s scatterers on the surface of the region. The slope of the plot in the range $R^{-1} < q < a^{-1}$ is $-(2D_m - D_s)$. From Oh and Sorensen (1999) (Copyright © 1999, Kluwer Academic Publishers). With permission of Springer.

$$\sim k_0 N_s \text{ for } a^{-1} < q \tag{5.13c}$$

It is useful to rewrite these equations because k_0 is dependent on the lattice geometry of the point scatterers and N_s is an unusual parameter. (Note that the product k_0N_s is not lattice dependent.) Use of equations (5.6) and (5.7) yields

$$I(q) \sim N^2 \text{ for } q < R^{-1} \tag{5.14a}$$

$$\sim N^2(qR)^{-2D_m + D_s} \text{ for } R^{-1} < q < a^{-1} \tag{5.14b}$$

$$\sim N^2(R/a)^{-2D_m + D_s} \text{ for } a^{-1} < q \tag{5.14c}$$

Figure 5.3 shows the generic behavior of equations (5.13). Equations (5.14) show that a graph of the structure factor $S(q) = N^{-2}I(q)$ vs qR is universal for all sizes R when $q < a^{-1}$. This fact and the simplicity of equations (5.14) are an advantage over equations (5.13). However, equations (5.13) explicitly show the volume and surface regimes and are more facile for considering complex examples. Also note that the predicted scattering is constant at the largest q values such that $q > a^{-1}$. This occurs because the length scale of the scattering, q^{-1}, is smaller than any length scale in the system, so making q^{-1} smaller has no effect; there is nothing left to resolve.

The scaling result is general and can be applied to diffractive scattering for a number of physical situations.

5.3.1 A uniform sphere

Let $a \to 0$ and $D_m = d$, $D_s = d - 1$ where d is the spatial dimension. By equation (5.13a) when $q < R^{-1}$, the forward scattering lobe, the intensity of the scattering goes as $N^2 \sim R^6$ and is not a function of q, hence not a function of the scattering angle. Because the scattering cross section must have dimensions of length squared, the wavelength dependence in this regime must be λ^{-4} (this dimensional argument will be made again in the chapter on Rayleigh scattering, chapter 6). Thus, this isotropic, R^6/λ^4, forward scattering regime is aptly termed the *Rayleigh regime* of scattering for *any* sized particle (remember, however, here we are only considering the diffraction limit).

Next, by equation (5.13b), when $R^{-1} < q < a^{-1}$, $S(q) \sim I(q) \sim N_s q^{-(d+1)} \sim N_s q^{-4}$ for $d = 3$, a result known as Porod's law (Porod 1951, Guinier and Fournet 1955). Note both the q functionality and the N_s term; the latter reminding us that when $q^{-1} < R$, only the surface (of q^{-1} depth) contributes to the scattering in this regime. If the solid sphere has a fractally rough surface, $d > D_s > d - 1$ and $I(q) \sim q^{-6+D_s}$ for d = 3 (Martin and Hurd 1987).

5.3.2 A uniform circular obstacle

For example, consider a circular disk. Let the disk be infinitely thin (or more realistically much thinner than its radius). Then the mass of the disk scales with its area, hence $D_m = 2$. With this perspective, its surface becomes the edge, the circumference of the disc, with surface scaling dimension $D_s = 1$. Let the light be incident perpendicular to the plane of the disk. Finally, let $a \rightarrow 0$. Then, following the same reasoning as for the uniform sphere, by equation (5.13a) for small q when $q < R^{-1}$, the intensity of the scattering goes as the mass squared, $N^2 \sim R^{2D_m} \sim R^4$, and is not a function of q, nor the scattering angle; this is the forward scattering lobe. Since the scattering cross section must have dimensions of length squared, the wavelength dependence in this regime must be λ^{-2}, and we again have a Rayleigh regime. Furthermore, by equation (5.13b) when $R^{-1} < q < a^{-1}$, $S(q) \sim I(q) \sim N_s q^{(2D_m - D_s)} \sim N_s q^{-3}$, a Porod regime for the circular obstacle. Now recognize that this reasoning applies to shapes other than circular.

5.3.3 A uniform linear obstacle

We follow the same reasoning as above. An idealized linear object is infinitely long yet has a mass linearly proportional to its half width R. Thus, $D_m = 1$. The surface of the obstacle, its edge, does not change with the width, hence $D_s = 0$. Now by equation (5.13a) when $q < R^{-1}$, the intensity of the scattering goes as the mass squared, $N^2 \sim R^{2D_m} \sim R^2$. Since the scattering cross section must have dimensions of length squared, there is no wavelength dependence, and we yet again have a Rayleigh regime. Furthermore, by equation (5.13b) when $R^{-1} < q < a^{-1}$, $S(q) \sim I(q) \sim N_s q^{-(2D_m - D_s)} \sim N_s q^{-2}$, a Porod regime for the linear obstacle.

Now return to figure 4.6 that shows the complete scattered intensity (envelopes and ripples), i.e. the diffraction patterns for one-, two- and three-dimensional 'pulses', better known as the single slit, circular aperture and sphere, respectively. Remember Babinet's principle (Hecht 2002) in that, in the Fraunhofer limit, the diffracted or scattered intensity from an obstacle (a pulse in Fourier transform lingo) and a same-shaped aperture are the same. There, along the top row, the classic representations of the single slit and circular aperture are seen, well known (I hope!) to all freshman or sophomore engineering and science majors. On the bottom row, each has plotted double logarithmically intensity versus the dimensionless parameter qR. Now the plots look like figure 5.3, the ripples notwithstanding. This bottom row shows the commonality between these differently dimensioned systems. Figure 4.6 also reminds us that the sphere, circular obstacle and linear obstacle above can be thought of as objects with dimensionality of $d = 3$, 2 and 1, respectively. With this perspective, equation (4.60) gives a general description of the Porod regime, consistent with the results above.

The $d = 3$ spherical diffraction pattern has become known as the Rayleigh–Debye–Gans (RDG) result[3]. It is valid when the internal coupling parameter $\rho' = 2kR[(m^2 - 1)/(m^2 + 2)]$ is small. This parameter is discussed extensively in section 6.1.2.2.

5.3.4 A long circular cylinder—orientationally averaged

Consider a right circular cylinder of length L and radius R with $L \gg R$, as drawn in figure 5.4. Diffractive scattering from a single cylinder orientationally averaged or an ensemble of cylinders at random orientations can be determined by the scaling approach. Unlike the 'spheres', above, the cylinder has two length scales, not one. Again, let $a \to 0$. For the smallest q, hence the largest q-regions, such that $q < L^{-1}$, equation (5.4) predicts a scattered intensity that goes as the square of the volume of the entire cylinder, thus $I \sim N^2 \sim R^4L^2$, and is not a function of q, nor the scattering angle. Proceeding to larger q, when $R^{-1} > q > L^{-1}$, one can construct q-regimes as shown on the left side of figure 5.4. The number of these q-regions is $n_q \sim L/(q^{-1}) = Lq$. Note how the q-regions divide the cylindrical rod into cylindrical segments q^{-1}

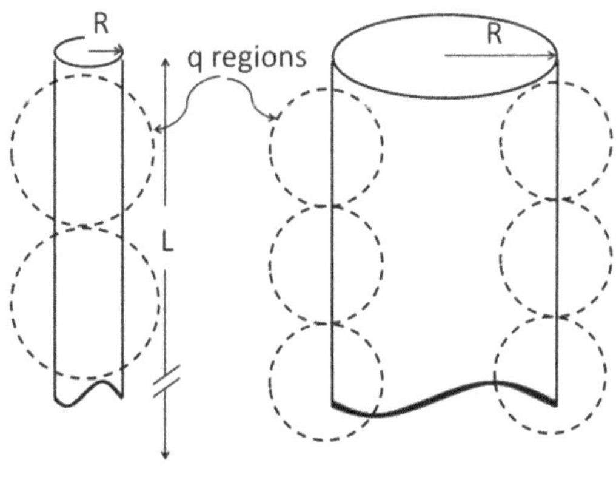

Figure 5.4. Scaling analysis of an orientationally averaged long cylinder, $L \gg R$. On the left are q-regions smaller than L but larger than R, $R < q^{-1} < L$; on the right are q-regions smaller than both L and R, $q^{-1} < R < L$.

[3] van de Hulst (1957) calls it 'Rayleigh–Gans', Kerker (1969) calls it 'Rayleigh–Debye', Bohren and Huffman (1983) provide an interesting discussion with their trenchant humor of possible terms and conclude in favor of 'Rayleigh–Gans', and Mishchenko *et al* (2002) acknowledge both these terms but use 'Rayleigh–Gans'. With the advent of fractal aggregate light scattering, the term 'Rayleigh–Debye–Gans' appeared with Koylu and Faeth (1994) and is now used throughout that literature.

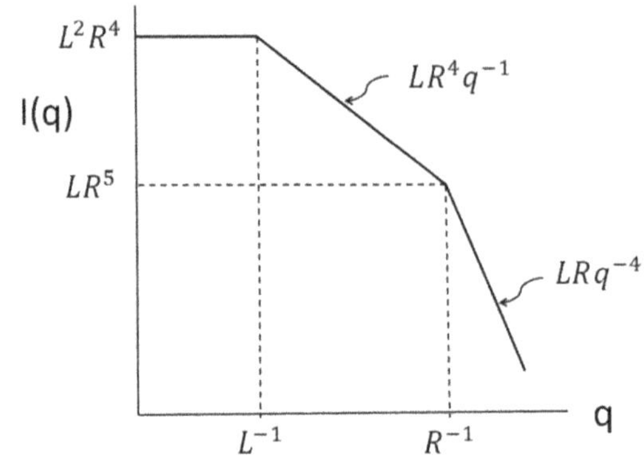

Figure 5.5. Log–log plot of the scattered intensity functionalities with q for an orientationally averaged long cylinder of length L and radius R.

long and R in radius. From this the number of scatterers in each q-region N_q is proportional to R^2q^{-1}. Application of equation (5.4) yields $I(q) \sim LR^4q^{-2}$. For yet larger q such that $q > R^{-1} > L^{-1}$ consider the right side of figure 5.4. Now the q-regions populate the cylindrical surface (the end caps are not considered because $L \gg R$). The number of q-regions is the ratio of the cylindrical surface to the area occupied by a q-region on the surface $n_q \sim LR/(q^{-1})^2 = LRq^2$. The number of scatterers in each q-region is proportional to the volume of the q-region, $N_q \sim (q^{-1})^3 = q^{-3}$. Application of equation (5.4) yields $I(q) \sim LRq^{-4}$. Note that this is a Porod law with both a −4 exponent and a surface area (LR) coefficient. The complete $I(q)$ versus q result is plotted in figure 5.5.

5.3.5 A disk (a short circular cylinder)—orientationally averaged

Consider a right circular cylinder of length L and radius R with $L \ll R$, i.e. a circular disk as shown in figure 5.6. Diffractive scattering from a single disk orientationally averaged or an ensemble of disks at random orientations can be determined by the scaling approach. Like the long cylinder, above, the disk has two length scales. Again let $a \rightarrow 0$. For the smallest q such that $q < R^{-1}$, equation (5.4) predicts a scattered intensity that goes as $I \sim N^2 \sim R^4L^2$, a volume squared dependence, and is not a function of q, nor the scattering angle. Proceeding to larger q, when $L^{-1} > q > R^{-1}$, one can construct q-regimes as shown in the upper part of figure 5.6. The number of these q-regions is $n_q \sim R^2/(q^{-1})^2 = (Rq)^2$. Note how the q-regions divide the disk into cylindrical segments L thick and q^{-1} in radius. From this the number of scatterers in each q-region N_q is proportional Lq^{-2}. Application of equation (5.4) yields $I(q) \sim L^2R^2q^{-2}$. For yet larger q such that $q > R^{-1} > L^{-1}$ and q^{-1} grows smaller, consider the lower part of figure 5.6. Now the q-regions populate the disk upper and lower surfaces. The number of q-regions is the ratio of the disk surface to the area occupied by a q-region on the surface $n_q \sim R^2/(q^{-1})^2 = R^2q^2$. The number of scatterers in each q-region is proportional to the volume of the q-region, $N_q \sim (q^{-1})^3 = q^{-3}$. Application of equation (5.4) yields $I(q) \sim R^2q^{-4}$. Note that this is a Porod law with

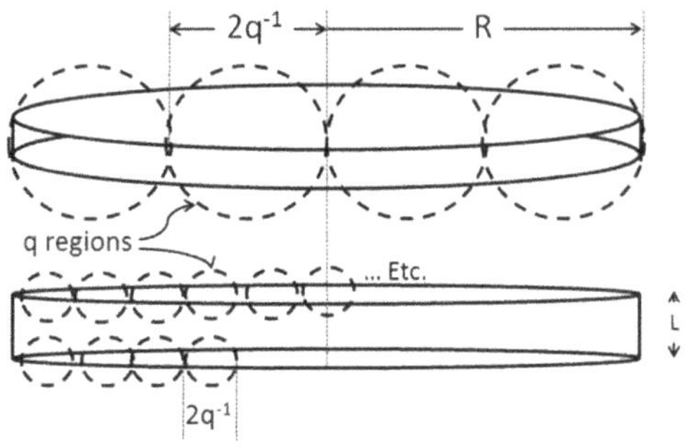

Figure 5.6. Scaling analysis of an orientationally averaged disk (short cylinder), $R \gg L$. Above are q-regions smaller than R but larger than L, $L < q^{-1} < R$; below are q-regions smaller than both L and R, $q^{-1} < L < R$.

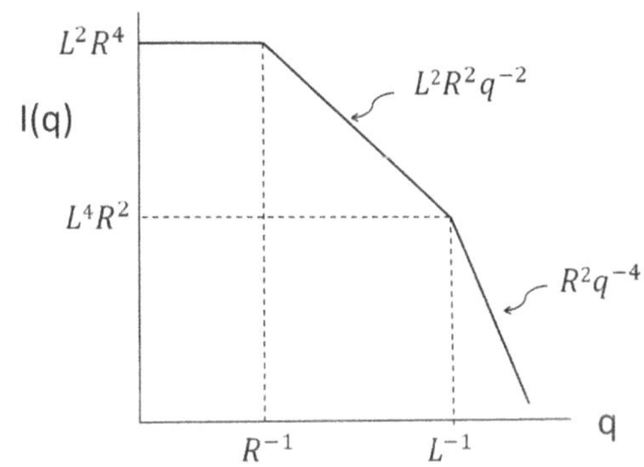

Figure 5.7. Log–log plot of the scattered intensity functionalities with q for an orientationally averaged disks (short cylinders) of thickness L and radius R.

both a −4 exponent and a surface area (R^2) coefficient. The complete $I(q)$ versus q results are plotted in figure 5.7.

It is valuable to compare the results in figures 5.5 and 5.7. The same shape and same volume dependence in the forward lobe, but different ordering of the width and the length changes the power law exponent in the intermediate regime between the two Guinier regimes. The switch also changes the coefficient of the q^{-4} power law in the large q regime. In all this, power laws describe the scattering. This happens because wave scattering depends on comparisons of the length scale carried by the wave, q^{-1}, and the length scales of the scatterer. Scaling and power laws are completely coupled.

5.3.6 A circular cylinder—aligned

The examples immediately above have all had spherical symmetry, often due to orientational averaging; hence q could be used as a magnitude and not a vector and the q-regions were spherical. To bring in the vector nature of $\boldsymbol{q}$ consider the situation of cylinders aligned perpendicularly to the scattering plane as depicted in figure 5.8. This figure shows two vectors $\boldsymbol{r}_1$ and $\boldsymbol{r}_2$, drawn such that $\boldsymbol{r}_1 - \boldsymbol{r}_2$ is in the vertical x direction, perpendicular to the horizontal $\boldsymbol{q}$. Then $e^{i\boldsymbol{q}\cdot\boldsymbol{r}_1} = e^{i\boldsymbol{q}\cdot\boldsymbol{r}_2}$, which means that the

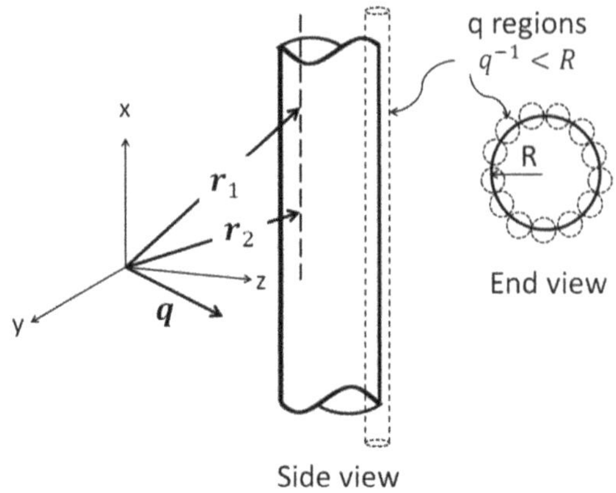

Figure 5.8. Scaling analysis of a cylinder aligned perpendicular to the yz scattering plane that contains the scattering wave vector $\boldsymbol{q}$. The q-regions are now cylinders with an axis perpendicular to the scattering plane. On the left is a demonstration that two vectors $\boldsymbol{r}_1$ and $\boldsymbol{r}_2$ differing solely by a vertical displacement perpendicular to $\boldsymbol{q}$ yield the same phase $\boldsymbol{q} \cdot \boldsymbol{r}$ at the detector.

phase at the detector doesn't change as the $\boldsymbol{r}$ vector changes vertically. No change in phase in turn means a q-region extends without limit in the vertical direction; consequently, the q-regions are infinitely tall vertical cylinders of radius q^{-1}. Infinitely tall, cylindrical q-regions are drawn in figure 5.8.

The scaling approach starts as usual with the magnitude q^{-1} larger than any object length scale. Then the scattering is proportional to $N^2 \sim L^2R^4$. However, because the q-regions are infinite in vertical extent regardless of the magnitude of q, the object's vertical length scale, in this case L, is never observed (hence the cylinders could be flat disks). Instead, only the horizontal scales are observed (because $\boldsymbol{q}$ is horizontal), which in this case is R. When $q > R^{-1}$, the cylindrical q-regions lie along the cylinder's surface as shown in figure 5.6. The number of q-regions is $n_q \sim R/(q-1) = Rq$. The number of scatterers in the q-region is $N_q \sim L(q^{-1})^2 = Lq^{-2}$. Then by equation (5.4) the scattering is $I(q) \sim L^2Rq^{-3}$. This result is sketched in figure 5.9.

5.3.7 A fractal aggregate

Fractal aggregates are described in chapter 9. In a fractal aggregate essentially all the monomeric particles that make up the aggregate are on the surface. Then the mass scaling dimension equals the surface scaling dimension, both of which we write as $D = D_m = D_s < d$. The scaling dimension D is usually called the *fractal dimension*. Orientationally averaged fractal aggregates would have a spherical symmetry. Then by equations (5.14)

$$I(q) \sim N^2 \text{ for } qR_g < 1 \tag{5.15a}$$

$$\sim N^2(qR_g)^{-D} \text{ for } qR_g > 1 \tag{5.15b}$$

In equations (5.15) N is the number of monomeric particles (or monomers or primary particles) in the aggregate. If N is the number of monomers, then $2a$ is the

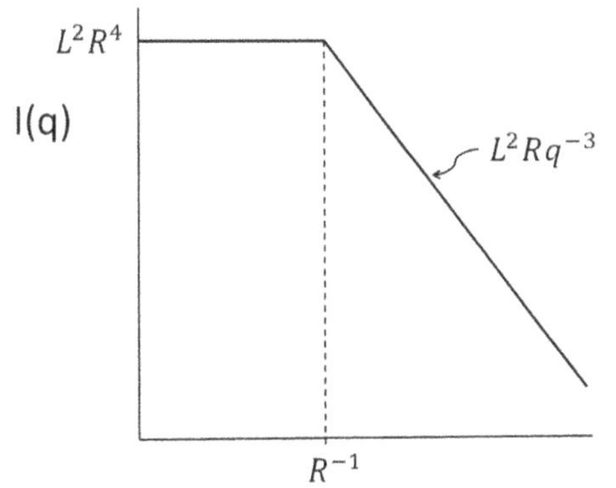

Figure 5.9. Log–log plot of the scattered intensity functionalities with q for a cylinder of length L and radius R with its axis perpendicular to the scattering wave vector $\boldsymbol{q}$.

monomer diameter. The radius of gyration R_g is also used here since this is a well-defined length scale for a random aggregate; it is the root mean square of the monomer distances from the center of mass of the aggregate. We do this because R_g comes naturally out of the rigorous discussion of the structure factor, see section 4.7, and because a geometric radius is ill defined for an aggregate, given its indefinite perimeter.

Equations (5.15) do not include the possibility of the monomer being observable with q, which can be achieved at large q such that $q > a^{-1}$. Said differently, large q means small q^{-1} such that $q^{-1} < a$ so that the light scattering length scale can resolve the monomers of size a. To include N dense spherical monomers with smooth surfaces, equations (5.14) can be applied twice, once for the fractal and once for the sphere, to yield a total intensity structure factor that is the product

$$I(q) \sim I_{\mathrm{FA}}(q) I_{\mathrm{sphere}}(q) \tag{5.16}$$

where FA means 'fractal aggregate.' This product is depicted in figure 5.10 and is consistent with the experiment. The crossover at $qa \sim 1$ was well described by Hasmy *et al* (1993).

5.4 The scaling approach for single particles

The development above was semi-quantitative in that it was largely concerned with the scaling exponents and functionalities of the coefficients on various length scales. Here, we include missing factors of π, 2, etc. We also normalize by the forward scattering to get the structure factor $S(q) = I(q)/I(0)$. Our results, which are published in Maughan and Sorensen (2019), are more quantitative than those above and find a universal description in terms of the aspect ratio of the scattering object, the particle.

To begin let N be the number of point scatterers in the system of scatterers of volume V. Then the number density of point scatterers is $n = N/V$. Let the q-regions have a radius of q^{-1} (the radius, not just the 'size'), and $D_m = 3$. Then reformulate equation (5.8) as

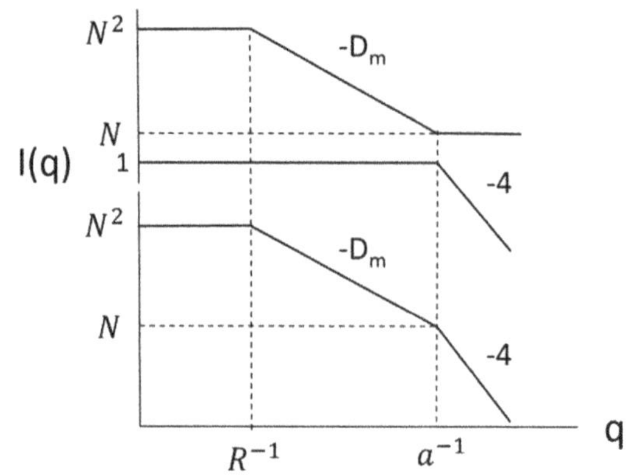

Figure 5.10. Log–log plot depicting generic scattering from a mass fractal aggregate of radius R_g with a mass fractal dimension D_m made up of spherical, smooth surfaced monomers of radius a. Upper curve is for a fractal of point particles spaced by $2a$, second curve is for spherical particles, lowest curve is the product of these, a representation of equation (5.16). From Oh and Sorensen (1999) (Copyright © 1999, Kluwer Academic Publishers). With permission of Springer.

$$N_q = n\frac{4\pi}{3}(q^{-1})^3 \tag{5.17}$$

Next let the surface area of the scattering system be S. The q-regions cut circles with radii q^{-1} in the surface the number of which is

$$n_q = S/\pi(q^{-1})^2 \tag{5.18}$$

Note the equalities in equations (5.17) and (5.18) compared to the proportionalities in the previous section. Finally, substitute these equations into equation (5.4) and use $n = N/V$ to obtain

$$I(q) \sim N^2\frac{16\pi}{9}\frac{S}{V^2}q^{-4} = N^2S(q) \tag{5.19}$$

This result applies for large q. Equation (5.19) reminds us that we need to distinguish S, which means surface area, from $S(q)$, which represents the structure factor. To within a factor of 8/9, this is the classic Porod (Porod 1951; Guinier and Fournet 1955) result.

$$S(q) = \frac{2\pi S}{V^2}q^{-4} = \frac{9}{2}(qR)^{-4} \tag{5.20}$$

Recognize that these equations represent the average behavior of the structure factor in the Porod regime, an average in which the ripples would be smoothed out. Because the ripples are typically due to a cosine squared term, and $\langle\cos^2\theta\rangle = ½$, the average is a factor of ½ smaller than the envelope of the structure factor plot. Then equation (5.20) agrees precisely with the large qR limit of the diffraction pattern for a sphere, equation (4.49), when the ripples are average out. Notice that the shape of the scattering system was not involved in the derivation of (5.19), hence this result applies to all solid, three-dimensional shapes. There are, however, subtleties that will not be described here but can be found in our published work (Maughan and Sorensen 2019).

Equation (5.19) can be reconfigured to look like equation (5.13b). Assume a cubic lattice for the point scatterers. Then $N = (V/2a)^3$ and $N_s = (S/2a)^2$. Solve these for V and S and substitute into equation (5.19) to find

$$I(q) \sim (\pi/8)N_s(qa)^{-4} \tag{5.21}$$

The factor π/8 is dependent on the lattice type (among other things) for the point scatterers and hence not significant. Equation (5.21) reinforces the fact that scattering at the end of the Porod regime is due to the surface of a homogeneous system.

It is often advantageous to plot the structure factor versus the dimensionless parameter qR. Furthermore, we seek an expression for any shape not just a sphere. To do this, equation (5.19) can be solved for $S(q)$ and then multiplied and divided by R_{veq}^4 where R_{veq} is the radius of a sphere with the same volume as the object, i.e. the *volume equivalent radius*. The structure factor for the power law region can then be expressed as

$$S(q) = \frac{16\pi S R_{\text{veq}}^4}{9V^2}\left(qR_{\text{veq}}\right)^{-4}. \tag{5.22}$$

Not only does this provide a dimensionless parameter as the independent variable but also yields a dimensionless coefficient for the power law. For a sphere equation (5.22) can also be reduced to

$$S(q) = 4\left(qR_{\text{veq}}\right)^{-4} \tag{5.23}$$

where we have used $S = 4\pi R_{\text{veq}}^2$. Remember, the exact result replaces the 4 with 9/2 in equation (5.23) as in equation (5.20).

Next we follow Maughan and Sorensen (2019) who applied a quantitative scaling approach to calculate the forward normalized structure factor, i.e. $S(q)$ when $S(0) = 1$. It was shown that the Guinier boundaries and the power law coefficients for a given shape are described by a single parameter, the aspect ratio of the scattering object. Recognize that a three-dimensional object could have three significant length scales. Here, we will assume that two of these three are equal, and we will refer to this as the width. The other dimension is the axial length, L. For example, a right circular cylinder has two equal diameters, $2R$ and $2R$; call this the width, W, and a length L. With this, we then define the *aspect ratio* as the ratio of the length divided by the width

$$\varepsilon = L/W \tag{5.24}$$

For example, a right circular cylinder has an aspect ratio of $\varepsilon = L/2R$. Note that ε can be both greater than and less then unity.

Four different particle shapes were considered in that work: cylinders, hexagonal columns, square columns, and spheroids. Results were completely consistent with figures 5.5 and 5.7. Two Guinier regimes occur when the aspect ratio moves away from unity, to become either much greater than, or much less than unity. The first Guinier regime (smallest q) will occur when q crosses roughly through the inverse of the largest length scale of the object, while the second Guinier regime occurs when q crosses roughly through the inverse of the smallest length scale of the object. Between the Guinier regimes and at q values larger than the second Guinier regime, the power laws of figures 5.5 and 5.7 ensue. The coefficients of these power laws were also calculated. For example, the coefficient in the q^{-4} regime can be calculated by expressing the surface area and volume in equation (5.19) and the volume equivalent radius in terms of the width and the aspect ratio. These coefficients are solely a function of ε, but the functionality depends on the shape. They are given in figures 5.11, 5.12 and 5.13. For more detail on this calculation, see the original publication. As with any particle, at small q, when q^{-1} is larger than the largest length scale, the scattering will go as N^2 and $S(0) = 1$.

Figure 5.11 shows the structure factors plotted versus qR_{veq} for the four particle shapes with $\varepsilon = 1/10$, i.e. short objects, disks and an oblate spheroid.

Figure 5.12 shows the structure factors plotted versus qR_{veq} for the four particle shapes with $\varepsilon = 1$, i.e. drums, cubes and spheres.

Figure 5.13 shows the structure factors plotted versus qR_{veq} for the four particle shapes with $\varepsilon = 10$, i.e. long cylinders, prisms and prolate spheroids.

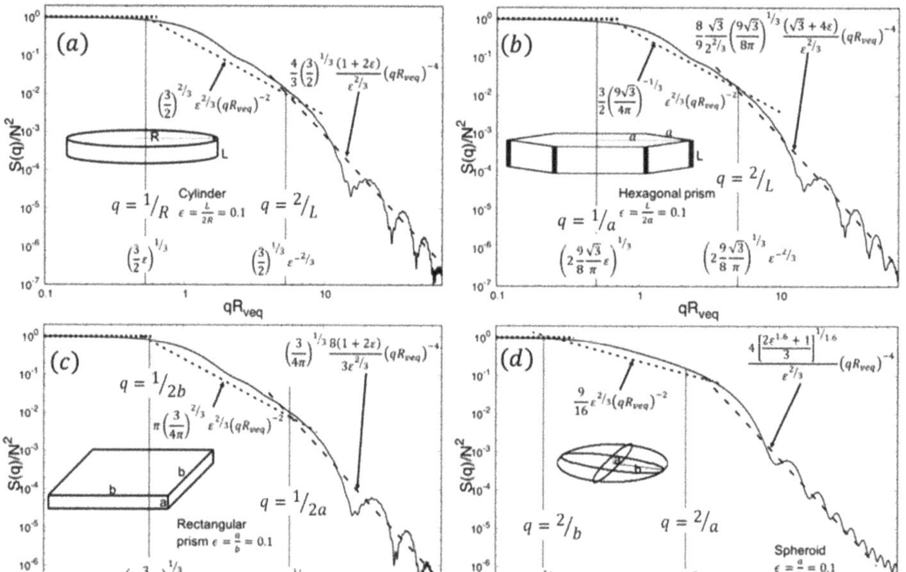

Figure 5.11. The normalized structure factor for a circular disc (a), hexagonal prism disc (b), rectangular prism disc (c), and an oblate spheroid (d), all with aspect ratio $\varepsilon = 1/10$. The solid black curves are the Fourier transforms calculated numerically using equation (5.1). The dashed lines and their dependence on ε were computed using the scaling approach. Reprinted from Maughan and Sorensen (2019), Copyright (2019), with permission from Elsevier.

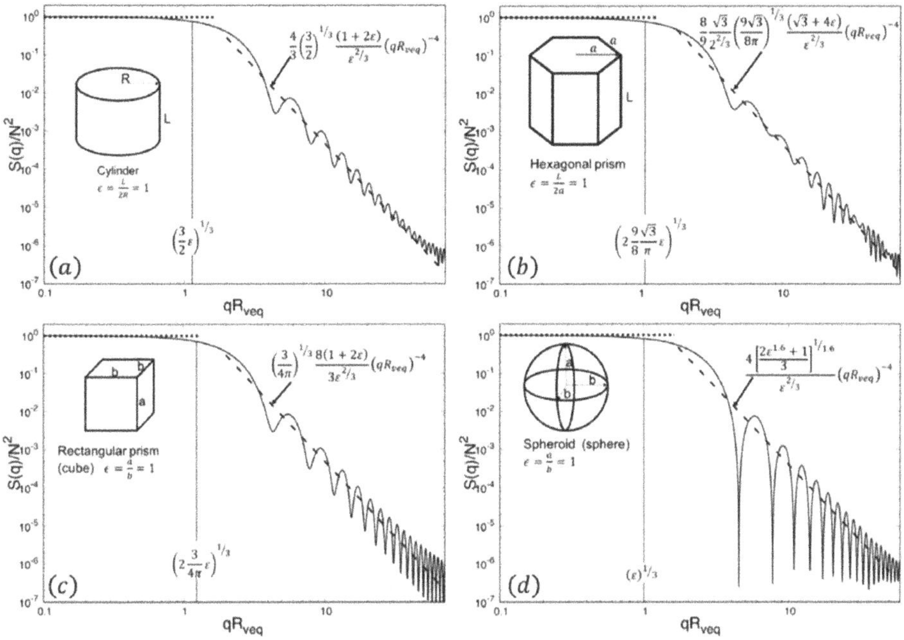

Figure 5.12. The normalized structure factor for a circular drum (a), hexagonal drum (b), cube (c), and a sphere (d), all with aspect ratio $\varepsilon = 1$. The solid black curves are the Fourier transforms calculated numerically using equation (5.1). The dashed lines and their dependence on ε were computed using the scaling approach. Reprinted from Maughan and Sorensen (2019), Copyright (2019), with permission from Elsevier.

Although the functionalities with ε in figures 5.11, 5.12 and 5.13 are complex, one can show that they are completely consistent with the proportionalities of figures 5.5 and 5.7, where $W = 2R$, a, b and b (see drawings in figures) for cylinders, hexagonal columns, square columns, and spheroids, respectively.

As advertised above, the Guinier boundaries and power law coefficients are solely dependent on the aspect ratio ε of the scattering object, but the functionality is shape dependent. There is considerable similarity between the plots for the different shapes in figures 5.11, 5.12 and 5.13. The Guinier crossovers are all at $q = 2/W$ and $2/L$, which merge when $\varepsilon = 1$. In practice, we find that they are hard to distinguish when $0.3 < \varepsilon < 3$. The coefficients of the Porod power laws $(qR_{\text{veq}})^{-4}$ vary widely with ε, but all shapes vary together as illustrated in figure 5.14. The coefficients of the -1 and -2 intermediate power laws are all essentially equal for a given power law as indicated by the same Guinier end points and slopes on the plots. Remember that these coefficients are averaged over a ripple structure as is depicted in the figures. The major difference is the intermediate power laws of -1 and -2, which occur for $\varepsilon > 1$ and $\varepsilon < 1$, respectively, independent of shape. With all this, one can conclude that diffraction scattering measurements, such as x-ray, neutron or quasi-index--matched light scattering, from rotationally averaged ensembles of particles of these shape types would not be able to differentiate the shapes but would be able to semi-quantitatively determine the aspect ratio ε.

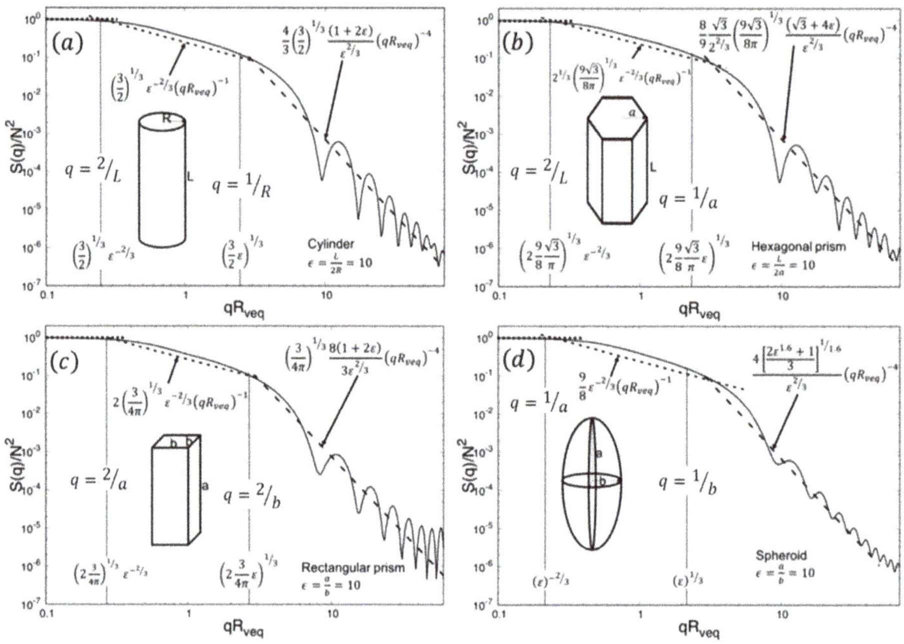

Figure 5.13. The normalized structure factor for a circular cylinder (a), hexagonal prism (b), square prism (c), and a prolate spheroid (d), all with aspect ratio $\varepsilon = 10$. The solid black curves are the Fourier transforms calculated numerically using equation (5.1). The dashed lines and their dependence on ε were computed using the scaling approach. Reprinted from Maughan and Sorensen (2019), Copyright (2019), with permission from Elsevier.

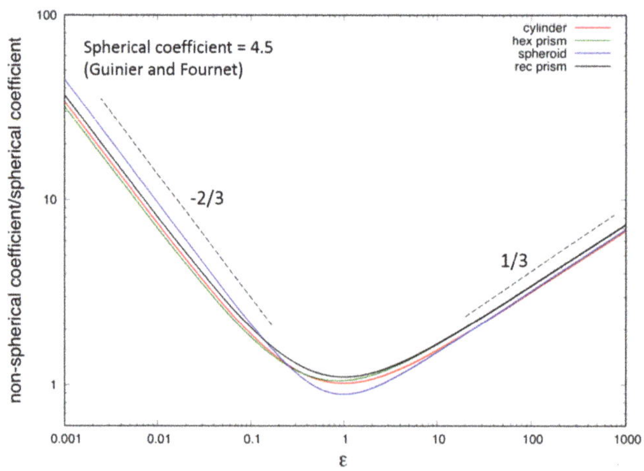

Figure 5.14. Ratio of the non-spherical coefficient of $(qR_{\text{veq}})^{-4}$ in the Porod regime to 9/2, the exact value for a sphere, equation (5.20), versus the aspect ratio ε for the four shapes studied here.

5.5 The scaling approach for ensembles of particles

Most experimental situations involve scattering from ensembles of particles. Such situations might invoke thoughts of dispersed aerosols or colloids, but also include dense particulate systems like gels, powders and atomic and molecular systems, such as liquids, glasses and crystalline solids. With regard to dense systems, much of the previous work in this area has endeavored to use scattering data, especially from x-rays and neutrons, to determine the spatial structure of the system (Guinier and Fournet 1955, Egelstaff 1967). Scattering of waves, like x-rays and neutrons, yields the structure factor, $S(q)$, which, as we have seen in chapter 4, is related via Fourier transformation to the system's density autocorrelation function, $g(r)$, which contains fundamental information about the real space structure of the system. Fourier inversion from $S(q)$ to $g(r)$, however, can be difficult given the limited range of q and the wide range of length scales involved in complex systems, as well as the great variety of structures. Given that an analysis is considered successful if key length scales and concepts of structural arrangements or shapes can be determined, we argue that there is good reason to stay in q-space, where the same structural information resides, so long as one is well equipped to understand and interpret q-space. We believe this simple economization allows one to interpret more clearly difficult features in the structure factor of more complex systems. With this view of q-space in mind, here we extend the scaling approach to any system from which waves can be scattered. In preparation for this extension, we first consider a situation in which density fluctuations exist in the volume.

5.5.1 Systems with volumetric density fluctuations

Consider a system of scatterers in a volume where the scatterers can be displaced randomly from their uniform lattice positions, as drawn in figure 5.15. Let ζa, where $\zeta > 0$ is a random multiplicative factor, be the average displacement. If $\zeta = 0$, small q-regions for which $q^{-1} < a$, hence $q > a^{-1}$, would only see fluctuations in the scatterer density at the surface of the system. This situation was studied above, and yielded a scattered intensity depending only on the surface, $I(q) \sim N_s$, equation (5.13c).

Now include volumetric density fluctuations by increasing ζ slightly above zero. Remember that q^{-1}, the inherent length scale of the scattering experiment, is the resolution of the scattering. As long as the length scale of the scattering, q^{-1}, is bigger than ζa, or equivalently $q < (\zeta a)^{-1}$, the density fluctuations will be invisible (unresolvable) to the scattering and again $I(q) \sim N_s$. However, once $q^{-1} < \zeta a$ (i.e. $q > (\zeta a)^{-1}$), the scattering can 'see' (resolve) the length ζa and then the density fluctuations contribute to the scattering. Then with $q > (\zeta a)^{-1}$, each scatterer contributes with random phase, so the total scattered intensity will be proportional to N. In total, for a system with randomized point scatterers representing density fluctuations, equations (5.13) modify to

$$I(q) \sim N^2 \text{ for } q < R^{-1} \tag{5.25a}$$

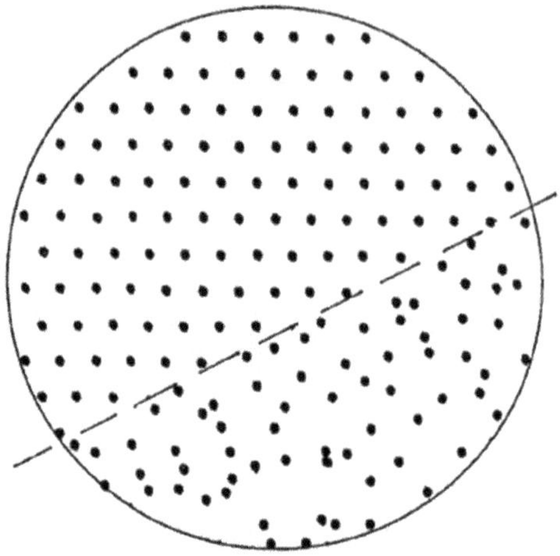

Figure 5.15. Drawing of a system of point scatterers in a spherical region. In the upper left the scatterers are on a uniform lattice and therefore have no fluctuations in their density within the volume of the sphere. In the lower right the scatterers are displaced randomly from their uniform lattice positions and thus the system has volumetric density fluctuations. From Oh and Sorensen (1999) (Copyright © 1999, Kluwer Academic Publishers). With permission of Springer.

$$\sim k_0 N_s (qa)^{-2D_m + D_s} \text{ for } R^{-1} < q < a^{-1} \tag{5.25b}$$

$$\sim k_0 N_s \text{ for } a^{-1} < q < (\zeta a)^{-1} \tag{5.25c}$$

$$\sim N \text{ for } (\zeta a)^{-1} < q \tag{5.25d}$$

This is our most general result. Figure 5.16 shows this result schematically. Note the functionalities on N^2, N and N_s. The transition regime at large q from surface scattering to volumetric density fluctuation scattering (from $\sim N_s$ to N) occurs because once q is big enough, q^{-1} is small enough to 'see' the small density fluctuations in the volume interior. This transition has a slope of approximately 2, hence a q^2 dependence. This can be understood by expanding the integral form of the structure factor in equation (4.17) at low q as described by Hasmy *et al* (1993). Finally, note that as $\zeta \to 0$ the fourth regime disappears and figure 5.16 becomes figure 5.3.

Figure 5.17 demonstrates equations (5.25) and figure 5.16 with a Fourier transform of a 3d spherical system of point scatterers. The system was created with a cubic point lattice with separation $2a$ inside a sphere of radius $R = 60a$ to yield $N = 112\,994$. Then, the positions of the point scatterers were randomized by moving them away from their lattice positions according to $\delta x = 2\zeta ar$ and $\delta y = 2\zeta ar$, where δx and δy are random displacements in the x and y directions, and r is a random number bounded as $-1 < r < 1$. Note that $\boldsymbol{q}$ is in the xy-plane. Each scatterer used the same ζ but had its own random number r. This system was created for various values of ζ, and the structure factors (Fourier transforms) were computed using equation (5.1). The results, given in figure 5.17 for a variety of ζ values, support the

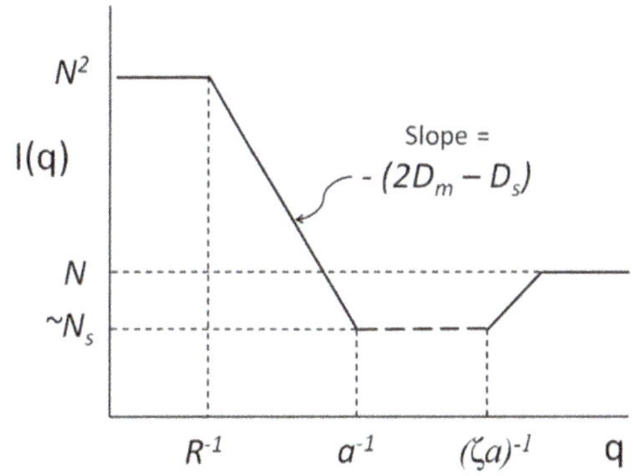

Figure 5.16. A schematic diagram of the scattered intensity, normalized to equal N^2 at small q, from a system of scatterers as a function of q as described by equations (5.25). The plot is log–log. This is also the structure factor as we define it multiplied by N^2 where N is the number of scatterers in the system. R is the linear size of the system, $2a$ is the spacing between the point scatterers, N_s is the number of point scatterers on the surface of the system, D_m and D_s are the mass and surface scaling dimensions, respectively, and $\zeta < 1$ (for this diagram) is the random multiplicative factor for the average displacement of the point scatterers. Finite ζ causes volumetric density fluctuations to occur leading to enhanced scattering for $q > (\zeta a)^{-1}$.

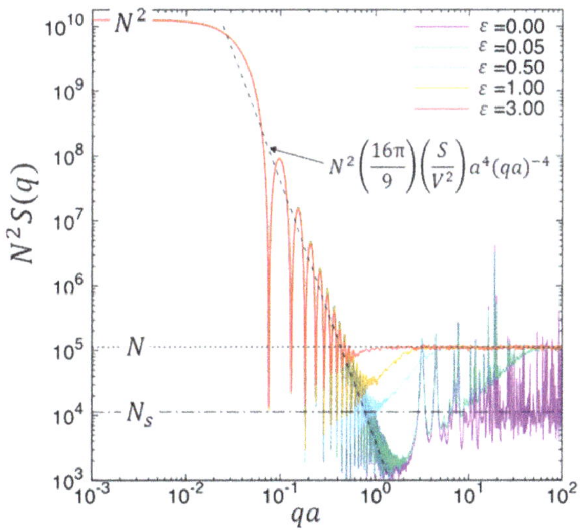

Figure 5.17. Structure factor multiplied by N^2 versus qa for a system of N = 112 994 point scatterers in a $d = 3$, spherical region of radius $R = 60a$ calculated using equation (5.1). N_s = 11 310 is the number of scatterers on the surface. The parameter ζ is a measure of the random spatial deviation of the point scatterers from a cubic lattice of spacing $2a$. When ζ = 0, the scatters all reside on the cubic lattice. Curves for increasing ζ show enhanced scattering for $q > (\zeta a)^{-1}$ due to volumetric density fluctuations in accordance with the scaling prediction of equation (5.21).

prediction of the scaling theory as given in equation (5.21) and drawn schematically in figure 5.16.

Figure 5.17 provides numerical verification of the scaling arguments above. When $qa \ll 1$, the intensity structure factor equals N^2. The Porod regime is well described by equation (5.20) especially when we remember that, as an average, equation (5.20) should lay a factor of two below the top of the ripples. When $qa \gg 1$, the limits of N and N_s are well verified, although a 'trough' occurs near $qa \simeq 2$, which is probably an anticorrelation between neighboring lattice sites. Semiquantitative agreement with the $(\zeta a)^{-1}$ crossover points is also achieved, and we speculate that this could be improved with a better randomization procedure.

The small random displacements from a crystal lattice we have used to describe density fluctuations are reminiscent of thermally induced displacements in a real crystal lattice, but they are not. Thermally induced fluctuations lead to a decreased scattered intensity of diffraction lines that remain sharp, the Debye–Waller factor (Sidebottom 2012). Thermal fluctuations, however, are random with time, and it is the time averaging of the detector that yields the decreased intensity. The fluctuations we introduce are stationary with time and hence are not described by the Debye–Waller factor.

Equations (5.25) and figure 5.16 describe the most general case for scattering of waves from a system of scatterers. As such, it can be used to describe nearly any given scattering situation.

5.5.2 An ensemble of particles in a scattering volume

Equations (5.25) and figure 5.16 can be usefully applied to the common situation of an ensemble of particles, either solid spheres or fractal aggregates (and with simple generalizations, any shape), in a finite sized scattering volume. This extension begins with the approximation that the total intensity structure factor of the entire ensemble is a product of individual intensity structure factors as

$$I_{\mathrm{TOT}}(q) \simeq I_{\mathrm{ens}}(q) I_{\mathrm{sc}}(q) I_{\mathrm{m}}(q) \tag{5.26}$$

In (5.26) subscript 'm' designates the monomer structure factor, 'sc' designates the single cluster structure factor representing the monomers of which are composed as points, and 'ens' designates the ensemble structure factor due to the entire ensemble of clusters in the scattering volume representing each cluster as a point. The equality in equation (5.26) is true only if the clusters are monodisperse. Polydispersity breaks the exact equality, but it still holds to a good approximation (Huang *et al* 1998); but with caution (Haw *et al* 1997).

Equation (5.26) is written for an ensemble of clusters of monomers. The concept embodied in (5.26) applies to any ensemble of particles. Recognize that in this picture the entire ensemble is a system of scatterers with a size and shape. Thus, a system of spheres would have a structure factor that is the product of the single sphere structure factor and the structure factor of the ensemble of these particles. We choose to study (5.26) because there are numerous examples for it in the literature.

Each of these three structure factors in equation (5.26) is described by equations (5.25) and figure 5.16 and each is represented in the example of figure 5.18. In this example, the monomers are uniform, compact spheres of radius a. The clusters each have N monomers, have a radius of R, and a mass fractal dimension of D_m. The clusters could be solid spheres ($D_m = d > D_s \geqslant d - 1$), fractal aggregates ($D_m = D_s < d$)

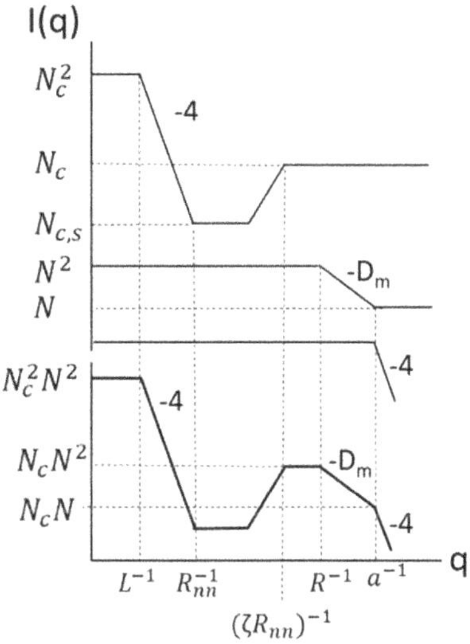

Figure 5.18. Prediction of equation (5.26) for a system of N_c fractal aggregates of mass fractal dimension D_m and radius R composed of N uniform spherical monomers of radius a. These aggregates reside in a scattering volume of size L and have a mean nearest neighbor separation of $\sim R_{nn}$. The upper curve is the structure factor for the system of clusters in the scattering volume taking them as points. The second curve is the structure factor for a single cluster taking its monomers as points. The third curve is the structure factor for the spherical monomers. The fourth, lowest curve is the product of the upper three to yield, by equation (5.26), the complete structure factor. The factor $\zeta > 0$ allows for variation in the ordering of the cluster ensemble. The plot is drawn for $1 \gg \zeta > 0$. If $\zeta = 0$ the cluster ensemble has long range order and the upper plot is constant at $N_{c,s}$ for large q, and the dip in the product, lowest plot, disappears. As ζ increases, the ensemble randomizes, the dip reappears until the randomization is $\zeta \gtrsim 1$ where it disappears again (Oh and Sorensen 1999). Compare this figure to figure 5.20. Adapted from Oh and Sorensen (1999) (Copyright © 1999, Kluwer Academic Publishers). With permission of Springer.

or something else; here, we choose fractal aggregates. The ensemble of clusters is contained in a scattering volume of size L. Because experiments are usually performed with a horizontal scattering plane with the vector $\boldsymbol{q}$ in that plane, this size is a horizontal width defined by both the incident light beam and the width of the beam observed from the detector. In the scattering volume, there are N_c clusters, $N_{c,s}$ of which are on the surface of the scattering volume. They are separated by average spacing of $2R_{nn}$ (the cluster nearest neighbor, center-to-center distance) but the fractional deviation from uniformity is ζR_{nn}. All three of these, the monomers, the clusters, the scattering volume are each a 'system of scatterers', and they each have structure factors in accordance with equation (5.26). Figure 5.18 shows all the structure factors drawn on the same q scale. Their product is the total structure factor and is shown as the lowest curve in figure 5.18.

The total structure factor in figure 5.18 can be described as follows. At smallest q, $q < L^{-1}$, the total scattering according to figure 5.18 is proportional to $N_c^2N^2$. This follows from the fact that the total number of monomers in the entire scattering volume is $N_m = N_cN$, and the total scattering for $q < L^{-1}$ is N_m^2. This, of course, follows from scaling approach rule No. 1 above; since q^{-1} is larger than the entire scattering volume, all the monomers scatter in phase regardless of the state of aggregation. Next, for $L^{-1} < q < R_{nn}{}^{-1}$ fluctuations in density only occur from the surface of the scattering volume; thus, in complete analogy to compact, uniform sphere scattering, a Porod regime for the scattering volume is encountered with slope $-(d+1)$, which is -4 in three dimensions. Note that we are taking the rather unusual perspective of treating the entire scattering volume as a scattering entity. Most often L is much too large, hence L^{-1} is much too small, to be detected by q in a real experiment. Indeed, such a q would require a scattering angle of a few hundredths of degree at optical wavelengths for a scattering volume of a few tenths of a millimeter. However, it is important to realize that this *scattering volume Porod regime* which begins at $q \approx L^{-1}$ can continue with increasing q for many orders of magnitude until it can be seen experimentally near $R_{nn}{}^{-1}$. Also note that as $L \to \infty$, it becomes the $\delta(\boldsymbol{q})$ of equation (4.33). Thus, from that perspective this regime is often ignored, neglected, or discounted in the literature. Despite this, the scattering volume Porod regime appears in experimental measurements, and its identification can be important for interpretation of structure factor data, see figure 5.20 (Sorensen *et al* 1998).

5.5.3 Scattering from dense systems

One can define a system of particles as dilute when the mean particle nearest neighbor separation is much greater than the particle size, $R_{nn} \gg R$, and dense when R_{nn} is within an order of magnitude or closer to R (Ebini and Sorensen 2019), which we write as $R_{nn} \gtrsim R$. These two length scales are related by

$$R_{nn} = R/f_{\mathrm{v}}^{1/d} \tag{5.27}$$

where f_{v} is the volume fraction of the particulate system (e.g. an aerosol or a colloid) and d is the spatial dimensionality, $d = 3$. Thus, if $f_{\mathrm{v}} = 10^{-3}$, the separation to size ratio is 10 to 1, an order of magnitude; if $f_{\mathrm{v}} = 10^{-6}$, two orders of magnitude. The latter can be considered dilute because being one or more orders of magnitude apart relative to the size implies that the particles rarely interact with each other. Then the particles will be dispersed randomly in space; their positions will be uncorrelated with each other. Our quantitative means of expressing this random dispersal is the parameter ζ. Random dispersal means ζ will be large. Thus, $(\zeta R_{nn})^{-1}$ in figure 5.18 will slide to the left with increasing ζ and obliterate the dip. Dilute systems have no dip.

The critical reader is likely to realize that the range of particle–particle interaction relative to R_{nn} is the dimensionless parameter for determining whether or not particles interact, cause correlations and thereby suppress ζ. The critical reader is correct. A good example would be a colloid in which long range electrostatic interactions are not screened.

If $f_{\mathrm{v}} = 0.1$, $R_{nn}/R \simeq 2.2$. This is dense even if the particle–particle interaction potential is only the hard sphere repulsion. The particles are in each other's face! With such a length scale ratio, steric interactions will occur if nothing else. The particles' positions will depend on their neighbors and our ζ parameter will be small and the dip will occur.

Having a dip and seeing it are two different issues. The dip region is always at smaller q than the Guinier regime for the particle near $q \simeq R^{-1}$. If the particle or cluster is big, small q is necessary to see it, and yet smaller q is necessary to see the dip. Small q needs small scattering angles, which can ultimately limit the measurement.

An example of a system with a dip is shown in figure 5.19, which displays light scattering measurements at various times from a colloid undergoing aggregation. The colloid consisted of 40 nm diameter polystyrene nanospheres in water at a volume fraction of 4×10^{-4} in water. The relative refractive index of these particles immersed in water is 1.20, which is low enough to use diffractive scattering analysis to a good approximation (it's better than doing nothing). Hence, we refer to the

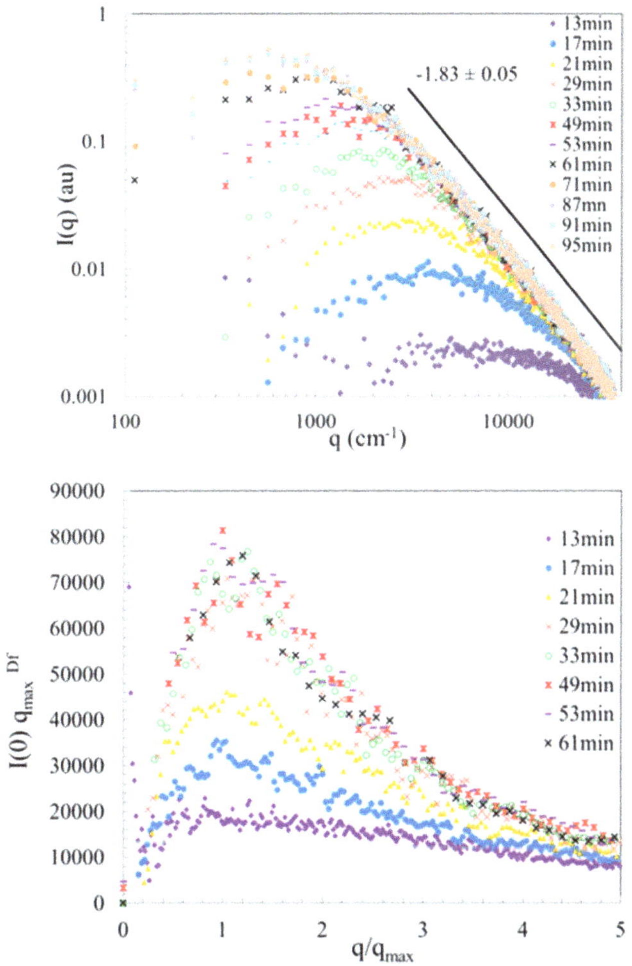

Figure 5.19. Upper: light scattering structure factor measurement for a destabilized colloid of 40 nm diameter polystyrene nanospheres at different times after the onset of aggregation. The aggregates so formed have a fractal dimension of 1.83 as indicated by the solid black line. Lower: scaling plot of much of the data above in accord with equations (5.28).

scattered intensity $I(q)$ as the structure factor. The colloid was destabilized by adding 0.2 M $MgCl_2$ which screens the electrostatic, repulsive potential between the particles. With this screening, the nanospheres can approach each other close enough so that the attractive van der Waals force will bind them together. Then the nanospheres become the monomers of growing fractal aggregates that have a fractal dimension of D = 1.83 as indicated by the power law with q at large q.

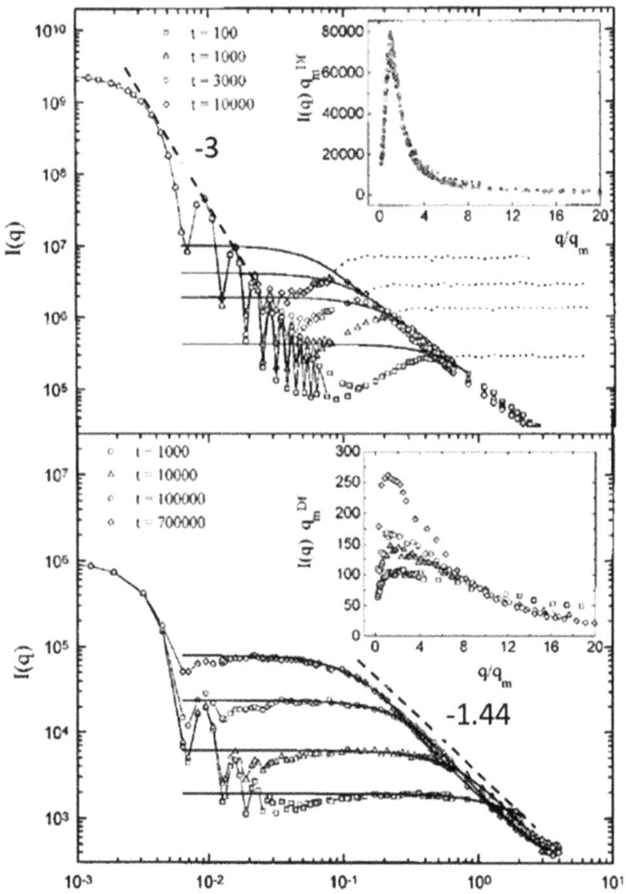

Figure 5.20. Structure factor intensity for simulated two-dimensional aggregating systems on a 1000 by 1000 square lattice. The monomer radius is $a = ½$ thus $q = 2 = a^{-1}$. The volume fraction of monomers is $f_v = 10^{-3}$ (left) and 50×10^{-3} (right). The aggregation mechanism is DLCA which yields $D = 1.44$ fractal aggregates in $2d$. On the left the mean aggregate nearest neighbor separation is $R_{nn} \simeq 32\ R$; on the right $R_{nn} \simeq 4.5R$ such that these length scales interfere with each other in this structure factor. Also shown are the single-cluster (solid line) and cluster–cluster (dotted line) structure factors. Dashed lines indicate slopes, hence power laws, with slopes of −3 and −1.44. Insets show attempts to scale the scattered intensities with one parameter, q_m the position of the peak. Also present at lowest q is the structure factor of the $d = 2$ system simulation box. Compare this figure to figure 5.18. Adapted figure with permission from Huang *et al* (1998). Copyright (1998) by the American Physical Society.

However, at small q, the light scattering structure factor shows a significant dip. This dip and the power law in scattering intensity at larger q create a peak in the scattering intensity. The position of that peak ranges from 500 to 5000 cm^{-1} to semi-quantitatively imply aggregate sizes ranging from approximately 20 to 2 microns, respectively, increasing with increasing aggregation time. Note also that the q range of this experiment could not detect the small monomers, which would require a large value near $q \simeq 5 \times 10^5$ cm^{-1}. Also, the R_{nn} scale at small q was either below the experimental q range or not intense enough to be measured.

Figure 5.19 is an example of a number of experiments on dense, aggregating colloids. Other workers have emphasized the peak not the dip (Carpineti and Giglio 1992, 1993, Bibette *et al* 1992, Robinson and Earnshaw 1993, Earnshaw *et al* 1996). The scattered intensity $I(q,t)$ in these experiments was found to exhibit a peak, a maximum, at a position q_m, which increased in intensity, while q_m decreased with time. For late stages of aggregation the scattered intensity was found to scale according to

$$I(q) \sim q_m^{-\alpha} F(q/q_m) \tag{5.28a}$$

$$F(q/q_m) \sim \left(\frac{q}{q_m}\right)^{-\beta} \quad (q/q_m) \gg 1 \tag{5.28b}$$

Early stages of aggregation did not display scaling. For colloids of non-compact, fractal aggregates $\alpha = \beta = D$, the mass fractal dimension of the aggregate. Remarkably, equations (5.28) have the same form for the structure factors measured for fluids undergoing spinodal decomposition except that then $\alpha = d$ and $\beta = d + 1$, where d is the spatial dimension (Furukawa 1985, Takenaka *et al* 1992, Butler *et al* 1995).

The lower part of figure 5.19 shows much of the data replotted according to the scaling of equations (5.28) with some success. There, we see that scaling does not hold early in the aggregation, times 13, 17 and 21 min, but does hold later at aggregation times of 29 through 61 min and more.

The implication of a single peak structure factor that can be scaled to a universal plot as happens in figure 5.19, and the numerous other situations that display scaling mentioned above, is that there is one, and only one, relevant length scale in the aggregating (or spinodally decomposing) system, and that length scale is q_m^{-1}.

Nevertheless, there are two length scales in the system. These length scales are R_{nn} and R, but they conspire in very dense systems that arise in the late stages of aggregation to look like one (Huang *et al* 1998, Cerda *et al* 2004). This will be explained below.

Figure 5.20 shows a simulation of an aggregating particulate system in $d = 2$ dimensions (Huang *et al* 1998). In this spatial dimension diffusion limited cluster–cluster aggregation (DLCA) leads to a fractal dimension of about $D = 1.44$. Compare these plots to figure 5.18. At smallest q, there are the Rayleigh and Guinier regimes of the system itself. This is the scattering volume of figure 5.18. As q increases, the Porod regime of the system appears with a slope of $-(d + 1) = -3$. The volume fraction in the lower part of the figure is $f_v = 10^{-3}$ to imply via equation

(5.27) $R_{nn}/R \simeq 32$ (remember $d = 2$). Notice that the Guinier regime, where $q = R^{-1}$, and the minor dip at the end of the system Porod regime (with slope $-(d + 1) = -3$) are about a factor of 30 apart, hence consistent with this ratio and the bottom plot in figure 5.18. At yet larger q, there is a flat Rayleigh regime of the aggregates followed by the aggregate Guinier regime and Porod regime with a power law of slope $D = 1.44$, the aggregate fractal dimension.

Now we address the two length scales and scaling issues. Notice that the inset of the lower plot of figure 5.20 demonstrates that scaling with one length scale, q_m^{-1}, does not work. It does not work for two reasons: first, there are two length scales, R_{nn} and R that are distinct because they are about a factor of 30 different; second, for fractal aggregates these two length scales have different time dependencies (see appendix B), hence are independent. On the other hand, in the upper plot the volume fraction is $f_v = 50 \times 10^{-3}$ to imply via equation (5.27) $R_{nn}/R \simeq 4.5$. This is close enough for the two length scales to intertwine as seen in the main plot of the upper figure. This leaves a single composite length scale. This blending of length scales apparently mixes the time scales and leads to one parameter scaling seen in the inset of the upper figure.

As an experimentalist, it is important to recognize that it is difficult to see all the regimes displayed in figure 5.18 with experiments because the length scales range at least four orders of magnitude, from ca. 10 nm to 0.1 mm. Simulations can be similarly limited due to restrictions on system size due to computer memory restrictions. However, for simulations a wider range of length scales can be obtained in two dimensions than three. That was the strategy involved in the generation of figure 5.20.

As an example of a measurement that has the right combination of system and scattering length scales to warrant analysis with the predictions of figure 5.18, consider figure 5.21 (Sorensen *et al* 1998). Therein are plotted small angle x-ray scattering (SAXS) data for bulk powders of soot. Because x-rays experience a near unity refractive index in condensed media, their scattering is diffractive and so the scattered intensity is proportional to the structure factor. *In situ* light scattering measurements with the flame source showed soot particles with a fractal dimension of 1.75 and mean size of a few hundred nanometers. The SAXS data in figure 5.21 are consistent with these data in the $0.005 < q < 0.1$ nm^{-1} region. When $q \geqslant 0.1$ nm^{-1}, a Porod regime with slope −4 on the log–log plot indicates monomers with a mean size of 10 nm and scaling dimensions $D_m = 3$ and $D_s = 2$, very likely well approximated as spheres. Indeed, spherical monomers are common for flame soot. Finally, another −4 Porod regime is visible just starting below $q < 0.005$ nm^{-1}. This is the large q end of the scattering volume Porod regime with an inverse length scale too small for the q values attainable with the SAXS apparatus.

5.6 Connections to other formulations

It is of value to compare the generic results of figures 5.16–5.18 to more detailed calculations of the structure factor for an ensemble of N scatterers, be they molecules in a fluid system or particles in a colloid, based on statistical mechanical theory, as

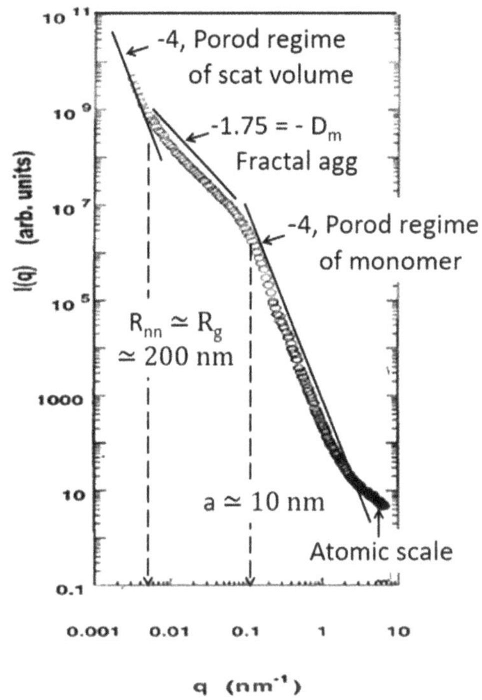

Figure 5.21. SAXS data for a bulk soot powder. Starting at smallest q one finds a fragment of the scattering volume Porod regime, a crossover at the soot aggregate size which is also the aggregate nearest neighbor separation because the aggregates are touching in the powder, the fractal scaling of the aggregate, a crossover at the monomer size, the monomer Porod regime, and finally, at largest q, scattering from the atomic scale. Reprinted figure with permission from Sorensen *et al* (1998). Copyright (1998) by the American Physical Society.

discussed in chapter 4 and Klein and D'Aguanno (1996). In this more traditional approach, a delta function at $q = 0$, i.e. $\delta(q)$, results, which is customarily omitted because it contributes only to scattering at zero angle. Moreover, what remains at $q \to 0$ is closely related to the isothermal compressibility of the system (Sidebottom 2012). Both these features reside in the scaling approach presented here. To see the similarity, one must take the thermodynamic limit, $L \to \infty$, $N \to \infty$ with N/L^d constant, of the scaling approach, a limit which is implicit in the statistical mechanical approach. Consider the scaling $L \to \alpha L$ with $\alpha > 1$. Then $V \to \alpha^d V$ so $N \to \alpha^d N$, where d is the spatial dimension. The thermodynamic limit occurs as $\alpha \to \infty$. Now consider $NS(q)$ as $\alpha \to \infty$. Then the low q peak of the scattering volume Porod regime will increase in magnitude proportional to α^d, i.e. $NS(q) \sim \alpha^d$ while its boundary at L^{-1} will decrease proportional to α. Since the magnitude of the slope of this scattering volume Porod regime is $(d + 1) > d$, the overall effect will be both to heighten and narrow this regime until, as $\alpha \to \infty$, it limits to $\delta(q)$, consistent with statistical mechanic theory and our equation (4.33). By omitting $\delta(q)$, previous approaches have in essence discarded the scattering volume term in $S(q)$ by assuming the scattering volume was infinite.

With the scattering volume term removed, $N^2S(q) = k_0N_s$ for $a^{-1} < q < (\zeta a)^{-1}$. Again normalize as $NS(q) = k_0N_s/N$. Scalings with α are $N \rightarrow \alpha^d N$, and $N_s \rightarrow \alpha^{d-1}N_s$; so for $a^{-1} < q < (\zeta a)^{-1}; NS(q) \rightarrow 0$ as $\alpha \rightarrow \infty$, the thermodynamic limit. This means the compressibility of our system is zero, which is correct because we have assumed no physics for our system that would lead to anything else than an incompressible system. Inclusion of fluctuations with various length scales would create various compressibilities, but this is beyond the scope of this discussion.

As mentioned above, for $(\zeta R_{nn})^{-1} < q < R^{-1}$ the scattering is due to every cluster in the ensemble scattering with phases random with respect to the rest. The regions $R^{-1} < q < a^{-1}$ and $a^{-1} < q$ were described above in reference to figure 5.16.

5.7 Assessment

The scaling approach is based entirely on wave physics, which includes linear superposition and the energy proportional to the amplitude squared. It applies to all waves. It and the concept of a system of scatterers are quite powerful in accurately describing the general features of wave scattering from a variety of systems. It also enhances physical intuition for the scattering process. However, its simplicity leaves it unable to describe any detail in the crossover regions as q^{-1} passes through a characteristic length scale of the system, hence figures 5.5, 5.7, 5.9 and 5.10 show unphysical sharp turns at the crossovers. It also does not predict the interference ripples present in scattering from objects with sharp surfaces. Nevertheless, its simplicity belies its power.

The scaling approach makes up for these deficiencies by yielding a general picture of power laws and length scale dependent crossovers for the structure factor for a system of scatterers, whatever that system might be. Perhaps the most attractive feature of the approach is that the concept of a system of scatterers is general enough to describe a wide variety of experimental situations including solid spheres with both smooth and fractally rough surfaces, fractal aggregates, and any ensemble of particulates in a scattering volume with or without correlations between the particulates. Thus, all these scattering phenomena are unified and physical insight is gained.

References

Bibette J, Mason T G, Gang H and Weitz D A 1992 Kinetically induced ordering in gelation of emulsions *Phys. Rev. Lett.* **69** 981–84

Bohren C F and Huffman D R 1983 *Absorption and scattering of light by small particles* (New York: Wiley)

Butler B D, Hanley H J M, Hansen D and Evans D J 1995 Dynamic scaling in an aggregating 2D Lennard–Jones system' *Phys. Rev. Lett.* **74** 4468–71

Carpineti M and Giglio M 1992 Spinodal-type dynamics in fractal aggregation of colloidal clusters *Phys. Rev. Lett.* **68** 3327–30

Carpineti M and Giglio M 1993 Transition from semiorder to disorder in the aggregation of dense colloidal solutions *Phys. Rev. Lett.* **71** 656

Cerda J J, Sintes T, Sorensen C M and Chakrabarti A 2004 Structure factor scaling in colloidal phase separation *Phys. Rev.* E **70** 051405
Earnshaw J C, Harrison M B J and Robinson D J 1996 Local order in two-dimensional colloidal aggregation *Phys. Rev.* E **53** 6155–63
Ebini R H and Sorensen C M 2019 Light scattering studies of the sol-to-gel transition in particulate systems *J. Colloid Interface Sci.* **556** 577–83.
Egelstaff P A 1967 *An Introduction to the Liquid State* (London: Academic)
Furukawa H 1985 A dynamic scaling assumption for phase-separation *Adv. Phys.* **34** 703–50
Guinier A and Fournet G 1955 *Small-angle Scattering of X-rays* (New York: Wiley)
Hasmy A, Foret M, Pelous J and Jullien R 1993 Small-angle neutron-scattering investigation of short-range correlations in fractal aerogels: simulations and experiments *Phys. Rev.* B **48** 9345–53
Haw M D, Poon W C K and Pusey P N 1997 Structure and arrangement of clusters in cluster aggregation *Phys. Rev.* E **56** 1918–33
Hecht E 2002 *Optics* (Reading, MA: Addison-Wesley)
Huang H, Oh C and Sorensen C M 1998 Structure factor scaling in aggregating systems *Phys. Rev.* E **57** 875–80
Kerker M 1969 *The scattering of light, and other electromagnetic radiation* (New York: Academic Press)
Klein R and D'Aguanno B 1996 ed W Brown *Light Scattering Principles and Development* (Oxford: Clarendon)
Koylu U O and Faeth G M 1994 Optical-properties of overfire soot in buoyant turbulent-diffusion flames at long residence times *J. Heat Transfer-Trans. ASME* **116** 152–59
Martin J E and Hurd A J 1987 Scattering from fractals *J. Appl. Crystallogr.* **20** 61–78
Maughan J B and Sorensen C M 2019 Application of the scaling approach to particles having simple, fundamental shapes, in the Rayleigh–Debye–Gans diffraction limit *J. Quant. Spectrosco. Radiat. Transf.* **222** 190–95
Mishchenko M I, Travis L D and Lacis A A 2002 *Scattering, absorption, and emission of light by small particles* (Cambridge: Cambridge University Press)
Oh C and Sorensen C M 1999 Scaling approach for the structure factor of a generalized system of scatterers *J. Nanoparticle Res.* **1** 369–77
Porod G 1951 Die Rontgenkleinwinkelstreuung von dichtgepackten kolloiden systemen 1 *Kolloid-Zeitschrift* **124** 83–114
Robinson D J and Earnshaw J C 1993 Long-range order in 2-dimensional fractal aggregation *Phys. Rev. Lett.* **71** 715–8
Sidebottom D L 2012 *Fundamentals of Condensed Matter and Crystalline Physics: An Introduction for Students of Physics and Materials Science* (Cambridge: Cambridge University Press)
Sorensen C M 2001 Light scattering by fractal aggregates: a review *Aerosol Sci. Technol.* **35** 648–87
Sorensen C M, Oh C, Schmidt P W and Rieker T P 1998 Scaling description of the structure factor of fractal soot composites *Phys. Rev.* E **58** 4666–72
Takenaka M, Izumitani T and Hashimoto T 1992 Slow spinodal decomposition in binary-liquid mixtures of polymers. 5. Molecular-weight dependence in the later stage of unmixing *J. Chem. Phys.* **97** 6855–63
van de Hulst H C 1957 *Light scattering by small particles* (New York: Wiley)

Chapter 6

Rayleigh scattering

Rayleigh scattering is the term used to describe elastic scattering from small entities, such as particles or thermodynamically driven fluctuations in the density of media. Of course 'small' is relative. In the discussion of particulate light scattering, a particle is small when compared to the only other length scale involved, the wavelength of light.

The quest to understand why the sky is blue led to the results of Rayleigh scattering. Many scientists, including Rayleigh, all of whom made significant contributions to the eventual understanding, followed that quest. A good review of this history is given by Kerker (1969).

For us, Rayleigh scattering represents an important reference point for light scattering by all types of particles. In this book up to this point, diffraction has been established as the foundation for scattering of all types of waves including light. Diffraction from a very small particle, a Rayleigh particle, is trivial because the diffraction, the scattering, is isotropic. However, with Rayleigh scattering, we encounter for the first time the electromagnetic part of the light wave. Thus, for us Rayleigh scattering is not only the limit for scattering when the particle becomes small, but it is the simplest limit that deals with light's electromagnetic nature. As we will see later in this book, the influence of Rayleigh scattering will remain as size, refractive index and shape evolve away from their small or simple limits.

6.1 Dimensional analysis

If the particle is small compared to the optical wavelength, the light completely penetrates the particle; the particle is uniformly illuminated by the incident light. Furthermore, if the particle is small compared to the optical wavelength, the phase of the incident wave across the particle is essentially constant. We can envision the particle as a system of point-like scatterers, as we defined in chapter 4, figure 4.1, so each scatterer within the particle sees the same wave magnitude and phase. Given the small particle, the scattered waves from each scatterer travel essentially the same

doi:10.1088/978-0-7503-5157-7ch6

distance to the detector to within a small fraction of the wavelength, hence they arrive at the detector essentially in phase. Recall equation (1.11) to see that if R is a characteristic spatial dimension of the particle, then the greatest change in phase across the particle is $\Delta\,\phi \simeq kR$, where $k = 2\pi/\lambda$. The argument above can be restated in terms of phase by stating that the condition for Rayleigh scattering is that $\Delta\phi \ll 1$ for any two waves encountering any two scatterers within the particle. Thus, we will (for now, see below) impose the following condition for Rayleigh scattering by a particle

$$kR \ll 1 \tag{6.1a}$$

The amplitude of the total wave is the in-phase sum of all the waves, which is proportional to the volume of the particle. Intensity goes as amplitude squared, so the scattered intensity is proportional to the square of the volume of the particle, viz.

$$I \sim V_{\text{part}}^2 \tag{6.2}$$

The scattered intensity is proportional to the scattering cross section, which must have dimensions of area, length squared. Equation (6.2) shows a length to the sixth power. We need another length scale to bring this down to cross section dimensions.

There are only two length scales involved in the scattering, the particle size and the wavelength of the light. Since we have already used the particle size in V_{part}, we only have the optical wavelength, λ, remaining. Thus, the only way to achieve the proper dimensions for the cross section is to divide by λ^4. Then

$$C_{\text{sca}} \sim V_{\text{part}}^2/\lambda^4 \tag{6.3}$$

This simple but powerful dimensional analysis has other implications beyond those in equation (6.3). While we were concerned about the size of the scattering particle, nowhere in the analysis was shape involved. Thus, the result (6.3) is independent of shape. Nowhere in the analysis did we consider scattering angle. Since angle is not involved, there is no functionality with angle; the scattering is isotropic. Quickly, however, we remind ourselves that we left out polarization effects. Polarization has direction and that breaks isotropic symmetry. Since our fundamental condition is small, polarization must affect the scattering in the same manner that it does for the smallest entity possible, the radiating dipole. Thus, we conclude Rayleigh scattering obeys the angular and polarization laws of dipole radiation, which includes isotropy in the plane perpendicular to the incident polarization. This will be described completely in figure 6.1.

Also missing in equation (6.3) are the optical properties of the particle as described by the real and imaginary parts of the refractive index, $m = n + i\kappa$. Below, when we describe the complete electromagnetic theory of Rayleigh scattering, we will see that the scattering magnitude has, not surprisingly, a strong dependence on refractive index. Before that, however, recall from elementary optics that the wavelength of light in a medium is the vacuum wavelength divided by the real part of the refractive index, λ/n. This implies that 'small' means small compared to the wavelength in the medium. Furthermore, the imaginary part could absorb the light before it uniformly fills the particle. The absorption length is $1/k\kappa$ (section 7.2.1),

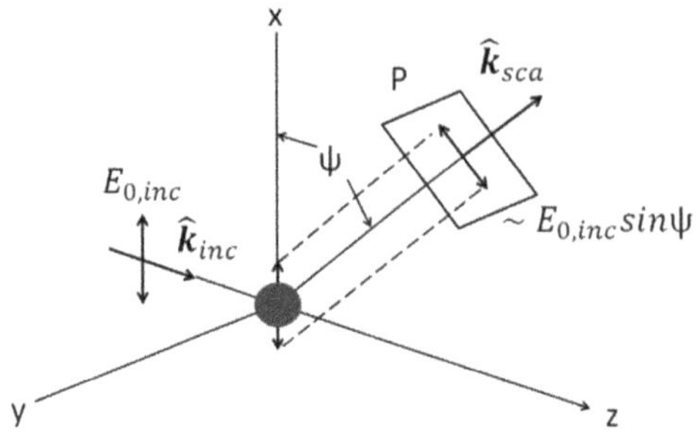

Figure 6.1. Light with an incident unit wave vector $\hat{k}_{\text{inc}}$ in the z-direction and amplitude $E_{0,\text{ inc}}$ scatters from a Rayleigh particle in the $\hat{k}_{\text{sca}}$ direction at an angle ψ from the incident vertical polarization direction. The polarization of the scattered light is the projection of the incident polarization onto the plane P that is perpendicular to the scattered direction. The magnitude of the scattered amplitude has the proportionality $E_{0,\text{ sca}} \sim E_{0,\text{ inc}}\sin\psi$.

so for the Rayleigh condition of uniform illumination to hold we require $R < 1/k\kappa$. These two needs combine to one additional condition for Rayleigh scattering, equation (6.1b). We conclude the conditions for Rayleigh scattering are

$$kR \ll 1 \tag{6.1a}$$

$$|m|kR \ll 1 \tag{6.1b}$$

These conditions are discussed again when we consider 'Rayleigh absorption' in section 7.2.2.

A similar argument for Rayleigh scattering can be made by recalling the lessons learned in chapter 5 that identified the length scale of the scattering experiment as the inverse of the scattering wave vector magnitude, q^{-1}. The largest q is $2k$; hence the finest resolution of the scattering is $1/2k$. If $1/2k > R$, i.e. if $2kR < 1$ (the Rayleigh condition, equation (6.1a) to within a factor of two), the light scattering can't resolve the particle. Then there is no detection of shape and no angular dependence for the scattering.

In the lab with visible light, $\lambda \simeq 500$ nm. Then $kR = 1$ when $R \simeq 80$ nm, a diameter of 160 nm. We conclude that Rayleigh scattering applies to particles only tens of nanometers in diameter.

6.2 The Rayleigh differential scattering cross section for a sphere: electromagnetic theory

The electromagnetic theory for Rayleigh scattering by a sphere assumes that the sphere experiences a uniform electric field across its volume, an assumption consistent with the notion that the wavelength of the light is much greater than the size of the sphere. This field polarizes the sphere in accordance with the Clausius–Mossotti equation (Jackson 1999). Since the incident light contains a time oscillating electric field, the polarization of the entire sphere oscillates in phase with the incident light to become an oscillating electric dipole. Electromagnetic theory predicts that an oscillating electric dipole radiates energy proportional to the dipole moment squared and the frequency to the fourth power. Rayleigh scattering is the result of the polarization of the sphere oscillating like a dipole hence radiating light. Thus, the Rayleigh differential scattering cross section for a spherical particle of radius R and relative refractive index of refraction m, possibly complex, is (Kerker 1969)

$$dC_{sca}/d\Omega = \frac{16\pi^4 R^6}{\lambda^4}\left|\frac{m^2-1}{m^2+2}\right|^2 \sin^2\psi \tag{6.4}$$

$$= k^4 R^6 \left|\frac{m^2-1}{m^2+2}\right|^2 \sin^2\psi \tag{6.5}$$

In equation (6.5), $k = 2\pi/\lambda$ and ψ is the angle between the incident polarization and the propagation direction of the scattered light, see figure 6.1. The electromagnetic theory provides, once again, the characteristic R^6/λ^4 functionality as found from the dimensionality argument above ($V_{part} = 4\pi R^3/3$). The R^6 term comes from the dipole moment being proportional to the square of the volume of the sphere, the λ^{-4} from the frequency to the fourth power, and the index of refraction dependence from the manner in which a static electric field polarizes a sphere as described by the Clausius–Mossotti equation. The dynamic incarnation of the Clausius–Mossotti term is the *Lorentz–Lorenz term*, the square of which is often written as

$$F(m) = \left|\frac{m^2-1}{m^2+2}\right|^2 \tag{6.6}$$

Then, simply

$$dC_{sca}/d\Omega = k^4 R^6 F(m)\sin^2\psi \tag{6.7}$$

Given this differential cross section, the scattered intensities can be found using equations (2.7) and (2.8) augmented for polarization effects. Because the polarization is oscillating like an electric dipole, the polarization laws for a dipole, displayed in figures 3.3 and 3.4, apply to the Rayleigh particle as well. This fact is embodied in figure 6.1. In a typical laboratory setting, one might have an incident laser beam with intensity I_{inc} traveling in the horizontal plane with vertical polarization, see figure 6.2. If the scattered light was also in the horizontal plane at an arbitrary angle θ from the incident direction (the scattering angle), it would be vertically polarized as well. Then the angle $\psi = 90°$ and by equation (2.8), the scattered intensity at a detector at a distance r due to a single particle is

$$I_V = \frac{k^4 R^6}{r^2} F(m) I_{inc,\,V} \tag{6.8}$$

where $I_{inc,V}$ is the incident intensity of the vertically polarized light, r is the distance from the particle to the detector, and I_V is the intensity of the detected light which is vertically polarized. Note that the scattering is isotropic, independent of the scattering angle, in the incident plane. If, with the same scattering geometry, the incident light is horizontally polarized, the scattered intensity would be the same as

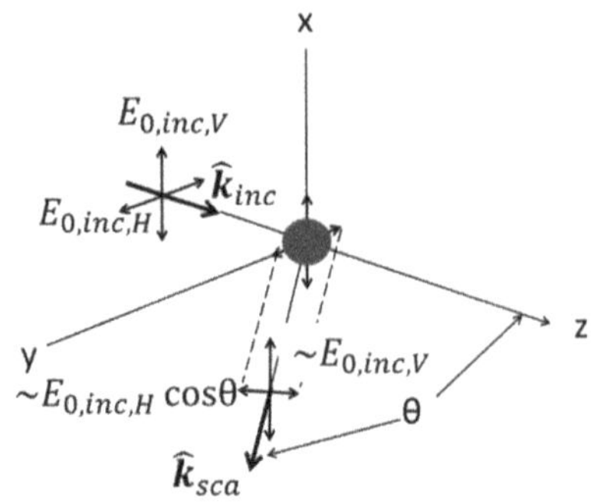

Figure 6.2. Light with an incident unit wave vector $\hat{k}_{\rm inc}$ in the *z*-direction with two polarization amplitudes, horizontal and vertical. The light scatters in the *yz* plane from a Rayleigh particle at an angle θ. The scattered polarizations are shown. The vertical polarization is angle independent; the horizontal has a $\cos\theta$ functionality. These functionalities are a result of the projection of the incident polarizations on a plane perpendicular to the scattered direction as indicated for the horizontal case by the dashed lines. Remember the scattered intensity is proportional to the field squared.

equation (6.8) but multiplied by a factor of $\cos^2\theta$ due to the $\sin^2\psi$ functionality of equation (6.7) with $\psi = 90° - \theta$. This situation is illustrated in figure 6.2 as well. Thus,

$$I_{\rm H} = \frac{k^4R^6}{r^2}F(m)\cos^2\theta I_{\rm inc,\,H} \tag{6.9}$$

where $I_{\rm inc,H}$ is the incident intensity of the horizontally polarized light, r is the distance from the particle to the detector, and $I_{\rm H}$ is the intensity of the detected light, which is horizontally polarized.

Finally, if, with the same scattering geometry, the incident light is unpolarized, as might occur with thermal light sources (e.g. the Sun, incandescent light bulbs), the scattered intensity would be

$$I = \frac{k^4R^6}{r^2}F(m)I_{{\rm inc},\,U}\frac{(1+\cos^2\theta)}{2} \tag{6.10}$$

where $I_{\rm inc,U}$ is the incident intensity of the unpolarized polarized light, r is the distance from the particle to the detector, and I is the intensity of the detected light that has a polarization described by the linear depolarization ratio described below. Equation (6.10) is a linear combination of equations (6.8) and (6.9). The factor of ½ accounts for the equal intensities of the two incident polarizations.

A consequence of all this is that the manner in which the scattered intensity varies with scattering angle depends only on the incident polarization. This is illustrated in figure 6.3.

The polarization can be quantified by the degree of linear polarization, equation (3.11), which for this situation is

$$P_{\rm l} = \frac{I_{\rm V} - I_{\rm H}}{I_{\rm V} + I_{\rm H}} \tag{6.11}$$

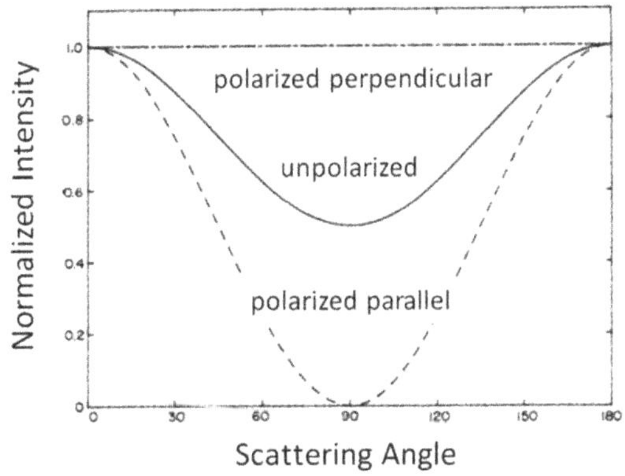

Figure 6.3. Plot of the angular distribution of the scattered intensities from a Rayleigh particle for incident light polarized perpendicular, unpolarized and parallel to the scattering plane.

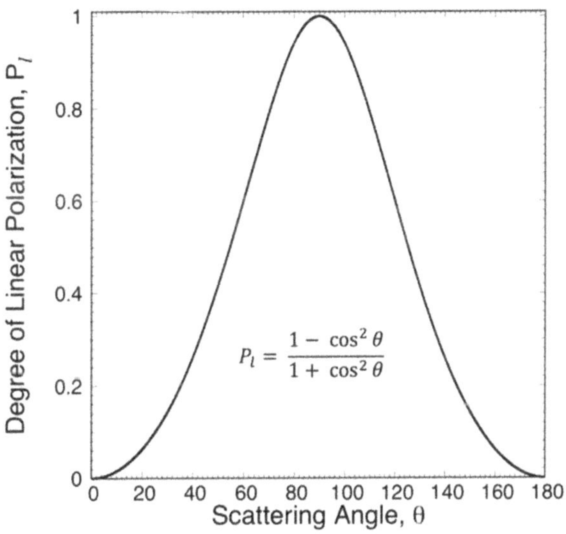

Figure 6.4. Degree of linear polarization of light scattered by a small particle in the Rayleigh regime scattering unpolarized light.

where I_V and I_H are the intensities of vertically and horizontally polarized scattered light, respectively. For Rayleigh scattering equations (6.8) and (6.9) lead to

$$P_l = \frac{1 - \cos^2\theta}{1 + \cos^2\theta} \tag{6.12}$$

This result is shown graphically in figure 6.4.

Equation (6.12) and figure 6.4 show that Rayleigh scattered light is completely linearly polarized, $P_l = 1$, as described in equations (3.11) and (3.12), in the vertical direction when $\theta = 90°$. .This is the polarization perpendicular to the plane of incidence, and there is no scattering of the polarization parallel to the plane of

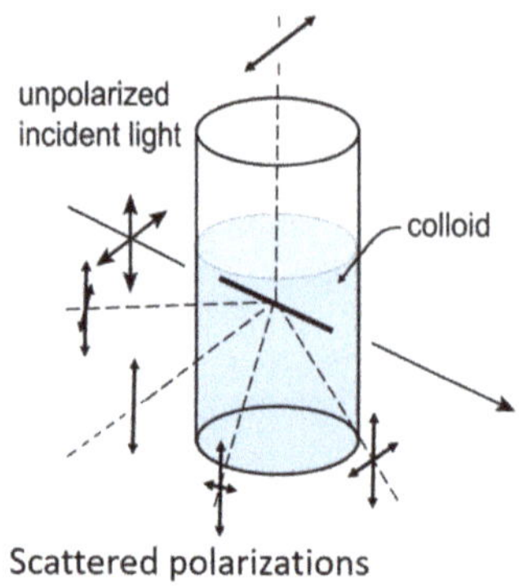

Figure 6.5. Unpolarized light incident on a colloid producing polarized scattered light. The plot in figure 6.4 applies to this drawing.

incidence. We shall see later, with figure 7.9, that as the particle size grows out of the small, Rayleigh limit, this maximum degree of polarization evolves to larger angles and decreases below unity. Figure 6.5 qualitatively demonstrates polarization upon Rayleigh scattering and reminds us that the scattering plane is not always horizontal.

6.3 The total Rayleigh cross section

The total scattering cross section is found by integrating the differential cross section over the complete solid angle of 4π steradians, equation (2.6). Notice that the Rayleigh differential cross section only has an angle dependence due to the polarization via the angle ψ. The solid angle integration of $\sin^2\psi$ yields a factor of $8\pi/3$. Thus, the total Rayleigh scattering cross section is

$$C_{\text{sca}} = \frac{8\pi}{3}k^4R^6F(m) \tag{6.13}$$

With this, the scattering efficiency, which is the ratio of the total scattering cross section to the geometric cross section, πR^2, is a model of simplicity

$$Q_{\text{sca}} = \frac{8}{3}\alpha^4F(m) \tag{6.14}$$

where $\alpha = kR$ is the size parameter. For Rayleigh scattering $\alpha < 1$ so the scattering efficiency is very small, which is to say that the scattering is much less than one would infer from the physical size of the sphere. This view emphasizes the weakness of Rayleigh scattering.

Summary

In summary, important characteristics of Rayleigh scattering are:

1. The scattering is proportional to the wavelength to the inverse fourth power, λ^{-4}.[1]

[1] If one includes refractive index dispersion, the power is typically larger than fourth, e.g. for small water droplets one would have an approximate $\lambda^{-4.12}$ dependence.

2. The scattering is proportional to the square of the particle volume, V_{part}^2. For a spherical particle this means that the scattering is proportional to the radius to the sixth power, R^6. For a non-spherical particle, R is the volume equivalent radius.
3. The $V_{\text{part}}^2/\lambda^4$ proportionality is independent of particle shape and index of refraction.
4. The scattering obeys the laws of dipole radiation with isotropy in the scattering plane when the incident polarization is perpendicular to the scattering plane.

6.4 Consequences of Rayleigh scattering

6.4.1 Blue skies, red sunsets

Perhaps the best-known consequence of Rayleigh scattering is the bluing of scattered light; the prime example being the blue color of our sky. The molecules in air, mostly N_2 and O_2, are much smaller than the optical wavelength, hence they are Rayleigh scatterers. Furthermore, they are randomly placed so that the scattered intensities from the molecules can be summed without concern to the wave nature of light, i.e. we need only sum their scattered intensities not their phase dependent amplitudes. Consider that the visible wavelength range is ca. 400 nm to 700 nm, purple to red. The inverse fourth power law then implies a factor of $(400/700)^{-4} \simeq 9.4$, essentially an order of magnitude greater scattering of the shorter visible wavelengths than the longer! This seems to imply a purple sky, however, there are other factors involved for the sky to look blue, most notably convolution with the optical response of the human eye. For an excellent description and story of this topic, I refer the reader to Pesic (2005).

If 'blue' light preferentially scatters out of a beam of white light, 'red' light must remain in the transmitted beam. This is the concomitant explanation of the red of the sunrise and sunset. The transmitted light is significantly reddened at sunrise and sunset because the path through the atmosphere is greatest at this time. However, don't forget this reddening happens even at noon to somewhat yellow the Sun. Indeed, I find the Sun significantly whiter when viewed from on top of a 14 000 foot Colorado mountain than from the Kansas prairie at 1100 feet.

The bluing of scattered and the reddening of transmitted light is readily apparent in the laboratory too, and the alert experimentalist can take advantage of this to qualitatively size the particles. Figure 6.6 shows two pictures of two colloidal suspensions of polystyrene particles with different sizes: radii $R = 40$ nm on the left of each picture and $R = 4800$ nm on the right. When viewed illuminated under normal room circumstances from a perspective that allows us to see the scattered light as in the left picture, we can see the small particle colloid has a bluish tinge indicating Rayleigh scattering and, importantly, telling us that the particles are small compared to the wavelength of light (which indeed they are). The white scattering of the larger colloidal particles on the right is due the fact that these particles are too large to obey Rayleigh scattering. In fact, as spheres, Mie scattering, see chapter 7, is appropriate to describe their scattering. The picture on the right shows the same two colloids but now from a perspective in which a white sheet of paper is behind each

Figure 6.6. Two pictures of two colloidal suspensions with different size polystyrene particles with radii $R = 40$ nm on the left of each picture and $R = 4800$ nm on the right in each picture. The left-hand picture is from a perspective that shows the scattered light. The smaller particle colloid shows bluish scattering indicating Rayleigh scattering to imply that they are smaller than the wavelength of light. The bigger particle colloid shows whitish scattering indicating they are not Rayleigh scatterers, hence are larger than the wavelength of light. The right-hand picture is from a perspective that shows the transmitted light. The smaller particle colloid shows reddish transmission indicating Rayleigh scattering removing some blue hence leaving some red for transmission, the bigger particle colloid shows whitish transmission because there was no preferential scattering of any color due to their relative largeness.

colloid. From this perspective, we see the transmitted light, and that light now appears reddened for the smaller, Rayleigh scattering colloid.

Although the subject of scattering by particles larger than the wavelength of light will be discussed extensively in later chapters, a dimensionality argument can be made now to explain the white scattering in figure 6.6. Consider a very big particle, a baseball for example. We would expect that its scattering cross section is its geometric cross section proportional to its radius squared. If so, unlike a Rayleigh particle that has a scattering cross section proportional to it linear size to the sixth power and needs an inverse wave length to the fourth power to get the scattering cross section the correct dimension, the baseball needs no wave length functionality; the geometric units are already correct. Thus, we conclude that for baseballs and other particles bigger than the wavelength of light, the scattering cross section is not a function of the wavelength so all colors scatter the same. White light in; white light out.

6.4.2 Polarization effects

Polarization consequences of Rayleigh scattering are also visible in the atmosphere. Recall that the key concept to polarization due to dipole radiation was that the

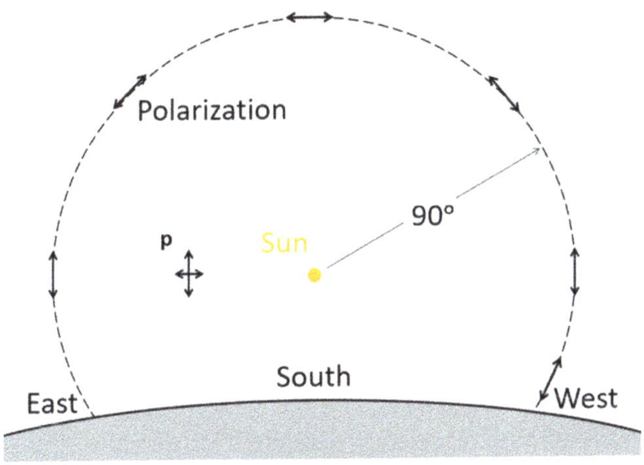

Figure 6.7. Sky polarization looking south at local noon on a cloudless day. Maximum polarization occurs along the dashed line circle 90° of arc from the Sun due to Rayleigh scattering from the molecules in the air. The polarization direction is tangential to the circle, as indicated by double headed arrows, hence perpendicular to a line from the Sun to the point of observation, as drawn. At other locations in the sky, such as point *p*, the tangential polarization persists, but a radial component is present as well to lead to partially polarized light. In fact, the sky light in not perfectly polarized along the dashed circle due to other, non-Rayleigh scattering effects.

projection of the dipole[2] vibration direction onto a plane perpendicular to the observation direction (the scattered direction) determines the polarization of the emitted radiation. Outside, when you look at the sky 90° of arc away from the Sun, the sky you see is due to light scattered by the air molecules at a 90° angle. See figure 6.7. On a clear day, the blue scattering is Rayleigh scattering. The unpolarized incident light from the Sun can be resolved onto two, equal perpendicular components. At a 90° scattering angle, one of these components sets the dipoles in the air molecules oscillating along the direction of observation. Such an oscillation has no projection on the plane perpendicular to the observation direction, hence no radiation is emitted, i.e. no scattering is caused by this polarization in that direction. On the other hand, the other polarization component, perpendicular to the first one, completely projects onto the plane perpendicular to the observation direction. With this second polarization, scattering ensues with a polarization perpendicular to the line in the sky from the Sun to the part of the sky being observed. The polarization typically is not complete due to depolarization effects from a variety of other sources, such as multiple scattering. Moreover, the polarization is only partial at angles other than 90°.

6.4.3 The Tyndall effect

The *Tyndall effect* is the increased scattering as a particulate system coarsens due to aggregation. For an ensemble of particles with volume V_{part} and number density n

[2] Dipoles can be either induced or permanent. In air they are induced.

the total scattering from the system is proportional to the scattering cross section per particle times the particle number density, viz.

$$I \sim nC_{\text{sca}} \sim nV_{\text{part}}^{\;2} \sim nV_{\text{part}} \cdot V_{\text{part}} \tag{6.15}$$

The last proportionality in equation (6.15) holds the Tyndall effect. The term nV_{part} is the total volume of particulate matter per unit volume of the system and, barring other processes such as settling, is conserved during aggregation. Thus, the total scattering involves this constant times the volume of the particles, which does increase with aggregation, hence the scattering increases as well. Of course, we must remember that this holds so long as the particles are in the Rayleigh regime, smaller than the wavelength of light.

A nice example of the Tyndall effect is the formation of a fog. A quantity of air can contain a given amount of gaseous, molecular water above the dew point. If the temperature falls below the dew point, the number density of water molecules in the air does not change significantly[3], but water's equilibrium phase state does. Below the dew point, some of the water becomes liquid and forms tiny droplets. Tiny as they are, they are still larger than the individual molecules of water that solely existed above the dew point. The amount of water in the air was conserved, but the size of the water particles grew from molecular dimensions to droplet sizes. By the last proportionality in equation (6.15), the system scatters more light and a fog is seen.

Another example of the Tyndall effect is precipitation of solids from solution. Add an aqueous solution of NaCl to an aqueous solution of $AgNO_3$ and a white cloud of precipitating AgCl will be seen. The amount of material during this reaction doesn't change, but the size of the entities does, from molecular to growing particles.

On might say an inverse Tyndall effect exists too. As a system of particles becomes more finely divided, it scatters less. This is the dilemma of scattering from small particles; the R^6 functionality is very rapid. This means a particle twice the diameter of another scatters 64 times more light! Simply said, small particles don't scatter much.

6.5 Rayleigh absorption

If the refractive index is complex, $m = n + i\kappa$, with a non-zero imaginary part κ, particles can absorb the light as well as scatter it. For small particles, we shall call this *Rayleigh absorption* although, as Bohren and Huffman point out (Bohren and Huffman 1983), Rayleigh did not explicitly consider absorption. As for scattering, a simple dimensionality analysis can be made for the absorption cross section. Since the particle is very small, the incident wave completely penetrates the volume of the particle. We can assume that this complete penetration occurs even though the particle has an imaginary component to the refractive index, which could cause absorption near the surface for a larger particle (when $\kappa kR \gtrsim 0.3$, see section 7.2.1). Absorption is not dependent on the phase of the light; it is an incoherent process

[3] The volume will contract hence the number density will rise approximately 0.3% K^{-1}.

independent of light's wave nature. Thus, the absorption cross section, C_{abs}, is proportional to the amount of stuff absorbing, the volume of the particle. Thus, $C_{abs} \propto V_{part}$. To get the cross section dimensions correct the cubic length of volume must be reduced to length squared, and this must be done with another length not involved in the size of the particle, but one still involved in the scattering process. That other length is (as above) the wavelength of light. Thus, we must have

$$C_{abs} \propto V_{part}/\lambda \tag{6.16}$$

Electromagnetic theory leads to a result consistent with (6.16)

$$C_{abs} = \frac{8\pi^2 R^3}{\lambda} Im\left(\frac{m^2 - 1}{m^2 + 2}\right) \tag{6.17}$$

$$C_{abs} = 4\pi k R^3 \, Im\left(\frac{m^2 - 1}{m^2 + 2}\right) \tag{6.18}$$

$$C_{abs} = 4\pi k R^3 \, E(m) \tag{6.19}$$

where

$$E(m) = Im\left(\frac{m^2 - 1}{m^2 + 2}\right), \tag{6.20}$$

and where *Im* means imaginary part. The absorption efficiency is simply

$$Q_{abs} = 4kRE(m) = 4\alpha E(m) \tag{6.21}$$

where $\alpha = kR$, the size parameter.

6.6 Rayleigh extinction

Extinction is the sum of scattering plus absorption. If the particle's refractive index is purely real, there is no absorption so extinction equals scattering. If there is an imaginary part to the particle's refractive index, the absorption will usually dominate the scattering in the Rayleigh regime where the size parameter α is small because from equations (6.13) and (6.19)

$$C_{sca}/C_{abs} = \frac{2}{3}\alpha^3 F(m)/E(m) \tag{6.22}$$

Equation (6.22) shows that when the size parameter $\alpha < 1$, as it is for Rayleigh scattering, $C_{sca}/C_{abs} \ll 1$ when $E(m)$ and $F(m)$ have similar magnitudes.

To get a feeling for the numerical values of $E(m)$ and $F(m)$ consider flame generated soot that has a refractive index of approximately $m = 1.6 + i0.6$. Then $E(m) = 0.27$ and $F(m) = 0.24$.

This difference between scattering and absorption for Rayleigh particles can be illustrated with the pictures of gold nanoparticle colloids in figure 6.8. Gold has a complex refractive index that has a strong wavelength dependence (highly

Figure 6.8. Two colloids of gold nanoparticles with diameters of 30 nm and 120 nm in water. The colloids have the same mass concentration. Absorption dominates scattering in the smaller colloid; scattering dominates in the larger.

dispersive). Two particle sizes are shown with diameters of $2R = 30$ nm and 120 nm. The gold concentrations are the same, 2 mg mL^{-1} in water. On the left, the 30 nm colloid shows no obvious scattering but a large absorbance. The size parameter in visible light is about $\alpha \simeq 0.36$, hence $\alpha^3 \simeq 0.05$. This small number substituted into equation (6.22) implies that the scattering is small compared to the absorbance, as is seen. On the other hand, the colloid pictured on the right has $2R = 120$, thus $\alpha \simeq 1.44$ and $\alpha^3 \simeq 3$. For these particles scattering should be comparable and even larger than absorption, which it is, even though for such a size the Rayleigh regime is passing out of validity.

Before we leave figure 6.8, note that the gold concentration is the same in each colloid but the scattering is greater for the larger particles (the absorption not withstanding). This is a demonstration of the Tyndall effect.

The different functionalities of absorbance and scattering for particles in the Rayleigh size range opens the possibility for light scattering measurement of the particle size and number density for small particles; so called scattering/extinction measurements (D'Alessio 1981). The scattered intensity at an arbitrary angle is proportional to the differential cross section and the number density of particles, n. Thus,

$$I_{sca} = CnR^6F(m) \tag{6.23}$$

In (6.23) C is a constant determined via calibration by scattering from a known scatterer such as a gas or liquid with known Rayleigh Ratio (see below). The problem with equation (6.23), however, is that it has two unknowns, n and R. Another independent measurement is needed with different functionality on n and R to give a system of two equations with two unknowns. That second equation is provided by an extinction measurement. Extinction is the sum of scattering and absorption, but, as equation (6.22) shows, absorption dominates scattering in the

Rayleigh regime (when $E(m)$ is comparable to $F(m)$) so that extinction is essentially entirely due to absorption. Extinction is measured with a transmission measurement to yield the turbidity as describe in section 2.5, equation (2.12), the Beer–Lambert law. The turbidity in the Rayleigh regime is given by

$$\tau = 4\pi nkR^3E(m) \tag{6.24}$$

Equations (6.23) and (6.24) represent two equations that can be solved for the two unknowns n and R. Note that this method requires knowledge of the refractive index of the particles.

6.7 Rayleigh albedo

The albedo for Rayleigh particles is found by combining equations (6.13), (6.19) and the definition of albedo, equation (2.10). Remembering that $C_{abs} \gg C_{sca}$ one obtains

$$\omega \simeq C_{sca}/C_{abs} = \frac{2}{3}\alpha^3F(m)/E(m) \tag{6.25}$$

where we have used the size parameter $\alpha = kR$. Since $\alpha \ll 1$ for a Rayleigh scatterer, $\omega \ll 1$ as well, and we conclude that small particles are very dark unless they have a perfectly real refractive index so that $E(m)$ is zero (Moosmuller *et al* 2021).

Both scattering/extinction measurements and the Rayleigh albedo approximation rely on $C_{sca} \ll C_{abs}$. Thus, it is worthwhile to know when this is true. Figure 6.9 plots the value of the size parameter as a function of the imaginary part of the refractive index, κ, for $C_{sca}/C_{abs} = 1.0$ and 0.1. This was calculated using the Mie equations for a sphere, which are valid for all sizes and refractive indices. The index of refraction is $m = 1.5 + i\kappa$. Below the $C_{sca}/C_{abs} = 0.1$ line, the condition $C_{sca}/C_{abs} \ll 1$ is quite accurate. Figure 6.9 tells us that the largest size parameters for which absorption dominates scattering and equation (6.25) holds are $\alpha = kR \lesssim 0.4$, and if the imaginary refractive index κ is different than ~0.5, the size must be smaller. $kR \simeq 0.4$ corresponds to diameters of $2R \simeq 70$ nm at optical wavelengths. For a more detailed description of the albedo in the Rayleigh regime, see Moosmuller and Sorensen (2018).

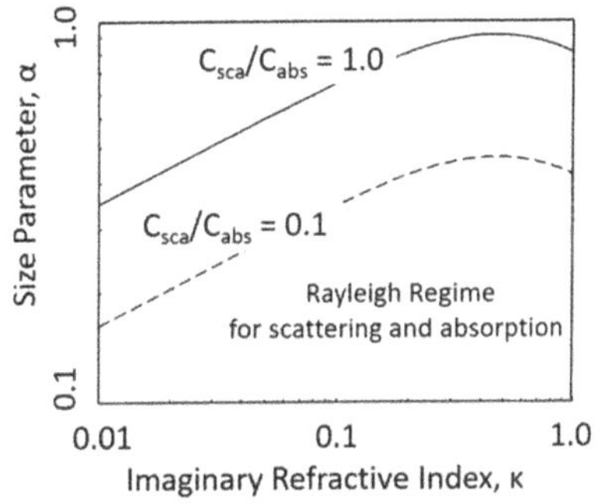

Figure 6.9. Plot of the size parameter as a function of the imaginary part of the refractive index, κ, under the conditions that $C_{sca}/C_{abs} = 1.0$ and 0.1. When C_{sca}/C_{abs} is small, the situation is in the Rayleigh regime for scattering and absorption. Note that when $\alpha = 0.4$ and $\lambda = 532$ nm, the diameter of the particle is $2R = 68$ nm.

6.8 The Rayleigh ratio

The Rayleigh ratio of a system of particles is the product of the particle scattering cross section and the particle number density of the system. There are two Rayleigh ratios, differential and total. For Rayleigh particles, i.e. those small enough for Rayleigh scattering to apply, these ratios are, respectively

$$\mathcal{R}_\theta = n dC_{\text{sca}}/d\Omega = nk^4R^6F(m)\sin^2\psi \tag{6.26}$$

and

$$\mathcal{R} = nC_{\text{sca}} = \frac{8\pi}{3}nk^4R^6F(m) = \frac{8\pi}{3}\mathcal{R}_\theta \tag{6.27}$$

The units of the differential Rayleigh ratio $\mathcal{R}_\theta$ are $\text{cm}^{-1}\ \text{sr}^{-1}$, the units of the total Rayleigh ratio $\mathcal{R}$ are cm^{-1}.

The utility of the Rayleigh ratio is at least three-fold: (1) it combines particulate number density and cross section into one single parameter that is readily measured, (2) in the absence of absorption the total Rayleigh ratio is the turbidity, a measure of optical transparency, and (3) it is a useful measure of scattering by liquids and gases.

Liquids and gases are not typically thought of as particulate media, their atomic nature notwithstanding, but they scatter light nevertheless, and the measure of their light scattering ability is the Rayleigh ratio. Table 6.1 contains Rayleigh ratios for a number of liquids and gases compiled previously by the author (Sorensen 2001).

Table 6.1. Differential Rayleigh ratios for some liquids and gases. The differential Rayleigh ratios below are for a temperature of 25 °C and a pressure of 1 atm; units are $10^{-6}\ \text{cm}^{-1}\ \text{steradian}^{-1}$. They are for scattering of polarized light in a plane perpendicular to the polarization. The literature reviewed for this table suggests that the uncertainty in these values is ca. ±5%. Gas Rayleigh ratios vary with λ^{-4} functionality. Liquid Rayleigh ratios are more complex and our cursory examination shows a rough $\lambda^{-4.7}$ variation. Adapted from Sorensen (2001).

Material	λ = 514.5 nm	λ = 488 nm	Reference
H_2	0.0038	0.0047	D'Alessio (1981)
Ar	0.0154	0.019	Rudder and Bach (1968)
O_2	0.0148	0.0183	D'Alessio (1981)
N_2	0.0173	0.0214	Rudder and Bach (1968)
CO_2	0.034	0.042	D'Alessio (1981)
CH_4	0.0386	0.0477	D'Alessio (1981)
Xe	0.0943	0.117	Rudder and Bach (1968)
C_2H_4	0.102	0.126	D'Alessio (1981)
C_2H_6	0.113	0.140	D'Alessio (1981)
CCl_4	13.5	17.1	Olivier *et al* (1992)
C_6H_6	23	29.1	Olivier *et al* (1992)
$C_6H_5CH_3$	24.2	31	Olivier *et al* (1992)
CS_2	82.2	111	Olivier *et al* (1992)

These Rayleigh ratios show a λ^{-4} dependence for the gases consistent with particulate Rayleigh scattering; atoms or molecules dispersed randomly and relatively widely separated. Liquid values are about three orders of magnitude greater than gas values, consistent with the ratio of liquid to gas densities, and roughly show a $\lambda^{-4.7}$ functionality.

The connection between liquid and gas media scattering and Rayleigh scattering by particles (composed of many atoms or molecules) can be made quite strong with the following calculation. A gas is, of course, a system of very small particles, atoms and molecules. Consider molecular nitrogen, N_2. A fit of pressure, volume and temperature data for N_2 to the van der Waals equation for gases (Atkins 2010) yields a molecular volume of 0.065 nm^3. Approximating the molecule as a sphere yields a radius of R = 0.25 nm. At T = 25 °C and p = 1 atm, the number density of N_2 molecules is $n = 2.47 \times 10^{22}$ cm^{-3}. For the refractive index of a N_2 molecule we will use the refractive index of liquid N_2 at 77 K, which is 1.199 (Johns and Wilhelm 1937). Then equation (6.26) with λ = 514.5 nm and $\psi = 90°$ yields $\mathcal{R}_\theta = 2.17 \times 10^{-8}$ cm^{-1} sr^{-1}. This compares quite favorably to the value for N_2 in table 6.1 (25% too big). The implication is that we can, to first order, understand scattering from gases (and liquids too) as Rayleigh scattering from the molecules behaving like particles.

The total Rayleigh ratio, as its units imply, can be inverted to get the *1/e* length for visual transparency. From equation (2.22) $\mathcal{R}^{-1}$ is also the photon mean free path. For air, which is a 4 to 1 mixture of N_2 and O_2, the data in table 6.1 and equation (6.27) imply $\mathcal{R} \approx 1.8 \times 10^{-7}$ cm^{-1} near the center of the visible spectrum, hence $\mathcal{R}^{-1}$ = 70 km. Thus, on a clear day, we can see about 70 km ≈ 40 miles. I personally recall many times first seeing the Rocky Mountains at about this distance when driving west along Interstate 70 on the high plains. And fortunately for astronomers, and all lovers of the night sky, this $\mathcal{R}^{-1}$ is much greater than the 1/*e* depth of the atmosphere (ca. 7 km) so we can see the stars at night!

6.9 Limits to the Rayleigh regime

Equations (6.1) define with certainty the Rayleigh regime. One might wonder if 'much, much less than one' is really necessary to fulfill the Rayleigh condition. In fact, it is valuable to know just how much we might cheat on these conditions and still use the Rayleigh results to good approximation. Figure 6.9 implies that at least $\alpha = kR \lesssim 0.4$ or smaller for the Rayleigh approximation regarding $C_{sca}/C_{abs} \ll 1$ to hold.

Considering scattering by itself, a fundamental characteristic of Rayleigh scattering is the isotropy of the scattering when the polarization is perpendicular to the scattering plane. Anisotropic scattering at small angles for any particle is well approximated by the Guinier equation, equation (4.43) in section 4.7. Using equation (4.43) the ratio of backscattering at an angle of θ = 180° to forward scattering and θ = 0° can be shown to be

$$I(180)/I(0) = 1 - (4/5)(kR)^2 \tag{6.28}$$

Equation (6.28) implies that a 10% reduction in the scattered intensity from θ = 0° to 180°, i.e. $I(180)/I(0) = 0.9$, occurs when kR = 0.35. This result is quite similar to that found from figure 6.9, both using a 10% error, but from rather different perspectives. The reader can continue this exercise for other estimates of anisotropy.

6.10 Epilogue

To conclude I now return to the forecast made in the Introduction to this chapter. Rayleigh scattering is the simplest limit that deals with light's electromagnetic nature. In the succeeding chapters, we will see remnants of Rayleigh scattering will remain as size, refractive index and shape become significant. In particular, Rayleigh scattering, or its straightforward modification by a universal parameter involving the size parameter and the Lorentz–Lorenz factor, will appear in the forward scattering of all particles regardless of size, refractive index or shape. When combined with diffraction, a universal description of light scattering by *all* particles will be achieved.

References

Atkins P W 2010 *Physical Chemistry* (New York: W. H. Freeman)

Bohren C F and Huffman D R 1983 *Absorption and Scattering of Light by Small Particles* (New York: Wiley)

D'Alessio 1981 Laser light scattering and flourescence diagnostics of rich flames, particulate carbon *Particulate Carbon* ed D C Siegla and G W Smith (New York: Plenum) pp 207–59

Jackson J D 1999 *Classical Electrodynamics* (New York: Wiley)

Johns H E and Wilhelm J O 1937 The refractive indices of liquid oxygen, nitrogen and hydrogen *Can. J. Res.* **15** 101–8

Kerker M 1969 *The Scattering of Light, and Other Electromagnetic Radiation* (New York: Academic)

Moosmuller H, Giri R, Sorensen C M and Berg M J 2021 Black metal nanoparticles from abrasion processes in everyday life: bicycle drivetrains and rock-climbing ropes *Opt. Commun.* **479** 126413

Moosmuller H and Sorensen C M 2018 Small and large particle limits of single scattering albedo for homogeneous, spherical particles *J. Quant. Spectrosc. Radiat. Transf.* **204** 250–5

Olivier B J, Sorensen C M and Taylor T W 1992 Scaling dynamics of aerosol coagulation *Phys. Rev.* A **45** 5614–23

Pesic P 2005 *Sky in a Bottle* (Cambridge, MA: MIT Press)

Rudder R R and Bach D R 1968 Rayleigh scattering of ruby-laser light by neutral gases *J. Opt. Soc. Am.* **58** 1260

Sorensen C M 2001 Light scattering by fractal aggregates: a review *Aerosol Sci. Technol.* **35** 648–87

Chapter 7

Light scattering and absorption by spherical particles

Light scattering by spheres is the central problem of light scattering. This is because spherical particles are common in nature and the lab. The spherical shape is the most symmetrical and economical of all shapes, and, as such, is, to some degree, an approximation to all other shapes. Moreover, the spherical symmetry is tractable to mathematical analysis. Thus, we desire to describe and understand scattering by spheres of arbitrary size and refractive index.

The solution for scattering of an electromagnetic wave by a sphere was obtained in various forms independently by Debye, Lorenz and Mie although Mie's paper in 1908 is usually the primary citation (Mie 1908)[1]. Thus, as is also customary, I will use the terms 'Mie theory' and 'Mie scattering' to mean the theory for and the scattering by a sphere of any size and refractive index. Mie theory solves the Maxwell equations for a plane wave encountering a spherical particle of arbitrary size and refractive index. The spherical shape imposes spherical boundary conditions. The theory yields a complex series expansion for both the scattered wave outside the particle and the internal field within the particle. This problem is analogous to the similar problem of an electromagnetic wave encountering a planar interface. This leads to the well-known Fresnel equations that describe reflection and refraction from a macroscopic interface (Hecht 2002); the reflected wave being analogous to the scattered wave and the refracted wave to the internal field. With the planar interface, the Cartesian boundary conditions, not surprisingly, lead to the much simpler Fresnel equations. It is, however, worthwhile to note that the process is the same: Maxwell's electromagnetic waves encountering a boundary of a given symmetry.

[1] Kerker gives a good historical account in Kerker M 1969 *The Scattering of Light, and Other Electromagnetic Radiation* (New York: Academic).

doi:10.1088/978-0-7503-5157-7ch7

The Mie solution is complex, and it wasn't until the advent of modern computers in the 1960s and 1970s that quick and accurate calculation of the solution was available. Now it is straightforward to quickly calculate the Mie solutions on a desk or laptop computer, or perhaps more conveniently, to use a Mie calculator found on the internet (Laven 2003) http://www.philiplaven.com/mieplot.htm. The previous great treatises on the subject of light scattering have typically described the complex mathematical path to the solution in detail. Here, with our modern setting (and given the author's limited abilities), we will use these wonderful calculational resources to study the Mie solutions systematically and extensively. Essentially, we will devote our energies not to the difficulties of obtaining the solution, but rather to digging deeply into the solutions themselves. We will act as empiricists by looking carefully at the solutions for patterns and generalizations.

7.1 The differential scattering cross section

7.1.1 The Rayleigh–Debye–Gans limit

Rayleigh scattering holds under the conditions of equations (6.1a) and (6.1b). With this foundation, it is conceptually useful to observe the behavior of scattering by any object with any shape as its size and refractive index grow. That is the general plan we will follow to gain an understanding of scattering (and absorption) by any particle.

The first step of our plan is to start with the simplicity of Rayleigh scattering, but relax the condition of small size, equation (6.1a), and allow the size parameter to be $kR > 1$. With this, condition (6.1b) will be broken as well because $|m|$ cannot be less than one. However, remember the index of refraction of the vacuum is $m = 1.0$ and the classical vacuum does not scatter light. Moreover, it is the contrast between the refractive indices of the particle and the medium in which it is immersed that leads to the particle's scattering. Most generally, m is the particle index normalized by the medium index. Thus, one can ad hoc propose a modification to condition (6.1b) as $|m - 1|kR \ll 1$. Since we wish to consider any size, our plan requires $|m - 1| \ll 1/kR$.

In the usual lab setting with incident light polarized vertically and scattering at an angle of θ in the horizontal plane, Rayleigh scattering has no angular dependence. An angular dependence will develop when the particle gets big enough so the waves emitted from different parts of the particle will have phase differences at the detector. These phase differences are accounted for by the structure factor, $S(q)$, the topic of chapter 4.

To complete our plan we combine the electromagnetic and wave natures of light to create the differential cross section for light scattering by a particle of arbitrary size. The simplest combination for any particle shape and any size is to multiply the Rayleigh differential cross section by the structure factor for that shape as

$$\text{Rayleigh–Debye–Gans (RDG)} = \text{Rayleigh Scattering } x\ S(q) \tag{7.1}$$

This combination is often called the *Rayleigh–Debye–Gans (RDG) limit (or approximation)*. It holds under the condition $|m - 1| \ll 1/kR$ where R is a length scale describing the overall size of the particle.[2]

The Rayleigh–Debye–Gans formulation combines the electromagnetic nature of light, embodied in the Rayleigh differential cross section, with the wave nature of light, embodied in the structure factor, via a simple multiplication.

Recall that the structure factor is derived using the situation where incident waves penetrate the object uninhibitedly and then scatter from the sub-volumes of the scattering object and then travel directly to the detector and do not interact with other sub-volumes of the scattering object. This situation is described as having no multiple scattering within the object, i.e. no internal multiple scattering, no internal coupling. It is a limiting case when a combination of small size and refractive index close to one is achieved. Thus, the RDG limit is one in which the interior field of the particle is the incident field. This field electrically excites sub-volumes of the particle to scattered radiation directly to the detector. There is no internal coupling between the sub-volumes.

7.1.2 Light scattering from soft spheres

To apply the RDG approximation to spheres we recall the Rayleigh differential scattering cross section for a spherical particle of radius R and relative refractive index of refraction m from equation (6.7)

$$dC_{\text{sca}}/d\Omega = k^4R^6F(m)\sin^2\psi \tag{7.2}$$

where $k = 2\pi/\lambda$, $F(m)$ is the square of the Lorentz–Lorenz term,

$$F(m) = \left|\frac{m^2 - 1}{m^2 + 2}\right|^2 \tag{7.3}$$

and ψ is the angle between the incident polarization and the scattering direction (see figure 6.1). Note that other than the $\sin^2\psi$ term Rayleigh scattering is isotropic, there is no dependence upon the scattering angle θ. The result of equation (7.2) is based on the electromagnetic nature of light.

The structure factor of a sphere was calculated in section 4.8 and is given in equation (4.47) as

$$S(q) = \left[\frac{3}{u^3}(\sin u - u\cos u)\right]^2 \tag{7.4}$$

where

$$u = qR \tag{7.5}$$

[2] This condition will be improved upon in section 7.1.3.2.

and

$$q = (4\pi/\lambda)\sin(\theta/2) \tag{7.6}$$

Recall that q is both the Fourier transform variable and the magnitude of the scattering wave vector. Also recall that $S(q)$ is known as the Fraunhofer diffraction of the 3d sphere. This result is based solely on the wave nature of light.

We now apply the RDG recipe of equation (7.1) with equations (7.2) and (7.4) to spheres to yield

$$\frac{dC_{\text{sca}}}{d\Omega}(\theta) = k^4R^6F(m)\sin^2\psi\left[\frac{3}{u^3}(\sin u - u\cos u)\right]^2 \tag{7.7}$$

Remember u is a function of θ via equations (7.5) and (7.6).

The Rayleigh–Debye–Gans (RDG) formula describes scattering of light from 'soft' spheres. The term 'soft' is usually meant to mean that the refractive index of the particle is not much different than the refractive index of the medium, i.e. $m \approx 1$. In fact, the Rayleigh conditions of equations (6.1) that have led us to this point imply that RDG holds for highly refractive but very small particles as well. The classic treatises indicate that RDG works well when the phase shift parameter ρ, equation (7.8) (van de Hulst 1957, Kerker 1969), is small. More recently, we have proposed a somewhat better parameter is the internal coupling parameter ρ', equation (7.9) (Heinson *et al* 2015). Thus, RDG holds when

$$\rho = 2kR\,|\,m - 1\,| < 1 \tag{7.8}$$

$$\rho' = 2kR\sqrt{F(m)} < 1 \tag{7.9}$$

Equation (7.8) is essentially our condition $|m - 1| \ll 1/kR$ from above. The internal coupling parameter will be derived in the next section.

The scattered intensity at a given angle is, of course, proportional to the differential cross section. Hence, we will often use the term 'intensity' when describing the differential cross section. We will use 'I_{Ray}' as the scattering that would come from a particle in the Rayleigh limit, proportional to equation (7.2). Then, for simplicity we can re-write equation (7.7) as[3]

$$I(q) = I_{\text{Ray}}[(3/u^3)(\sin u - u\cos u)]^2 \tag{7.10}$$

We will often use a '*Rayleigh normalized*' scattered intensity which is the scattered intensity divided by the scattered intensity for that particle assuming the Rayleigh formula holds even if the conditions (6.1) are broken. In the RDG limit the Rayleigh normalized scattered intensity is the structure factor.

Figure 7.1 displays plots of the Rayleigh normalized scattered intensity for spheres in the RDG limit with a variety of size parameters kR. The incident polarization is perpendicular to the scattering plane, i.e. $\psi = 90°$. This Rayleigh

[3] To be precise, according to equation (2.8), these scattered intensities are obtained when the incident light has a value of one and the detector is one unit away.

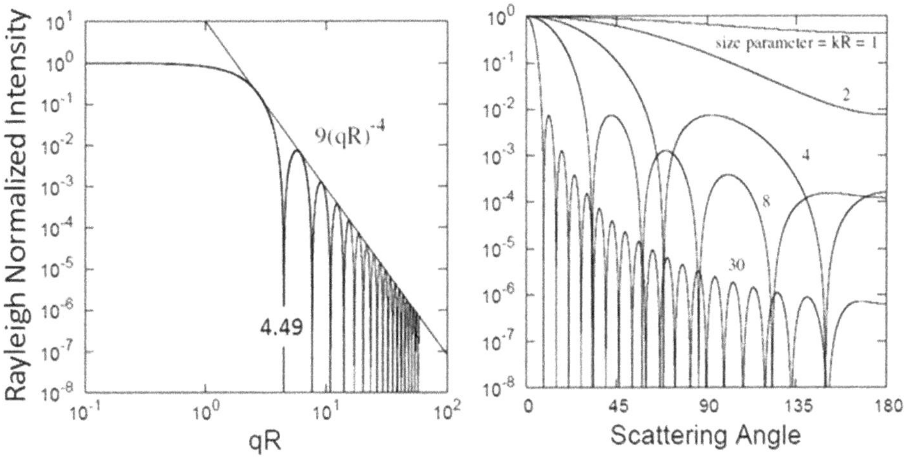

Figure 7.1. The Rayleigh normalized scattered intensity for spheres in the RDG limit with a variety of size parameters kR. The incident light polarization is perpendicular to the scattering plane, which contains the scattering angle. On the left, the plot is versus the dimensionless qR. A power law envelope and a first minimum are indicated. This plot applies to all size parameters; a given size parameter ranges in the interval $0 \leqslant qR \leqslant 2kR$. On the right, the plot is versus the scattering angle.

normalized intensity is equivalent to the structure factor for a sphere, equation (7.4), except, of course, we know the magnitude of the scattering. Both plots contain the same data. On the right, these data are plotted linearly versus the scattering angle. This plot demonstrates how small sizes are fairly isotropic in their scattering and that as size increases, the anisotropy increases to the forward direction.

The log–log plot on the left shows several important features:

1. *Universality.* Rayleigh normalization and plotting with the dimensionless qR, equation (7.5), yields one, universal plot for all sizes. Note that for a given size, the maximum qR is $2kR$.
2. *The forward scattering lobe.* When $qR < 1$ the scattering is not a function of q, hence not a function of θ. Moreover, the scattering cross section in this regime equals the Rayleigh differential cross section for the particle as given in equation (7.2). Hence, the forward scattering lobe can also be called the *Rayleigh regime* (also see section 5.3.1) for a particle of any size. This forward lobe ends near $qR \simeq 1$ which corresponds to an angle of $\theta \simeq \lambda/\pi D$.
3. *The Guinier regime.* Near $qR \simeq 1$ is a Guinier regime which obeys equation (4.43) and hence can be used to size the particle. Expansion of both trig functions for small $u = qR$ in the spherical particle structure factor, equation (7.4), yields

$$I(q) = k^4R^6F(m)[1 - (q^2R^2/5)] \tag{7.11}$$

which is consistent with the Guinier equation, equation (4.43), because for a sphere $R_g = \sqrt{3/5}\,R$.

4. *The ripples at large q*. Interference ripples appear for $qR > 4$. The first three minima are at $qR \simeq 4.493...,7.85...$, and 11.0 ..., respectively. At large qR the cosine term in the spherical particle structure factor, equation (7.4), dominates to yield

$$I(q) \sim 9k^4R^6F(m)(qR)^{-4}(\cos^2 qR) \tag{7.12}$$

We see that the spacing at large qR is determined by the cosine term hence the spacing is $\delta(qR) = \pi \simeq 3.14...$. This fact can be used for particle sizing.
5. *The large q envelope*. The envelope at large q is the coefficient of the $\cos^2$ term in equation (7.12). Rayleigh normalization is equivalent to normalizing the scattering by the scattering in the forward direction, $I(0)$. This leads to an equation for the large qR envelope as

$$I(q)/I(0) = 9(qR)^{-4} \tag{7.13}$$

This normalized envelope is shown in figure 7.1. In many practical situations scattering occurs from an ensemble of particles with a finite size distribution, or the angular resolution of the detector cannot resolve the ripples. Then the ripples smooth out to yield a uniform power law and the $\cos^2$ term in equation (7.12) averages to ½. Then the factor of 9 becomes 9/2 in equation (7.13), i.e. $I(q) \sim (9/2)(qR)^{-4}$.

Another useful perspective for the envelope combines the R terms in equation (7.12) to find

$$I_{\text{sca envelope}} \sim R^2/q^4 \sim S/q^4 \tag{7.14}$$

This envelope regime is also called the Porod regime (Guinier and Fournet 1955), and for dense, three-dimensional objects, it has a characteristic q^{-4} functionality. The third term in equation (7.14) emphasizes that scattering in the Porod regime is due to the surface S of the particle. We now see that the small qR, Rayleigh regime is dependent on the volume squared, whereas the large qR, Porod regime is dependent on the surface area of the particle. We have seen that this is a general result for all shapes in the diffraction limit, chapters 4 and 5, and also holds in the RDG limit because of the simplicity of equation (7.1).

7.1.3 The limits of Mie scattering and the internal coupling parameter

Mie scattering has two limits. First, as discussed immediately above, is the weakly refractive, RDG limit as $m \rightarrow 1$ in which the scattering is described by three-dimensional Fraunhofer diffraction multiplied by the Rayleigh differential cross section for the sphere. At various times, we have and will call this either the *RDG* or 3d *diffraction limit*. Second is for particles that are large size and have a significant refractive index. In this limit, the scattering is described by two-dimensional Fraunhofer diffraction from the projected geometric cross section of the object. We will call this the 2d *diffraction limit*. Here we derive the *internal coupling parameter* to describe how Mie scattering transforms between these limits of 3d diffraction and 2d diffraction.

7.1.3.1 The 2d diffraction limit

As the refractive index increases, the internal electromagnetic coupling begins, and the RDG formulae equations (7.7) and (7.10) no longer hold; then the full Mie formalism must be used. However, ultimately, as the sphere becomes a macroscopic object, a baseball for example, simplicity is regained. For then the sphere becomes an obstacle, the projection of which is a two-dimensional circle, and the scattering in the far field becomes the two-dimensional diffraction pattern independent of the refractive index. In this 2d *diffraction limit* we are back to waves, there is no electromagnetism overtly involved! However, recognize that because the Mie equations for scattering by an arbitrary sphere are based on the Maxwell equations, which are fundamental equations of nature, they should contain this two-dimensional diffraction limit.

For a circular obstacle of radius R and 'sufficient' (a condition that will become clear below) refractive index, experiencing an incident intensity of unity, the 2d Fraunhofer diffraction pattern (the Airy pattern) is (Hecht 2002)

$$I(\theta) = (k^2R^4/4)\left[\frac{2J_1(u)}{u}\right]^2 = I(0)\left[\frac{2J_1(u)}{u}\right]^2 \tag{7.15}$$

In (7.15) $J_1(u)$ is the first Bessel function of the first kind, and, again, $u = qR$. The second equality follows from $J_1(u) \to u/2$ *as* $u \to 0$ (Abramowitz and Stegun 1972). We expect that this is also the scattering from a large, refractive sphere in the 2d diffraction limit. Notice that there is no dependence on the refractive index.

7.1.3.2 The internal coupling parameter

Two limits of Mie scattering have been identified above; briefly, the weak and strong refractive limits. In the course of our research we found that for spheres when the phase shift parameter was $\rho \geqslant 10$, the forward scattering was fairly well described by the Rayleigh scattering divided by $3\rho^2$. We also showed that in the limit $\rho \to \infty$ the Mie equations yielded the two-dimensional Fraunhofer limit, which is refractive index independent, as they should because they are based on the Maxwell equations. However, dividing Rayleigh forward scattering by ρ^2 will never yield the two-dimensional Fraunhofer limit because it cannot cancel the refractive part of the Rayleigh scattering, viz. the Lorentz–Lorenz term $F(m)$. This led us to propose a modified phase shift parameter.

Figure 7.2 diagrams the two limits represented by the forward scatterings. Needed now is a parameter similar to ρ that can allow the Rayleigh limit to become the 2d diffraction limit when divided by its square.

The parameter we seek is the square root of the ratio of the scattering in the forward direction for the two limits. It is

$$\rho'^2 = \frac{k^4R^6\left|\frac{(m^2-1)}{(m^2+2)}\right|^2}{(k^2R^4/4)} \tag{7.16}$$

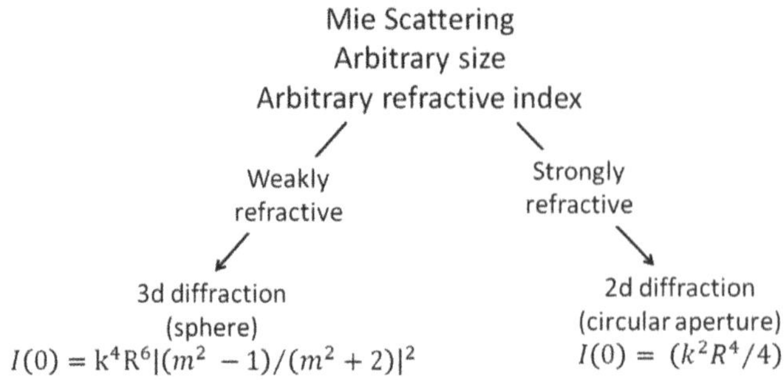

Figure 7.2. Scheme for the limits of Mie scattering.

Thus,

$$\rho' = 2kR\,|(m^2 - 1)/(m^2 + 2)| \tag{7.17}$$

Does the parameter ρ' in equation (7.17) make sense? To address this question consider the fact that the sub-volumes dV of the particle are infinitesimally small; thus, they are Rayleigh scatterers. Then each sub-volume of the particle scatters a field proportional to $k^2\,|(m^2 - 1)/(m^2 + 2)|\,dV$. From this it is expected that the most accurate description of the refractive index functionality for the internal coupling is $|(m^2 - 1)/(m^2 + 2)|$, the Lorentz–Lorenz term. This is precisely what we have achieved with ρ' in equation (7.17). Therefore, we will henceforth call ρ' the *internal coupling parameter* (Heinson *et al* 2015).

The internal coupling parameter is very similar to the phase shift parameter ρ, equation (7.8), in magnitude, functionality on m and kR, and significance. For magnitude as $m \to 1$, $\rho'/\rho = 2/3$; when $m = 1.5$, $\rho'/\rho = 10/17$. Functionality similarity is obvious. Significance lies in the fact that both describe whether or not the sphere can be regarded as being in the RDG limit as demonstrated in equations (7.8) and (7.9). Below we will demonstrate that ρ' is superior to ρ as a parameter to unify the scattering by spheres with different sizes and refractive indices. The phase shift parameter is typically used in the limit $m \to 1$, and it is in this limit that ρ and ρ' both have the $(m - 1)$ functionality. The phase shift parameter is equal to the phase difference between a wave traveling through a diameter of the sphere and one traveling the same distance outside the sphere. However, this has nothing to do with internal coupling and hence the internal field. The phase shift parameter cannot negate the Mie scattering refractive index dependence to yield refractive index independent, large, refractive particles.

7.1.4 Scattering by arbitrary spheres with real refractive index

7.1.4.1 Angular functionality

As the sphere size and/or refractive index increase, the internal coupling between different sub-volumes of the sphere increases. The combined effect of size and refractive index are contained in ρ' so that when either of these drives ρ' to become comparable to or greater than one, the RDG formulation breaks down. Then to determine the scattering the full Mie formulation must be used.

Examples of the Mie solution for spherical particle scattering, 'Mie scattering', are given in figures 7.3 and 7.4. These figures show the Mie calculated results for two different real refractive indices and a variety of size parameters kR. The plots are normalized by the scattered intensity at $\theta = 0$; they are 'forward normalized'. The incident light is vertically polarized perpendicular to a horizontal scattering plane containing the scattering angle θ. When compared to RDG in figure 7.1, similarities are seen with the scattering anisotropy increasing to the forward direction as kR increases, and the ripple structure with spacing about π/kR. However, beyond those

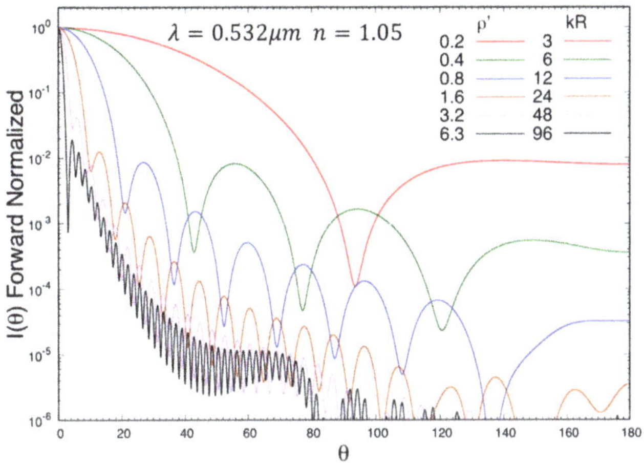

Figure 7.3. Forward normalized Mie scattering curves as a function of scattering angle θ for spheres of refractive index $m = 1.05$ and a variety of size parameters kR and consequently internal coupling parameters ρ'. The incident light is vertically polarized perpendicular to the scattering plane containing the scattering angle.

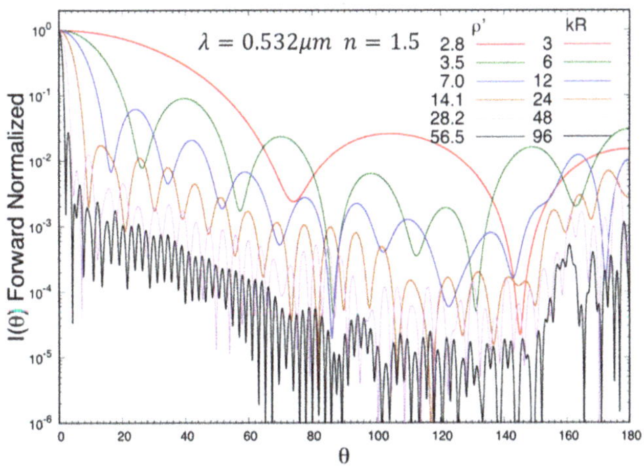

Figure 7.4. Forward normalized Mie scattering curves as a function of scattering angle θ for spheres of refractive index $m = 1.50$ and a variety of size parameters kR and consequently internal coupling parameters ρ'. The incident light is vertically polarized perpendicular to the scattering plane containing the scattering angle.

trends, the curves are much more complex than the RDG limit, especially for the more refractive situation of $m = 1.5$ in figure 7.3. There, for $\rho' \geqslant 7$ one finds the scattering begins to increase for $\theta \geqslant 140°$, and the ripple structure is then much more complex.

7.1.4.2 Forward scattering

In figures 7.3 and 7.4, the effects of the forward scattering, $I(0)$, were normalized out in order to illustrate the effects of the scattering angle. However, $I(0)$ is very important and has strong functionalities on size and refractive index. Figures 7.3 and 7.4 show a narrow angular range of intense scattering in the forward direction especially for large particles. This invites the question of what is the result of the narrow angular range combined with the huge increase in relative intensity. In section 7.3 we will find that they conspire to contain approximately half of all the scattered light across the entire $\theta = 0°$ to $180°$ range. Before we demonstrate that, however, we need to study the scattering at exactly $\theta = 0$.

Figure 7.5 plots the forward scattered intensity as the differential cross section at $\theta = 0$ versus the sphere radius R for a variety of refractive indices. Two major regimes are seen. When $R \leqslant \lambda$, the scattering goes as the sixth power of R and depends on the refractive index. It then crosses over near $R \simeq \lambda$ to a fourth power of R dependence for $R \gtrsim \lambda$ and becomes independent of the index of refraction. The sixth power occurs when the spheres are in the RDG limit, consistent with equation (7.7); the fourth power is consistent with the 2d diffraction limit in equation (7.11).

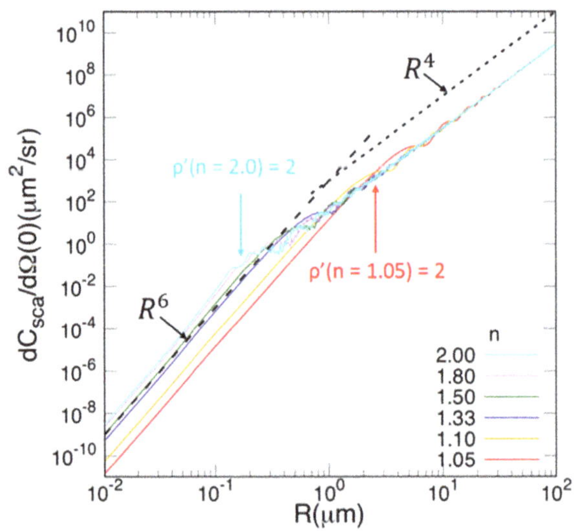

Figure 7.5. Forward scattering by spheres. The differential cross section at $\theta = 0$ as a function of the spherical particle radius calculated from the Mie equations for a variety of real refractive indices. The light wavelength is $\lambda = 0.532$ μm. Arrows mark the radii for which the internal coupling parameter $\rho' = 2$ for a given refractive index. Note the two functionalities on R, hence by a dimensional analysis, two functionalities on λ (not shown). Also note with the R functionality crossover, the scattering becomes independent of the refractive index when $R > \lambda$; more precisely, when $\rho' > 1$.

Figure 7.5 is very useful to emphasize the strong functionality on size and the regimes where refractive index is and is not important. However, the strong functionality can be tamed by normalization by the Rayleigh scattering cross section equation (7.1) and unified by changing the independent variable to ρ'. This is done in figure 7.6.

We first demonstrate that the internal coupling parameter ρ' does a better job than the phase shift parameter ρ in unifying the behavior of the forward scattering magnitude. Figure 7.6(a) shows plots of the forward scattered intensity, $I(0)$, normalized by the Rayleigh scattered intensity, equation (7.1)[4], (a ratio which we will refer to as the 'Rayleigh normalized forward scattering intensity') versus the phase shift parameter ρ (Berg *et al* 2005). A quasi-universal dependence on ρ is seen despite the wide variation on the refractive index. In contrast, figure 7.6(b) plots the

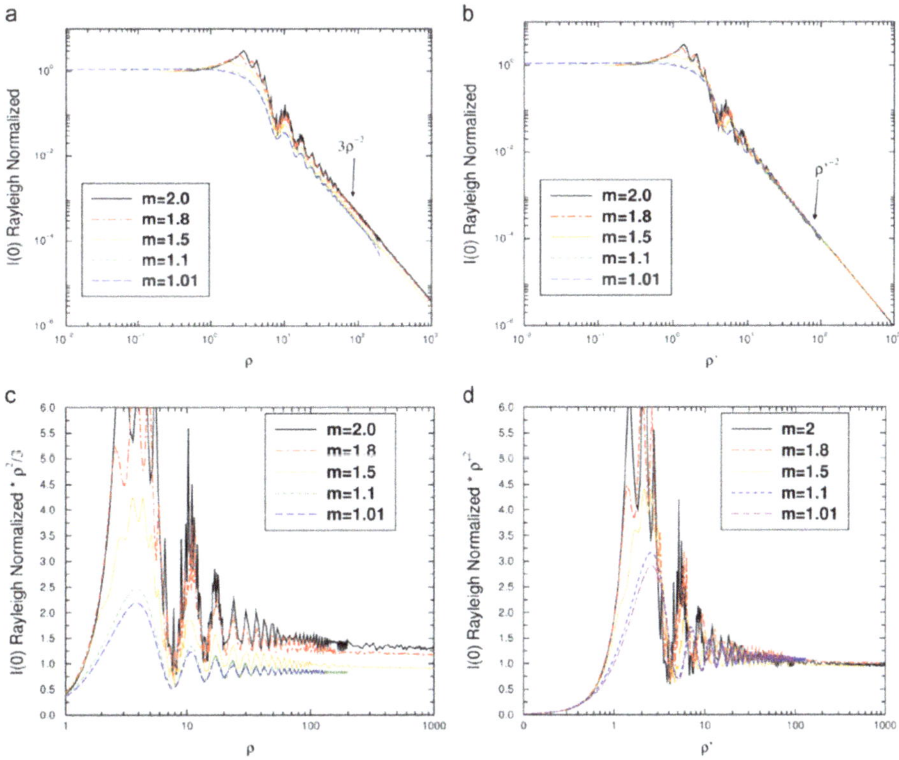

Figure 7.6. (a) Rayleigh normalized Mie forward scattering intensity for spheres with various refractive indices, m (a) versus ρ, (b) versus ρ'. (c) Data from (a) multiplied by $\rho^2/3$. (d) Data from (b) multiplied by ρ'^2. As spheres reach the 2d diffraction limit, ρ or $\rho' \to \infty$, plotting with ρ still shows refractive index dependency while plotting versus ρ' does not. Note, however, on the other hand, the ripple structure is more coherent with ρ than ρ'. Reprinted from Heinson *et al* (2015), Copyright (2015), with permission from Elsevier.

[4] Note that at $\theta = 0$, $\psi = 90°$.

Rayleigh normalized forward scattering intensity versus the internal coupling parameter ρ'. Now, at large ρ', all the plots fall together on the same line. Higher resolution views of this comparison are shown in figure 7.6(c) and (d), where the Rayleigh normalized scattering intensity is multiplied by either ρ^2 or ρ'^2, respectively. While figure 7.6(c) shows differences on the order of a factor of 2 at large ρ, the plots using ρ' in figure 7.6(d) fall precisely together at large ρ'. This property was built into ρ' as seen in figure 7.2. Nevertheless, these results show the power that ρ' has as an accurate and unifying parameter to describe Mie scattering intensity in the forward direction. Finally, note that ρ has its own merit in providing a more coherent functionality for the ripple structure than ρ'.

The properties of spherical particle Mie scattering in the forward direction is summarized as:

(i) When $\rho' < 1$, the forward scattering for all spheres regardless of size equals the Rayleigh scattering for that sphere.

(ii) When $\rho' > 1$, the forward scattering for different refractive indices m begin to separate and oscillate. Near $\rho' \approx 2$ the oscillations are largest and then damp out with increasing ρ'. When $\rho' \gtrsim 5$, the overall trend for all m is the forward scattering equals the Rayleigh scattering decreasing by a factor of ρ'^2 until for $\rho' > 100$, $I(0) = k^2R^4/4$. These features are summarized in equations (7.18)

$$I(0) = k^4R^6F(m) \qquad \rho' < 1 \tag{7.18a}$$

$$I(0) \simeq k^4R^6F(m)/\rho'^2 \qquad \rho' \gtrsim 5 \tag{7.18b}$$

$$I(0) = k^4R^6F(m)/\rho'^2 = k^2R^4/4 \quad \rho' \gtrsim 100 \tag{7.18c}$$

In the next chapter it will be shown that the scattering is equal to $I(0)$ for the range $u = qR \ll 1$, not just $\theta = 0$.

Equations (7.18) show that the forward scattering regime of a sphere transforms from Rayleigh scattering with the classic λ^{-4}, R^6 and strong refractive index functionalities for small or weakly refractive spheres ($\rho' < 1$) to weaker λ^{-2}, R^4 functionalities and a total obliteration of the refractive index dependency for the combined large and/or strongly refractive spheres ($\rho' > 10$). This k^2R^4 functionality is that of two-dimensional, circular aperture diffraction. With this, we see that the circular, as opposed to the spherical, nature of the sphere's scattering appears for $\rho' \simeq 10$ and greater. For $m = 1.33$ this corresponds to $kR \simeq 25$, and if $\lambda = 0.532$ micron, a particle diameter $2R \simeq 4$ microns (only).

Another way to look at figure 7.6 and equations (7.18) is that only when $1 \lesssim \rho' \lesssim 30$ does one need to use the nasty Mie equations. To a good approximation when ρ' is small, the Rayleigh formula works, and when ρ' is large, $I(0) = k^2R^4/4$.

7.1.4.3 The Guinier regime, qR $\simeq$ *1*

In the RDG limit, $\rho' \ll 1$, the Guinier result for diffraction will hold for spheres. Here we examine how well the Guinier equation works for an arbitrary sphere. Mie scattering curves for known values of the sphere radius hence known radius of gyration $R_g^{\ 2} = (3/5)R^2$ were fit to the Guinier equation, equation (4.43), using the restriction $qR \leqslant 1$ (Sorensen and Shi 2000). This generated a fit value for the radius of gyration that we called the 'Guinier inferred radius of gyration, $R_{g,G}$'. Thus, the Mie curves for $qR \leqslant 1$ were fit to

$$I(q) = I(0)(1 - q^2 R_{g,\,G}^{\,2}/3) \tag{7.19}$$

The comparison is shown as the ratio $R_{g,G}/R_g$, the inferred divided by the true, in figure 7.7. Figure 7.7 shows that the ratio $R_{g,G}/R_g$ starts at unity when the internal coupling parameter $\rho' = 0$. This is expected since this is the weak scattering limit and where none of the assumptions leading to the Guinier equation are violated. With increasing ρ', the ratio increases and oscillates. The large-scale oscillations appear universal with ρ' and have a period of about $\Delta\rho' \simeq 4$. Shorter period oscillations and spikes set in with increasing m. The large scale oscillations damp with increasing ρ and have an asymptotic limit $\rho' \to \infty$ of $R_{g,G}/R_g = 1.12$. Remarkably, values of $R_{g,G}/R_g > \sqrt{5/3} = 1.29$ exist to imply $R_{g,G} > R$, i.e. the measured radius of gyration is larger than the geometric radius of the sphere, a physically impossible condition for a sphere.

The asymptotic limit $\rho' \to \infty$ of $R_{g,G}/R_g = 1.12$ occurs because in this limit the sphere acts like an opaque obstacle with circular cross section, the 2d diffraction limit. The incident light is then subjected to Fraunhofer diffraction described by equation (7.11), above. Expansion of the first Bessel function in equation (7.11) and using $R^2 = 5/3R_g^2$ yields

$$I(q) \simeq I(0)[1 - (5/12)q^2 R_g^2] \tag{7.20}$$

Comparing equations (7.19) and (7.20) implies $R_{g,\,G}^{\,2} = (5/4)R_g^2$, thus $R_{g,G}/R_g \simeq 1.12$ as found in figure 7.7.

From a practical point of view equation (7.20) and figure 7.7 provide a recipe to adjust the Guinier analysis for refractive spheres with $\rho' > 1$ in order to determine

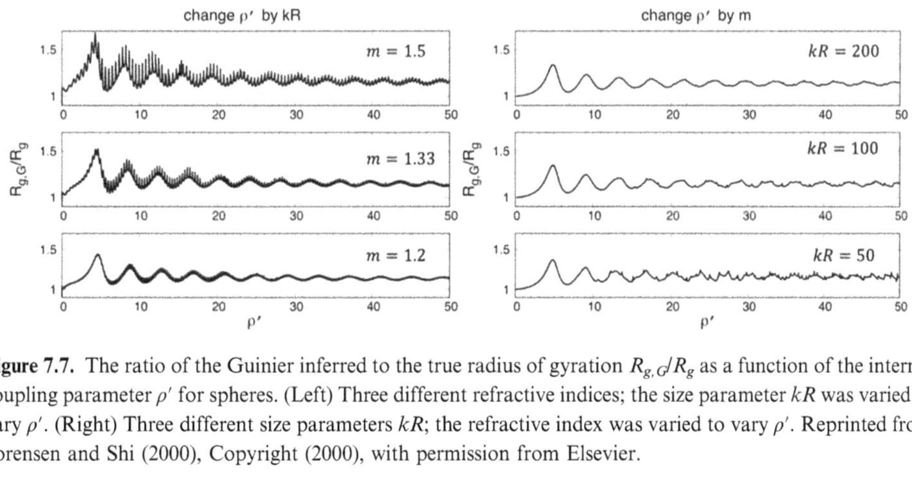

Figure 7.7. The ratio of the Guinier inferred to the true radius of gyration $R_{g,G}/R_g$ as a function of the internal coupling parameter ρ' for spheres. (Left) Three different refractive indices; the size parameter kR was varied to vary ρ'. (Right) Three different size parameters kR; the refractive index was varied to vary ρ'. Reprinted from Sorensen and Shi (2000), Copyright (2000), with permission from Elsevier.

their size by light scattering. There are, however, other details as discussed in Sorensen and Shi (2000). Becoming even more practical, if the experimenter can accept some uncertainty (and who can't?), the Guinier analysis provides a reasonable size measurement for particles of any refractive index or shape. Said differently, the Guinier analysis for size is useful beyond the domain of small angle x-ray diffraction. I find it remarkable that light scatterers rarely, if ever, take advantage of this. A measurement with caveats is better than no measurement at all.

7.1.4.4 The ripples

Interference ripples have been studied as a function of the phase shift parameter ρ (Sorensen and Shi 2002) and in prior work as a sizing method (Kerker *et al* 1964). In the $\rho = 0$, RDG limit the ripples are equally spaced in qR-space, the spacing being $\delta(qR) = \pi$ once past the first maximum throughout the entire angular range to 180° (see the discussion associated with equation (7.12), above). As ρ increases, the ripple spacing evolves with some chaos until for $\rho \gtrsim 5$ ($\sim\rho' \gtrsim 3$), the spacing is better described in θ-space with the spacing being $\delta\theta = \pi/kR = \lambda/2R$. Note that this is equivalent to $\delta(qR) = \pi$ but only for small θ. Furthermore, it is convenient to recognize that this is equivalent to $\delta\theta = 180°/kR$. A spacing of $\delta\theta = \pi/kR = \lambda/2R$ is truest for angles past the forward scattering regime but before ca. 90°, although this statement needs much qualification. Inspection of figure 7.4 supports this. For large ρ' 2d Fraunhofer diffraction dominates the forward scattering and in this regime equation (7.12) holds with a ripple periodicity of $\delta(qR) = \pi$. Ripple spacing can be used for particle sizing if the system has a narrow size distribution with geometric width less than 1.2 or so; wider distributions wash out the ripples. The reader is invited to return to figures 7.3 and 7.4 and verify these results with the ripples seen there. Figure 7.8 gives an example for scattering from a single polymer drop.

In the extreme back direction, $\theta \sim 170°$ to 180°, the realm of the glory, the angular spacing is well described by $\delta\theta = \pi/kR = \lambda/2R$.

7.1.4.5 Polarization

When the incident light is polarized in the scattering plane (typically horizontally polarized, H), the light scattered at 90° is zero in the RDG limit. Slightly beyond this limit, the angle at minimum grows larger than 90° and the scattering intensity at the minimum grows from zero. This is illustrated in figure 7.9 for a sphere with an index of refraction of $m = 1.5$. One can see that the RDG result is a good approximation up to size parameter $kR \simeq 1$ but fails significantly by $kR = 3$. Thus, at $kR \simeq 3$ and beyond, the lack of a finite projection of the H polarization in the $\theta = 90°$ is not significant.

7.1.4.6 Rainbows

Rainbows are well known and beautiful phenomena often seen when the Sun shines on falling rain. Given their beauty, a great deal has been written about them, e.g. Minnaert (1993). Often two bows are seen with the arc centers at the anti-solar point, usually the shadow of the observer's head. The primary bow is near $\theta = 138°$ with a width of about 1.6°, and the fainter secondary bow is near $\theta = 129°$ with a width of about 3°. The spread in scattering angle is due to the dispersion of the index

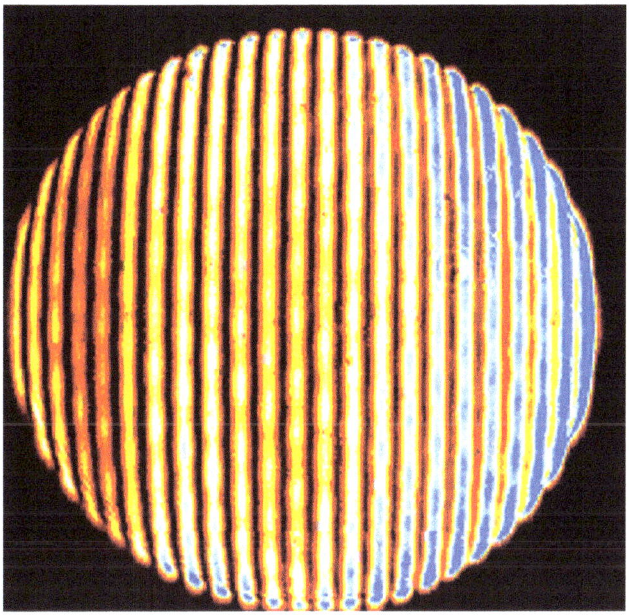

Figure 7.8. False color image of scattering of $\lambda = 0.532$ micron light from a polymer drop with a diameter of 23 microns. The center of the field is at a scattering angle of $\theta = 90°$. The angular spacing is $\delta\theta = 1.3°$.

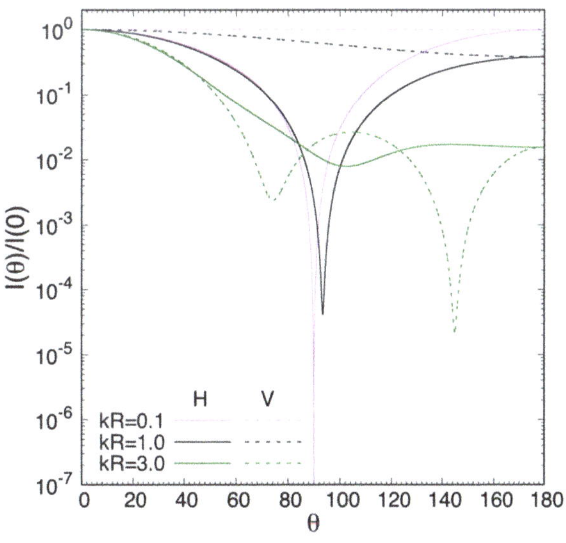

Figure 7.9. Normalized scattering intensity versus the scattering angle in the horizontal plane for both incident light polarized horizontally, H, solid lines and vertically, V, dashed lines for three different size parameters, kR.

of refraction of water across the visible spectrum. The primary bow has red furthest from the anti-solar point; the secondary bow has red closest. Between the bows lies Alexander's Band, a relatively dark region of fainter scattering. Both bows are strongly, but not completely polarized tangentially to the bow.

Rainbows occur with millimeter size water drops hence ray optics can be used to explain the bow. From this perspective, the primary bow is a result of light rays refracting into the sphere, reflecting once on the inside and then refracting out of the sphere; the secondary bow is the result of rays having two interior reflections before refracting out. These are the $p = 2$ and $p = 3$ terms of the Debye series for the primary and secondary bows, respectively, where $p - 1$ is the number of internal reflections (see section 8.2.8).

Figure 7.10 shows a series of monochromatic, $p = 2$ rays to form the primary bow (Nussenzveig 1977, Adam 2002). The purpose of figure 7.10 is to show how parallel, incident rays of increasing impact parameter (the distance from a ray on the diameter), rays 1 through 7, yield, after refraction into the sphere and reflection once at the 'back' of the sphere, re-emerging via refraction as rays of deceasing scattering angle θ as the impact parameter increases. This decreasing θ-dependence ceases at ray 7 and for rays 8 through 12, θ increases again. At this turn around near ray 7, the rays are compacted in θ, and this causes the intensity of the scattered light to be larger as these rays conspire. This creates the rainbow and ray 7 is called the 'rainbow ray' and has the smallest scattering angle (from the forward direction) of all the $p = 2$ rays with various impact parameters from a central ray. Notice how a significant number of other rays are tightly spaced in angle with this ray. This confluence of rays explains the brightness of the rainbow.

The ray tracing scattering angle θ for a p-ray can be related to the incident angle θ_i by (Laven 2003)

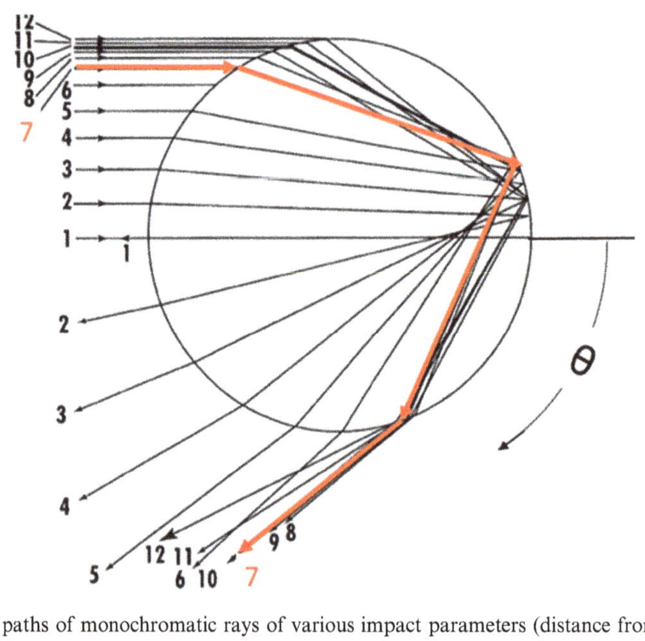

Figure 7.10. The paths of monochromatic rays of various impact parameters (distance from the central ray, ray 1, along the diameter). Ray 7 is the rainbow ray having the smallest scattering angle of all the rays. Many of the other rays are close to the rainbow ray hence the scattering is strong in this direction. Reprinted with permission from Greenler (2020).

$$\theta = (p - 1)180° + 2\theta_i - 2p \sin^{-1}(\sin \theta_i/n). \quad (7.21)$$

The incident angle is related to the radius normalized impact parameter, b, see figure 7.11, by

$$b = \sin \theta_i \quad (7.22)$$

Figure 7.11 plots these equations as θ versus b for a variety of real refractive indices.

Figure 7.11 affords another attempt to use the $p = 2$ ray to explain the rainbow, which occurs for water droplets with real refractive index of $n = 1.33$ and to generalize the bows for other spherical particles with different refractive indices. When the curves in figure 7.11 pass through a region of zero slope, there will be a range of impact parameters that all scatter at the same scattering angle. The grouping of a range of impact parameters into a single scattering angle causes a spike in the scattering, these spikes are the generalized rainbows. Use of $p = 3$ in equation (7.21) yields similar plots with points of zero slope to indicate the generalized secondary rainbow.

Maxwell's equations portrayed in the Mie calculation contain the rainbows as shown in figure 7.12. There the differential cross section for $\lambda = 500$ nm, unpolarized, incident light scattering from water droplets of various sizes is plotted for a range of scattering angles to include the rainbows. For a radius of $R = 1000$ μm = 1.0 mm distinct peaks in the scattering are seen near $\theta = 138°$ and 130° corresponding to the primary and secondary rainbows, respectively. Notice that the primary rainbow is about an order of magnitude more intense than the secondary rainbow and the region between the two, Alexander's band, is two orders of magnitude less intense.

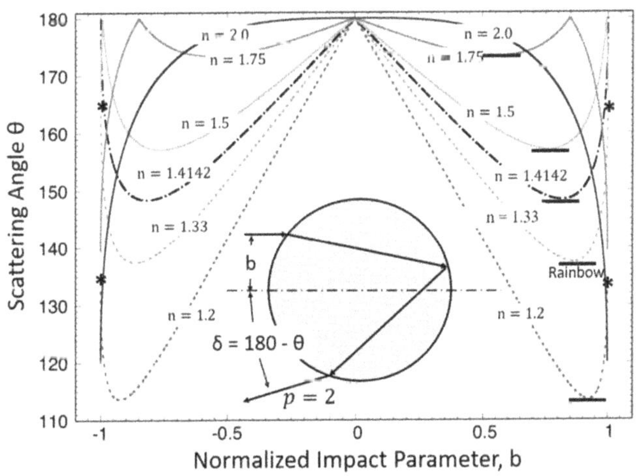

Figure 7.11. Scattering angle versus the radius normalized impact parameter for the Debye series $p = 2$ ray interacting with a non-absorbing sphere with a variety of refractive indices, n. Short, thick horizontal lines on right side, $b > 0$, indicate regions where a rainbow or bow occurs. These bow regions also occur symmetrically on the left side of the figure but are not marked. Asterisks for $n = 1.2$ and 1.33 indicate where $b = \pm 1$. When > $\sqrt{2} = 1.4142\ldots$, the scattering passes through $\theta = 180°$ to smaller angles on the opposite side of the optical axis, dot-dash line.

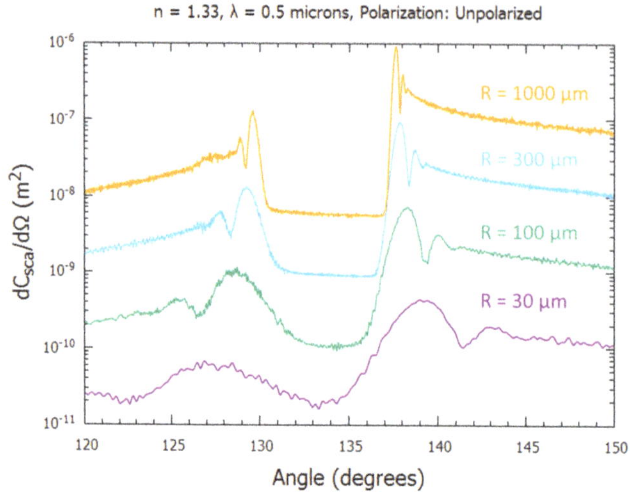

Figure 7.12. The differential cross section calculated from the Mie equations for λ = 500 nm, unpolarized incident light scattering from water droplets of various sizes with refractive index n = 1.33 (n = m) plotted for a range of scattering angles to include the rainbows.

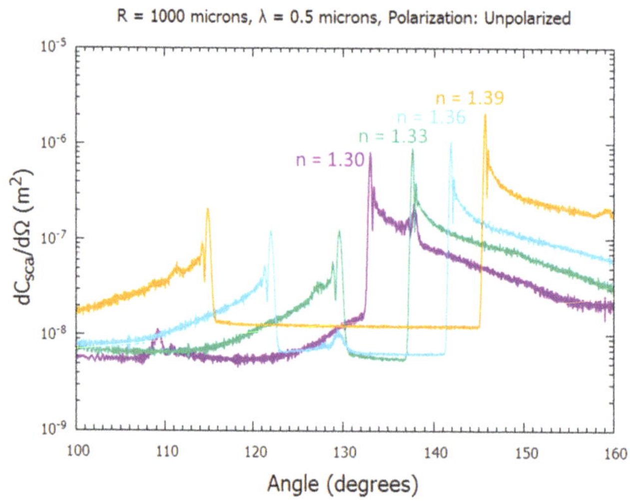

Figure 7.13. The differential cross section calculated from the Mie equations for λ = 500 nm, unpolarized incident light scattering from R = 1.0 mm spheres of various refractive indices plotted for a range of scattering angles to include the rainbows.

Also notice that the rainbow peaks fuzz out as the radius falls below ~100 μm. This fuzzing causes the rainbow colors to blend together to make a pastel and eventually, with yet smaller droplets, creates a white bow known as a fog bow.

Figure 7.13 demonstrates the effect of refractive index on the bows. Note that the secondary bow is nearly gone for n = 1.30.

7.1.4.7 The glory

The glory is an atmospheric phenomenon visible as a glorious, colorful display of rings subtending a few degrees of arc centered on the shadow of the observer cast upon a fog of water droplets (Minnaert 1993). A common occurrence might be observed when looking out an airplane window at the plane's shadow on a cloudbank. The droplets of the fog that produce the glory are typically on the order of 20 microns in diameter within half an order of magnitude. The glory is an enhancement of the scattering in the backward direction, $\theta \simeq 180°$, or in terms of the phase angle $\delta = 180° - \theta$, near $\delta = 0$. Mie theory can calculate scattering in the back direction and hence simulate the glory, but it does not give a physical explanation. van de Hulst (1957) suggested surface waves cause the glory. Nussenzveig (1979, 2002) formalized the surfacc wave model with his complex angular momentum (CAM) theory, but the theory while successful is complex.

Laven (2003) has presented a useful physical picture of the glory. He showed that the $p = 2$ term of the Debye series (see section 8.2.8) is dominant over other Debye terms for most refractive indices and comparable to the full Mie result in forming the glory. This term corresponds to light that refracts once into thc sphere and then experiences one internal reflection before refracting back out in the backward direction, see drawing in figure 7.11. He also found that the $p = 2$, low impact parameter, axial rays only serve to enhance the white spot at the center, hence $\theta \approx 180°$, of the glory. Thus, the task of forming the major part of the glory is left to large impact parameter, $p = 2$ rays. However, for a water droplet with a refractive index of $n = 1.33$, these large impact rays cannot scatter at angles great than 165°. This fact is illustrated in figure 7.11 with asterisks at points $(b, \theta) = (\pm 1, 165°)$. Nevertheless, it is here that the wave nature of light kicks in. As proposed by van de Hulst (1947) and Laven (2003), surface waves can propagate along the outer circumference of the sphere beyond the exit point of refraction, and these surface waves shed light tangentially as they proceed to allow $p = 2$ light to reach 180° and beyond. Then if two rays enter the droplet at diametrically opposite points on the sphere and follow the $p = 2$ trajectory, they can exit the droplet at the same phase angle $\delta = |180° - \theta|$ if one ray follows a 'short' surface wave path and the other follows a 'long' surface wave path. This is illustrated and described in figure 7.14 (left).

This physical scenario can be used to predict the phase angle at which monochromatic light will experience constructive interference of the two $p = 2$ rays in the back direction. Figure 7.14 (right) shows detail of the left diagram; points a, a′, b and b′ are the same. The rays A and B have traveled the same optical distance to points a and b and then via a surface wave to points a' and b′. It can be shown that this is independent of the impact parameter of the incident rays. Note, however, that ray A has to travel an extra distance a′ to a′ before it sheds at a phase angle of δ, parallel to ray B. From figure 7.14 (right) it is apparent that a'a′ = $2\delta R$. When this path difference a′ to a′ is an integer number of wavelengths, rays A and B will constructively interfere in the back direction. This condition is

$$2R\delta = n\lambda \tag{7.23}$$

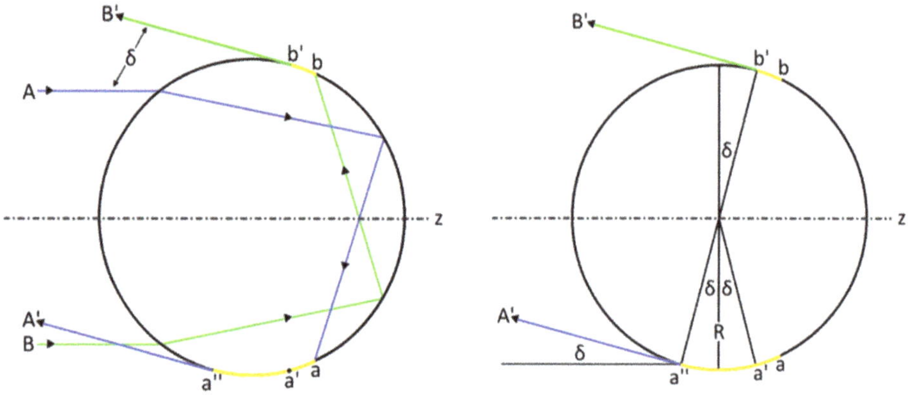

Figure 7.14. (Left) Two rays, A and B, enter a spherical particle parallel to the *z*-axis at diametrically opposite points. After this initial refraction, they partially reflect once off the inside front surface of the sphere. They then propagate to points a and b, respectively, where they refract out of the sphere (rays not shown); they are Debye $p = 2$ rays. At a and b, they also launch surface waves along the sphere's circumference, paths aa′ and bb′ (yellow), respectively. These surface waves shed light tangentially as they propagate. Rays that shed at a′ and b′ are parallel and thus can meet in the far field. However, they have a path difference equal to the arc length a′a′. If this path difference is an integer number of wavelengths, the two rays constructively interfere to give rise to the glory. (Right) Following the diagram on the left, two rays, A and B, have shed off the surface waves, the orange sectors a to a′ and b to b′, to travel parallel at phase angle δ into the far field.

where n is a positive integer and λ is the wavelength of light in the medium surrounding the droplet. Equation (7.23) leads to the phase angle for the monochromatic glory rings to be

$$\delta = n(\lambda/2R) = n(\pi/kR) \tag{7.24}$$

Figure 7.15 shows the prediction of equation (7.24) for $m = 1$, the first maximum in intensity, versus radius of a water droplet with real refractive index of $n = 1.33$ compared to the Mie prediction. The agreement with the overall trend is quite good.

Before we leave figure 7.11, notice that once the refractive index $n > \sqrt{2} = 1.414...$, $p = 2$ ray refraction can occur out to 180° and beyond. Then surface waves are not needed, and one finds the glory to be much brighter.

Figure 7.15 also reveals many bumps, wiggles and spikes in the Mie plot. Given the polychromatic nature of visible light, these features are not seen in the natural glory; rather simply the trend of equation (7.24). Figure 7.16 shows a higher resolution portrayal of these features along with a plot of the scattering efficiency (the total scattering cross section divided by the geometric cross section πR^2). An obvious strong correlation between the features of these two plots is seen. These features are complex and a result of many factors. A viable physical interpretation is that they are due to internal resonances inside the sphere. When the path length for a ray reflecting along the inner circumference of the sphere is an integer number N of wavelengths, a resonance can occur. If the sphere radius is much greater than the wavelength, which it has to be for this ray picture to work, the path length is almost equal to the circumference and then the resonance condition is

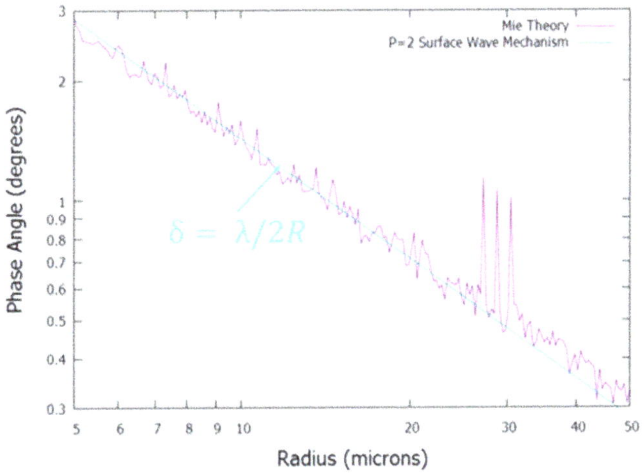

Figure 7.15. Phase angle δ of the first glory maximum versus radius R for water droplets interacting with light of wavelength $\lambda = 500$ nm. The index of refraction of water is $m = 1.33$. Blue line is the prediction of the $p = 2$ surface wave mechanism, equation (7.24); purple line is the prediction of Mie theory. In the Mie theory calculation, numerous spikes and smaller amplitude erratic structure occur due to internal wave resonances. Only a small fraction of the spikes are capable of being plotted because they are very narrow and the plotted point resolution is such that they often do not match the location of the spikes.

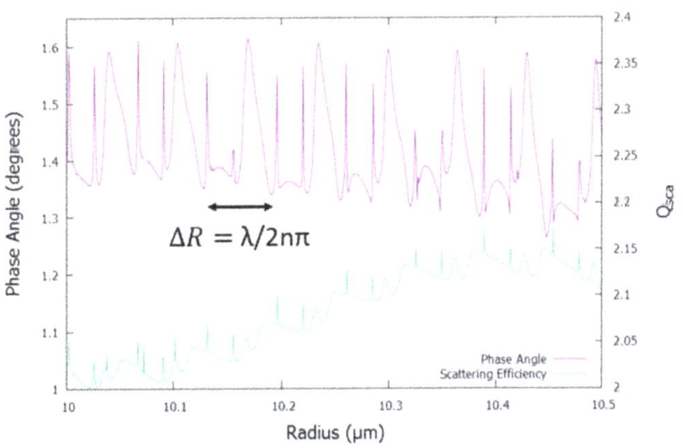

Figure 7.16. Glory phase angle, left axis, and scattering efficiency, right axis, as a function of sphere radius. The incident light is unpolarized with wavelength of 500 nm. The refractive index of the sphere is $n = 1.33$. The double arrow indicates ΔR as calculated with equation (7.26).

$$N\lambda = 2\pi R \tag{7.25}$$

At constant λ, N can change from one integer to the next, i.e. $\Delta N = \pm 1$, when R changes by

$$\Delta R = \lambda_{\text{vac}}/2\pi m \tag{7.26}$$

where λ_{vac} is the wavelength in vacuum and m is the real refractive index of the sphere. Figure 7.16 includes a double arrow for ΔR that indicates a good agreement with the length scales of the bumps and spikes of the glory and scattering efficiency.

Surprisingly, there are not many experimental studies of the glory in the literature. However, our group built an apparatus to measure scattered light in the range of $\theta = 180°\pm6°$ (hence $\delta = \pm6°$) and used it to study backscattering by micron sized polystyrene latex spheres (PLS) colloidal particles in water (Heffernan *et al* 2016). The incident light was either polarized vertically, v, or horizontally, h, with a horizontal scattering plane and detected with no polarizer (unpolarized, u). The light had a wavelength of 532 nm in air. Figure 7.17 gives an example for $R = 5.055$ μm particles.

Figure 7.17 shows a good agreement with theory for the peaks of the scattering as determined by both the Mie calculations in the figure and equation (7.24), above. Note, however, the difficulty in getting a good fit for the relative intensity. The manufacturer of the PLS specified the refractive index to be 1.59, however this did not yield a good fit to the data. Also included in the figure are Mie calculations for slightly different refractive indices and then one can see the extreme sensitivity of the relative intensity to the refractive index. On the other hand, note that the angles of intensity maxima are not sensitive to the refractive index, although the refractive index changes are small.

This sensitivity of backscattering on particle properties is explored further in figure 7.18, which shows the intensity ratio $I(180°)/I(178°)$ versus radius for spheres with a refractive index of $m = 1.59$ immersed in water with $m = 1.33$. There one can see order of magnitude variations of this ratio with small changes in radius.

Figure 7.19 shows the same ratio $I(180°)/I(178°)$ but now plotted versus refractive index. Again large variations to indicate that the relative scattering intensities are very sensitive to the particle properties.

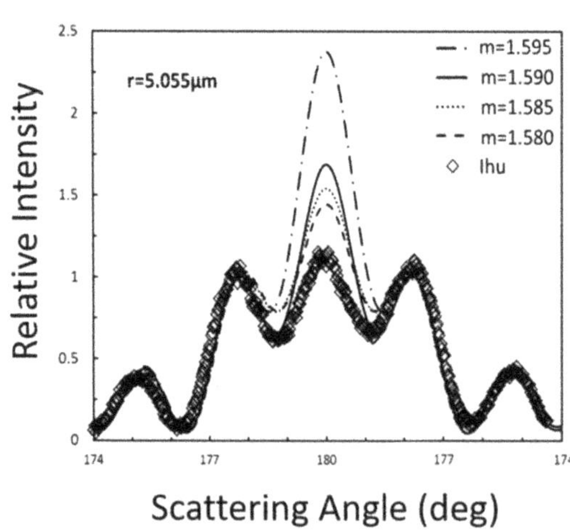

Figure 7.17. Backscattered intensity (normalized at 178°) versus $\theta = 180° - \delta$ for PSL spheres with average radius of 5.055 μm. Reproduced from Heffernan *et al* (2016). © The Optical Society.

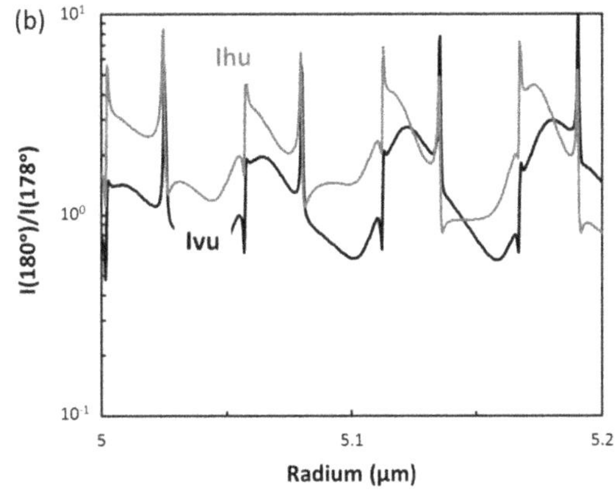

Figure 7.18. $I(180°)/I(178°)$ versus radius over the range of 5–5.2 μm. Reproduced from Heffernan *et al* (2016). © The Optical Society.

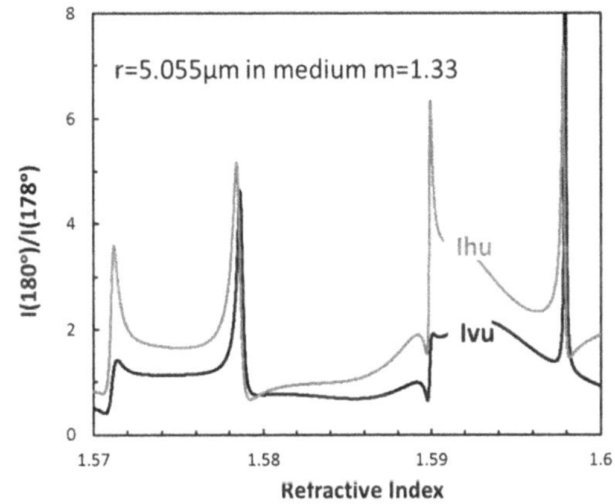

Figure 7.19. $I(180°)/I(178°)$ versus refractive index over the range of 1.57 to 1.60 for radius of 5.055 μm in a medium with a 1.33 refractive index. Reproduced from Heffernan *et al* (2016). © The Optical Society.

We may conclude from these studies that scattering in the extreme back direction has strong functionalities on the particle properties superimposed on the regular functionalities written in equation (7.24).

7.1.4.8 Stationary angles

Figure 7.20 shows the angular dependence of the differential scattering cross sections for an $R = 10$ μm sphere scattering polarized light with a wavelength of $\lambda = 532$ nm calculated using the Mie equations. These plots have been averaged over an angular range of 10° at each angle. Thus, if the scattering angle is 37°, the plotted differential cross section is the result of averaging the cross sections calculated from angles 32° to 42° at one-degree intervals. This removes the ripple structure with spacing ~$\lambda/2R$.

Figure 7.20 demonstrates what we have called a *stationary angle* for the real part of the refractive index at 37° (Waez *et al* 2018, 2021). The magnitude of the

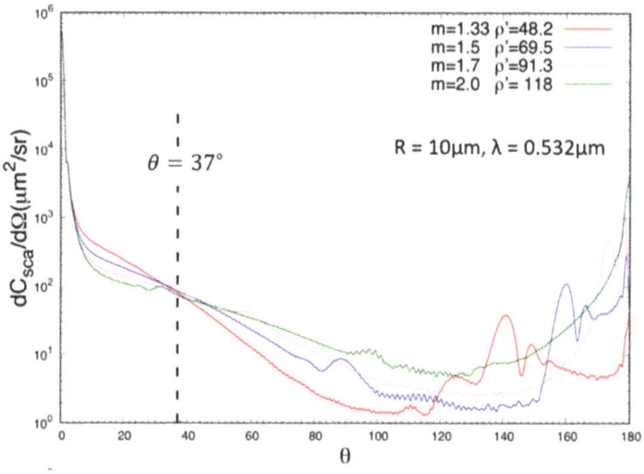

Figure 7.20. Light scattering differential cross section for R = 10 μm spheres of four different real refractive indices and nearly the same total cross section. The incident light has a wavelength of λ = 532 nm and is polarized perpendicular to the scattering plane. The cross section has been averaged over 10° scattering angle ranges (i.e. ±5°) and plotted versus the middle angle. A stationary angle at 37° is marked.

differential cross section does not change, it is stationary, when the real part of the refractive index changes. Remember this includes the angle range ±5°. We will soon learn below (section 7.1.5) that for particles this large with $\rho' \gg 1$ and a real refractive index, the total scattering cross section is very nearly $2\pi R^2$, independent of the refractive index. Thus, one might surmise that since the total scattering is the integral over all space, i.e. over all 4π steradians, the stationary angle results from scattering shifting between angles larger and smaller than the stationary angle while the total scattering remains constant. Indeed, figure 7.20 shows just that. The stationary angle at 37° is found for all spheres with real refractive index such that $R \geqslant$ ~3 μm.

Figure 7.21 demonstrates the stationary angle for the imaginary part of the refractive index characterized by the skin depth parameter, κkR. As for figure 7.20, this includes averaging over the angular range ±5°. One can see that this stationary angle is more of a stationary region and not particular stationary. It lies in the region of ~100° to 140° the average of which is a nominal value of 115°. While imperfect, the functionality on κkR abates in this angular range.

We have demonstrated that these stationary angles make the inverse problem of light scattering, which is to determine R and $m = n + i\kappa$ from the scattering properties, simpler and more accurate for spheres (Waez *et al* 2018, 2021).

7.1.5 The spherical particle total scattering cross section

The total scattering cross section is the integral of the differential cross section over the entire 4π steradians available in three-dimensional space, equation (2.6). For scattering of light polarized perpendicular to the scattering plane from particles in the Rayleigh regime where $kR < 1$ and $\rho' < 1$ the differential cross section is uniform

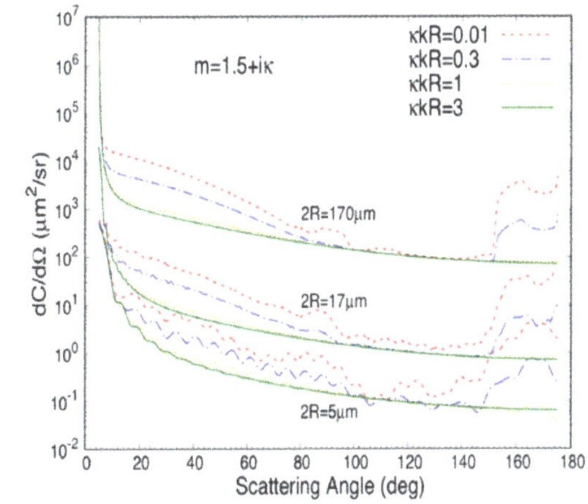

Figure 7.21. Light scattering differential cross section for spheres of three different radii and four different skin depth parameters, κkR. The incident light has a wavelength of $\lambda = 532$ nm and is polarized perpendicular to the scattering plane. The cross section has been averaged over 10° scattering angle ranges (i.e. ±5°) and plotted versus the middle angle. A stationary region spans roughly from 100° to 140°. Reproduced from Waez *et al* (2021). © The Optical Society.

but this integral is affected by the polarization to lead to a factor of 8π/3, as we have seen in section 6.3. The total scattering Rayleigh cross section is (equation (6.13))

$$C_{\text{sca}} = (8\pi/3)k^4R^6F(m) \tag{7.27}$$

where $F(m) = [(m^2 - 1)/(m^2 + 2)]^2$, the square of the Lorentz–Lorenz term.

The total scattering cross section beyond the Rayleigh regime is shown in figure 7.22. This figure shows the total scattering cross section for light polarized perpendicular to the scattering plane for spheres as a function of the radius for a variety of real refractive indices and a wavelength 532 nm. There one can see the strong dependence on size and refractive index when $\rho' < 2$ (see arrows); this is the Rayleigh regime described by equation (7.27). Note the dramatic crossover with increasing radius in the region near $\rho' \approx 2$ regardless of the refractive index. When $\rho' > 2$, the refractive index dependence eventually disappears leaving only a dependence on the geometric cross section, which is proportional to R^2.

The crossover occurs in the region where the two length scales of the system, the sphere size and the light wavelength, are the same. However, it is readily apparent that the refractivity plays an important role in more precisely determining where the crossover occurs. The internal coupling parameter ρ', which contains both the relative size and the refractivity, provides a quantitative indication of the crossover. The value near $\rho' \simeq 2$ appears to describe the crossover best. This indicates that the scattering has left the Rayleigh regime and significant internal multiple scattering is at hand.

The crossover can be viewed conceptually as a change from electromagnetic wave optics to geometric ray optics. This is true except for the factor of two in the total cross section $C_{\text{sca}} \to 2\pi R^2$ when $\rho' \gg 2$. Shouldn't a baseball, for example, scatter

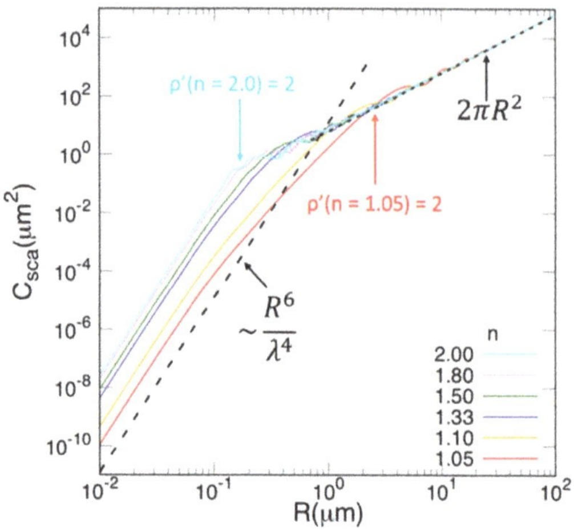

Figure 7.22. Total scattering cross section as a function of spherical particle radius for a variety of real refractive indices m and an incident wavelength of $\lambda = 532$ nm. Arrows mark the radii for which the internal coupling parameter $\rho' = 2$ for a given refractive index. Also shown are the small particle Rayleigh functionality R^6/λ^4 and the large particle limit for the cross section, $2\pi R^2$. The light was linearly polarized perpendicular to the scattering plane.

only the light that hits its geometric cross section, πR^2? The fact that the total scattering cross section is twice as big as it 'should be' for a non-absorbing particle is called the *extinction paradox*, which is definitely a wave phenomenon. The extinction paradox will be discussed in detail in section 8.4. Nevertheless, it can with certainty be stated that near and above $\rho' \simeq 2$, the material dependence, as embodied in the refractive index, is significantly depreciated.

Rayleigh scattering tells us that blue light is scattered more than red, and the light transmitted through a system of non-absorbing Rayleigh scatterers is thereby reddened. Now we see that for large particles with $\rho' > 2$, a dimension analysis implies that there is no wavelength dependence for the scattering; all colors scatter the same. Thus, a system of large particles when illuminated by white light will scatter that white light with no color preference, and therefore appear white. Moreover, the transmitted light through a system of non-absorbing large particles, having uniformly lost all colors equally, will appear grey.

Figure 6.6 in the chapter on Rayleigh scattering illustrates this difference between small and large particles. There we saw two colloids one with small particles (40 nm diameter) and another with large particles (9.8 micron diameter). Under room light illumination, which is essentially white light, the smaller particle colloid appears bluish when viewed via what it scatters, but yellowish when viewed via what it transmits. On the other hand, the bigger particle colloid appears white or grey, both of which imply a lack of functionality with wavelength for the scattering.

Figure 6.6 illustrates blue and white scattering in the lab. However, a sunny but partly cloudy day does a more pleasing job. The blue sky is due to Rayleigh scattering from the molecules that compose the air and the fluctuations in their collective density, which are smaller in extent than the wavelength of light; hence,

Rayleigh scattering ensues. In contrast, the whiteness of the clouds is a result of scattering by water droplets that are much larger than the light wavelength hence have scattering from the geometric scattering size range, which has no wavelength dependence. And when the sky is overcast on a cloudy day, the sky is grey.

A simple argument can be made to relate the total scattering cross section to the forward scattering lobe. We have seen above that when ρ' is large, the intensity in the forward lobe is uniform and has a R^4/λ^2 functionality, equation (7.18c). This is the behavior of the differential cross section, $dC_{sca}/d\Omega$, in the forward lobe as well. The angular size of the lobe is ~λ/R radians, so the solid angle the lobe subtends is $\Omega \approx (\lambda/R)^2$ steradians. To find the total scattering cross section one integrates the differential cross section over the entire 4π steradians of 3d space. However, the differential cross section falls off rapidly beyond the forward lobe, so we can, to a good approximation, include only the forward lobe in the integration[5]. Moreover, since the differential cross section is constant within the forward lobe, it can come out of the integral. The result is that the total scattering cross section is approximately the scattering differential cross section in the forward lobe times the solid angle size of the forward lobe. That is

$$C_{sca} = \int \frac{dC_{sca}}{d\Omega} d\Omega \approx \left(\frac{dC_{sca}}{d\Omega}\right)_{lobe} \Omega_{lobe} \approx (R^4/\lambda^2)(\lambda^2/R^2) = R^2 \qquad (7.28)$$

Equation (7.28) shows that while the forward lobe gets brighter with size, it also get narrower, but the brightening dominates. It also shows that while the forward lobe gets brighter with smaller wavelength, it also get narrower, and the two exactly cancel to leave no dependence on wavelength.

7.2 The spherical particle absorption cross section

Absorption occurs when the imaginary part of the refractive index κ is non-zero. The absorption cross section is the difference between the extinction and scattering cross sections calculated from the Mie equations. Here we discuss results from our study (Sorensen *et al* 2019) and follow the discussion there closely.

The functional dependence of the absorption cross section, C_{abs}, on κ is explored in figure 7.23 for a wide range of size parameters kR. For a given kR, different regimes of functionality occur as κ is varied. All the functionalities are linear on this log–log plot to imply power laws. Crossovers define four regimes of functionality and most of the crossovers are defined by a new parameter, κkR, the product of the refractive index imaginary part and the size parameter. We have named and specified the four regimes as follows:

1. A Rayleigh Regime when $\kappa kR < 0.3$.
2. A Geometric Regime when $kR > 1$ and $\kappa kR > 1$ and $\kappa < 1$.
3. A Reflection Regime when $\kappa > 3$ and $\kappa kR > 3$.
4. A Crossover Regime when $0.3 < \kappa kR < 3$ and $kR < 1$.

[5] In fact, approximately one-half of the scattered light is in the forward scattering lobe. See section 7.4.1.

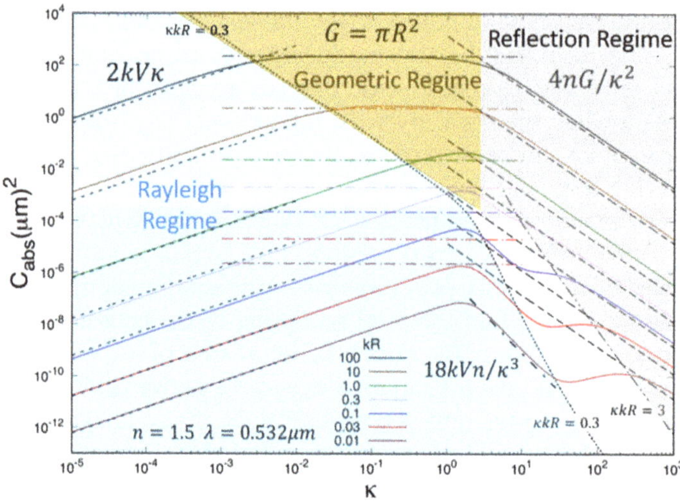

Figure 7.23. The total absorption cross section for spheres of seven different size parameters kR plotted as a function of the imaginary part of the refractive index κ. The refractive index is $m = 1.5 + i\kappa$ and the light wavelength is 0.532 μm. Dashed lines show the functionalities $2kV\kappa$ and $4nG/\kappa^2$ as marked; horizontal dot-dash lines indicate $C_{abs} = G = \pi R^2$, and the short dot-dash line shows the functionality $18kVn/\kappa^3$ as marked. Dotted lines and dash-double-dot lines mark the boundaries of the crossover regime at $\kappa kR = 0.3$ and 3.0, respectively. Different shading indicates different regimes. Reprinted from Sorensen *et al* (2019), Copyright (2019), with permission from Elsevier.

Figure 7.23 is for a real refractive index part of $n = 1.5$. Plots for $n = 1.1$ and 3 are similar to figure 7.23, but deviations from the Rayleigh regime prediction of $C_{abs} = 2kV\kappa$ are less for $n = 1.1$ and greater for $n = 3$.

Before we study the four regimes, we determine the physical significance of the new parameter, κkR that has emerged.

7.2.1 The skin depth parameter κkR

Consider the spatial dependence of a plane wave traveling in the x-direction. In complex notation, the wave is described by equation (1.17)

$$E(x) = E_0 e^{ikx} \tag{7.29}$$

In (7.29) $k = 2\pi/\lambda$ and $\lambda = \lambda_{vac}/m$ where λ_{vac} is the wavelength of light in the vacuum, the subscript 'vac' implying 'vacuum', $m = n + i\kappa$ is the complex refractive index, and we ignore the time dependent part. Thus, $k = mk_{vac} = (n + i\kappa)k_{vac}$. Now substitute this into equation (7.29) and recall the simple fact that $i^2 = -1$ to find

$$E(x) = E_0 e^{\text{ink}_{vac}x} e^{-\kappa k_{vac}x} \tag{7.30}$$

Equation (7.30) has both an oscillatory wave term, the first term on the right hand side, and an exponentially decaying term, the second right hand side term. Equation (7.30) demonstrates that non-zero κ causes an exponential decay of the wave as it propagates through a medium. The characteristic decay length of the wave indicated by equation (7.30) is often called the *skin depth* (Hecht 2002) or the penetration depth and designated by

$$\delta = 1/\kappa k_{\text{vac}} = \lambda_{\text{vac}}/2\pi\kappa \tag{7.31}$$

When the wave travels a distance δ, the wave amplitude is damped by a factor of $1/e \simeq 0.37$.

Equation (7.30) describes a plane wave, but we expect that the concept of damping of the wave over distances of δ to apply at least semi-quantitatively as a plane wave encounters a spherical particle. Then the significance of the damping depends on how δ compares to the particle's size, R. The comparison is achieved by a dimensionless ratio

$$R/\delta = \kappa k_{\text{vac}} R \tag{7.32}$$

Equation (7.32) shows that κkR (we have dropped the subscript 'vac') is the relative extent that the incident wave penetrates into the sphere. When κkR is small, $\delta \gg R$, thus the light thoroughly penetrates the particle so that absorption does not impede the incident light from affecting the entire sphere. Hence, the effects of κ are insignificant. When κkR is large, $\delta \ll R$, thus the incident light is confined to the incident surface, and the effects of κ are large. Given these results, we shall call κkR the *skin depth parameter*, or very often simply refer to it as κkR.

This physical insight implies that it is not the magnitude of κ that matters, but rather the magnitude of the dimensionless κkR. It also forecasts a volumetric to surface dependence of both absorption and scattering based on κkR and not κ directly. In fact, we will see in section 8.2.7 that κkR has significant importance for scattering (Wang *et al* 2015). Finally, the quantity $2\kappa k = 2/\delta$ is often refered to as the *absorption coefficient* (Kerker 1969, Mishchenko *et al* 2002). Thus, $2\kappa kR$ is the radius times the absorption coefficient.

Figure 7.24 draws an idealized, schematic representation for what the illumination of a large sphere might look like for $\kappa kR \simeq 3$. There the light wave only penetrates to a depth of about 1/3 radius into the incident side of the sphere.

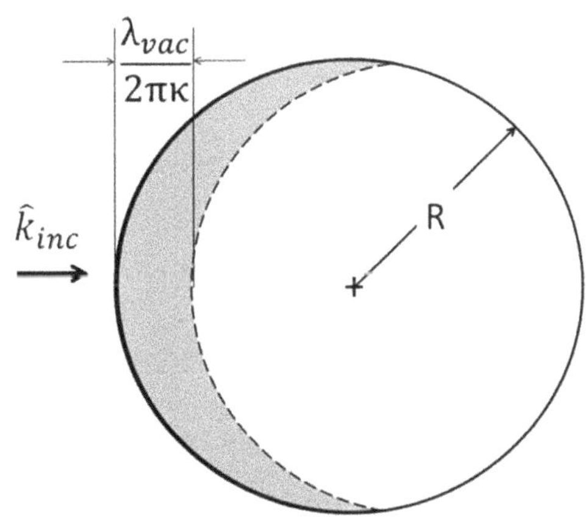

Figure 7.24. Drawing of the illuminated 'incident cap' of a sphere when $\kappa kR \simeq 3$. $\hat{k}_{\text{inc}}$ indicates the direction of the incident light.

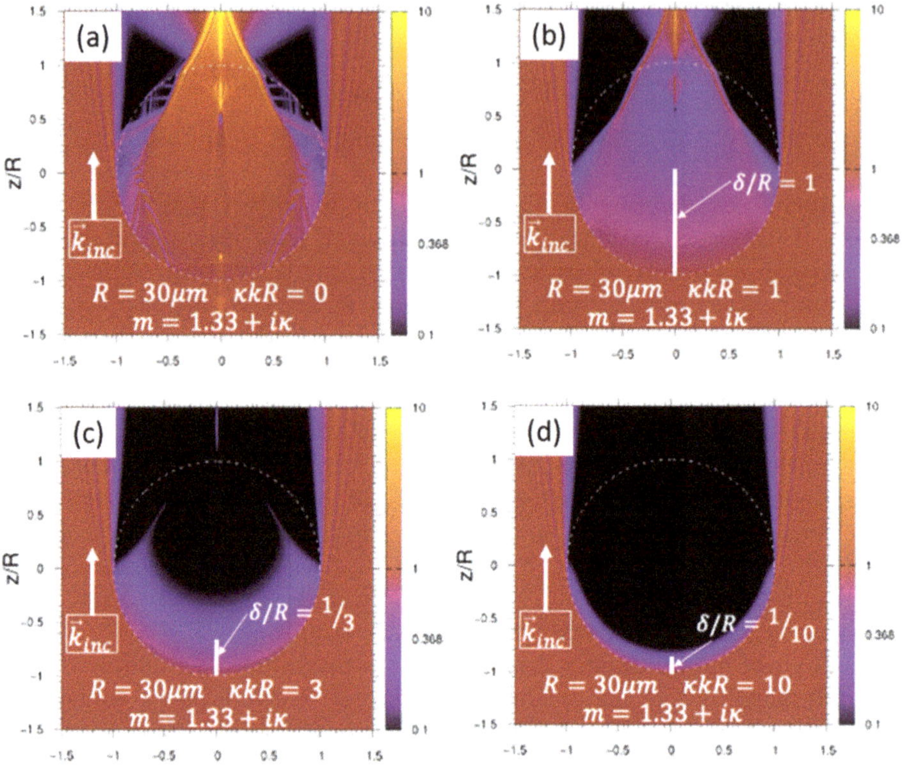

Figure 7.25. The magnitude of the internal fields and the fields near the surface of R = 30 μm spheres, white dashed lines, calculated with the Mie equations normalized by the incident field for four different values of the skin depth parameter, κkR. The view is looking down the y-axis onto the xz-plane. The light is incident from below. The incident polarization is in the y-direction. Color-coded field scale bars are to the right of each sub-figure. White lines marked with δ/R indicate the relative skin depth (Maughan and Sorensen 2020).

Figure 7.25 shows exactly what happens to the magnitude of the internal field and the field near the surface of the spheres when κkR increases from zero. The sphere is large enough, R = 30 μm, that ray optics focusing is apparent. With increasing κkR, the interior field falls to zero on the non-incident side of the sphere and ultimately confined to a thin layer on the incident side.

We now return to figure 7.23, which indicates various functionalities within the different regimes. We next attempt to derive and explain these functionalities.

7.2.2 Rayleigh regime absorption

The Rayleigh regime is usually discussed in reference to scattering. Then the conditions for Rayleigh scattering are expressed as when $kR \ll 1$ and $|m|kR \ll 1$, equations (6.1). These conditions arise from the expansion of the Mie series equations using the first condition to yield terms of third, fifth and higher order in kR. The third order term yields the Rayleigh limit, but one of the fifth order terms is persistent as $kR \to 0$ if $|m|kR \ll 1$ is large. Hence the second condition. A physical explanation of the role of refractive index was given by Bohren and Huffman (1983) who used an electrostatics approach in which the entire sphere acts like an oscillating dipole. This requires that the internal field be uniform in both magnitude and phase. The discussion section 7.2.1, above, indicates that the magnitude of the wave will be uniform if $\kappa kR \ll 1$. Phase uniformity is needed for the scattered waves from the sub-volumes of the sphere to reach the detector in phase. Phase uniformity can be achieved if the sphere's size is much less that the wavelength inside the sphere, i.e. $R \ll \lambda_{vac}/n$. This is equivalent to $nkR \ll 1$. Because the complete refractive index is $m = n + i\kappa$, the combined condition for Rayleigh scattering is $(m)kR \ll 1$. Now we emphasize that the Rayleigh condition for scattering is that the internal field is homogeneous in phase *and* amplitude.

In contrast to scattering, absorption is not significantly affected by the real part of the refractive index because absorption is not dependent upon in-phase addition of waves at the detector. Now we propose that the Rayleigh condition for absorption is that the internal field is homogeneous in amplitude independent of phase. Then the Rayleigh condition for absorption would best be stated as $\kappa kR \ll 1$ not $|m|kR \ll 1$. This implies that Rayleigh absorption can occur for all size parameters kR because κ can be very small. It is now necessary to justify this proposal.

The Rayleigh theory absorption cross section is, from equation (6.19)

$$C_{\text{abs Rayleigh}} = 4\pi R^2(kR)E(m) \tag{7.33}$$

where

$$E(m) = Im\left(\frac{m^2 - 1}{m^2 + 2}\right) \tag{7.34}$$

The expression Im(x) means to take the imaginary part of x. With $m = n + i\kappa$, equations (7.33) and (7.34) combine to yield

$$C_{\text{abs Rayleigh}} = 3kV\left(\frac{6n\kappa}{(n^2 - \kappa^2 + 2)^2 + 4n^2\kappa^2}\right), \tag{7.35}$$

where $V = 4\pi R^3/3$ is the sphere volume. There are a few different limits that one could take to simplify this equation. Here we start with $\kappa < 1$. Then

$$C_{\text{abs Rayleigh small}\,\kappa} \simeq 18kVn\kappa/(n^2 + 2)^2. \tag{7.36}$$

A simpler expression comes with using $n = 1.0$

$$C_{\text{abs Rayleigh small}\,\kappa}(n = 1.0) \simeq 2kV\kappa \tag{7.37}$$

This result is included in figure 7.23 as a dashed line. It is seen to have the correct functionality on all three parameters and to be in semi-quantitative magnitude agreement with the Mie result for all size parameters as long as $\kappa kR < 0.3$ and $\kappa < 1$.

On the other hand, when $\kappa \gg 1$, the Rayleigh absorption cross section, equation (7.35), leads to

$$C_{\text{abs Rayleigh large }\kappa} \simeq 18kVn/\kappa^3. \tag{7.38}$$

This result is also included in figure 7.23 as a short dot-dash line and seen to be in semi-quantitative agreement with the Mie result for all small size parameters ($kR \ll 1$), $\kappa > 3$ and $\kappa kR < 0.3$. Note that although κ is large ($\kappa \gg 1$), the combined parameter is $\kappa kR < 1$, so the entire volume of the sphere is still nearly uniformly illuminated to yield a volume dependence in equation (7.38).

The facts that the Rayleigh theory-derived equations (7.37) and (7.38) have the correct functionalities and apply semi-quantitatively for spheres of all sizes when $\kappa kR < 0.3$ implies that the *sole condition for Rayleigh absorption is* $\kappa kR < 0.3$ as proposed above.

The magnitude of the deviations of Rayleigh theory from the Mie result is studied in figure 7.26, which plots the Mie calculated absorption cross section divided by the Rayleigh theory prediction of equations (7.33) and (7.34). As expected, when $kR \lesssim 1$, the Rayleigh and Mie theory predictions agree. In the other extreme when $kR \gtrsim 100$, Rayleigh theory under-predicts the absorption cross section by about a factor of two as long as $\kappa kR \lesssim 0.01$. Nevertheless, the two are proportional. When $1 \lesssim kR \lesssim 100$ and $\kappa kR \lesssim 0.01$, Rayleigh theory not only under-predicts, but it misses a strong ripple structure in the Mie result. For all kR the ratio of Mie and Rayleigh theory calculated absorption declines rapidly after $\kappa kR \gtrsim 0.1$. When $\kappa kR \gtrsim 1$, the ratio falls off linearly

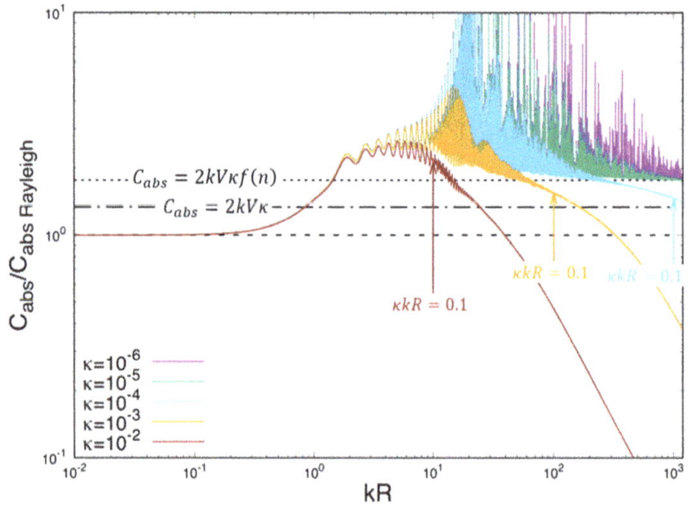

Figure 7.26. The ratio of the absorption cross section calculated with the Mie equations for spheres divided by the exact Rayleigh theory predictions of equations (7.33) and (7.34) versus the size parameter kR. The light wavelength is $\lambda = 0.532$ μm. The refractive index is $m = 1.5 + i\kappa$. The range of imaginary refractive indices is given in the key. The dashed line indicates a ratio of 1.0. The dot-dashed line indicates the ratio of $C_{\text{abs}} = 2kV\kappa$ (also used in figure 7.23), equation (7.37), to the Rayleigh theory prediction and the short-dashed line indicates the ratio of $C_{\text{abs}} = 2kV\kappa f(n)$, equation (7.39), to the Rayleigh theory prediction. Note that some of the spike-like structure may be affected by aliasing. Reprinted from Sorensen *et al* (2019), Copyright (2019), with permission from Elsevier.

with kR, consistent with the transition of the absorption cross section from a volumetric to the geometric cross sectional area dependence seen in figure 7.23.

Bohren and Huffman (1983) used a geometric optics approach to derive a formula for the absorption cross section valid for large size parameters and weakly absorbing spheres. Their result is

$$C_{\text{abs B\&H}} = 2kV\kappa\left\{n^{-1}[n^3 - (n^2 - 1)^{3/2}]\right\} = 2kV\kappa f(n). \tag{7.39}$$

Note that the same product $kV\kappa$ included in this equation also appears in equations (7.36), (7.37) and (7.38). When $n = 1.0$, the bracketed term in equation (7.39), i.e. $f(n)$, equals 1. Then equation (7.39) becomes identical to equation (7.37), which was derived from the Rayleigh theory absorption cross section in the $n \to 1$ limit. For $n = 1.5$ the bracketed term $f(n) = 1.318$ and inclusion of this factor yields perfect agreement between equation (7.39) and the ratio $C_{\text{abs}}/C_{\text{abs Rayleigh}}$ plotted in figure 7.26 for large kR and $\kappa kR < 0.01$.

These results show that in the regime where $\kappa kR < 0.3$ the Rayleigh theory equations for absorption yield the correct functionality of $kV\kappa$, give magnitudes in semi-quantitative agreement with Mie theory and, geometric optics calculations even when kR is small. These facts apply regardless of the size parameter, large or small, so long as $\kappa kR < 0.3$, i.e. $\kappa < 1/3kR$. Under this condition, the absorbance does not significantly affect the light propagation into the volume of the sphere.

7.2.3 The geometric regime when $kR > 1$ and $\kappa kR > 1$ and $\kappa < 1$

The shaded geometric regime in figure 7.23 is determined by when the absorption cross section C_{abs} is approximately equal to the geometric cross section, $G = \pi R^2$. It occurs for large spheres such that $kR > 1$. When the size is large, the geometric regime evolves out of the Rayleigh regime, where the absorption is volume dependent, and crosses over to a surface dependence as κkR grows larger than one. This is consistent with our physical picture that the light interacts with the entire volume of the sphere when $\kappa kR \ll 1$, but becomes confined to the surface when $\kappa kR \gtrsim 1$ as shown schematically in figure 7.24. Ultimately, when $\kappa > 1$, the absorption cross section falls below the geometric limit of G due to reflection from the surface as soon to be described below.

Figure 7.27 explores the accuracy of $C_{\text{abs}} = G$ in this regime by plotting the absorption efficiency $Q_{\text{abs}} = C_{\text{abs}}/G$ versus the size parameter kR. For $\kappa kR > 1$ and large kR, $Q_{\text{abs}} \simeq 1$ to within ±20%. However, when κ increases through unity, absorption gives way to reflection and Q_{abs} decreases rapidly. Arrows indicating where $\kappa kR = 1$ mark the crossover from volumetric dependent Rayleigh regime, when $\kappa kR < 1$, to the surface dependent geometric regime that results in the geometric cross section.

Figures 7.26 and 7.27 both show a broad hump of enhanced absorption. Figure 7.26 also shows a high frequency ripple structure superimposed on the hump. There is some ripple in two of the plots in figure 7.27. These features are usually referred to as the interference and ripple structures, respectively, and are readily seen in plots of the scattering and extinction efficiencies (Bohren and

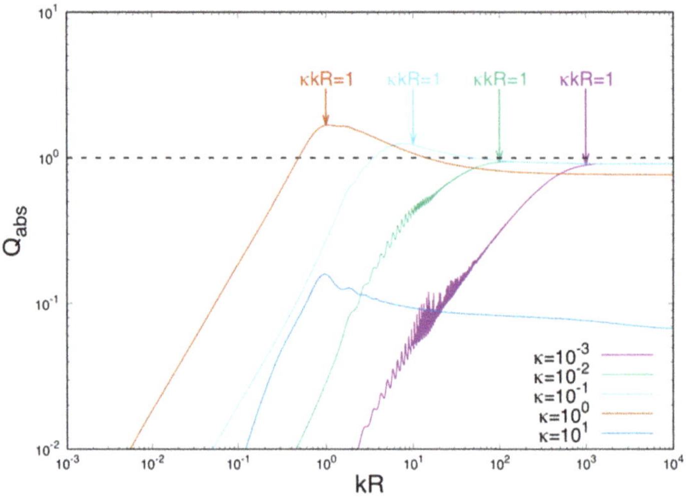

Figure 7.27. The absorption efficiency $Q_{abs} = C_{abs}/G$ calculated with the Mie equations for spheres versus the size parameter kR. The light wavelength is $\lambda = 0.532$ μm. The refractive index is $m = 1.5 + i\kappa$. The range of imaginary refractive indices is given in the key. Plots are made for a wide range of imaginary refractive indices, κ. Arrows indicate where $\kappa kR = 1$ for different κ. Reprinted from Sorensen *et al* (2019), Copyright (2019), with permission from Elsevier.

Huffman 1983, Mishchenko *et al* 2002). In figure 7.26, the broad hump interference structure begins near $kR \simeq 1$ and dies away as κkR increases through ~0.1. The interference structure has been explained as being due to interference between light that has been diffracted by the particle with that which has passed through the particle (van de Hulst 1957, Chylek and Zhan 1989, Lock and Yang 1991, Moosmuller and Sorensen 2018). The diminishment of the broad hump with increasing κkR is consistent with a picture of light no longer passing through the particle at large κkR. The ripple structure has been explained as due to resonances of internal rays propagating along the inner circumference of the sphere. In figure 7.27, the ripples begin to disappear when $\kappa kR \simeq 0.01$ and are essentially gone when $\kappa kR \simeq 0.1$. These numbers are rather small, but if we take the relevant scale comparison of the 1/e propagation length δ as not the radius but rather the circumference, the appropriate parameter to describe the loss of the ripples would be $2\pi\kappa kR$. Therefore, we can restate that the ripple structure begins to disappear when $2\pi\kappa kR \simeq 0.06$ and is essentially gone when $2\pi\kappa kR \simeq 0.6$; numbers sufficiently large to support the contention.

7.2.4 The reflection regime when $\kappa > 3$ and $\kappa kR > 3$

Figure 7.23 shows that a third regime occurs whenever $\kappa > 3$ and $\kappa kR > 3$. In this regime, the absorption cross section falls quadratically with increasing κ from the geometric limit. Considering for the moment only large particles, it is well known that then the extinction cross section, which is the sum of the absorption and scattering cross sections approaches $2G = 2\pi R^2$ as $kR \to \infty$, i.e.

$$C_{\text{ext}} = C_{\text{abs}} + C_{\text{sca}} = 2G. \tag{7.40}$$

Then if C_{abs} decreases from G with increasing κ, C_{sca} must increase from G with increasing κ. The conclusion is that as κ passes from 1 to 10 or more, absorption gives way to specular reflection to keep the extinction cross section equal to $2G$. One can envision that a smooth, black sphere with $C_{\text{abs}} = \pi R^2$ becomes shiny like a steel ball as κ increases beyond ~3. Note that reflection is a form of scattering so, ironically, the parameter that causes absorption, κ, ultimately causes scattering and quenches absorption!

Here we construct a geometric optics description of absorption by large spheres with $kR > 1$ when $\kappa > 3$ and $\kappa kR > 3$. Because it appears that large κ will favor scattering to the detriment of absorption, we consider the scattering. For large spheres in the geometric optics regime, the scattering cross section can be written as the sum of scattering cross sections for diffraction (D), transmission (T), and surface reflection (R) (ignoring internal reflections) as (Moosmüller and Arnott 2003)

$$C_{\text{sca}} = C_{\text{sca}\_D} + C_{\text{sca}\_T} + C_{\text{sca}\_R}. \tag{7.41}$$

As $kR \to \infty$, C_{sca_D} approaches G, and as $\kappa kR \to \infty$, C_{sca_T} becomes 0. Then with the help of equation (7.41), equation (7.40) becomes

$$C_{\text{abs refl}} = G - C_{\text{sca}\_R} \tag{7.42}$$

Equation (7.42) essentially claims that if the light reflects, it doesn't get absorbed, and when it reflects in the geometric limit it diminishes the absorption, as argued above.

C_{sca_R}, the scattering cross section due to reflection from the sphere, can be calculated using geometric optics by integrating the Fresnel reflection coefficients over a spherical surface as (Moosmüller and Arnott 2003, 2019)

$$\begin{aligned} C_{\text{Sca}\_R} = \frac{G}{4}\int_0^{\pi} d\theta \sin(\theta)\Bigg\{&\left(\frac{[\sin(\theta/2) - u_R]^2 + v_R^2}{[\sin(\theta/2) + u_R]^2 + v_R^2}\right) + \\ &\left(\frac{[(n^2 - \kappa^2)\sin(\theta/2) - u_R]^2 + [2n\kappa\sin(\theta/2) - v_R]^2}{[(n^2 - \kappa^2)\sin(\theta/2) + u_R]^2 + [2n\kappa\sin(\theta/2) + v_R]^2}\right)\Bigg\}, \end{aligned} \tag{7.43}$$

where the two fractions in the large, round brackets contain the s- and p-polarized contributions, respectively, and u_R, v_R, and a are defined as

$$u_R = \sqrt{\frac{\sqrt{a^2 + (2n\kappa)^2} + a}{2}} \tag{7.44}$$

$$v_R = \sqrt{\frac{\sqrt{a^2 + (2n\kappa)^2} - a}{2}} \tag{7.45}$$

$$a = n^2 - \kappa^2 - \cos^2(\theta/2). \tag{7.46}$$

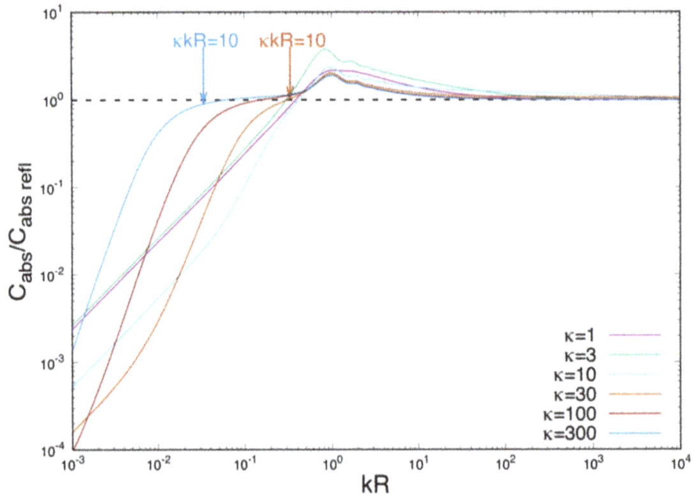

Figure 7.28. The ratio of the absorption cross section calculated with the Mie equations for spheres divided by the reflectance absorption cross section of equations (7.42)–(7.46) versus the size parameter kR. The light wavelength is $\lambda = 0.532$ μm. The refractive index is $m = 1.5 + i\kappa$. The range of imaginary refractive indices is given in the key. The dashed line indicates where the ratio is one. Plots are made for a wide range of imaginary refractive indices, κ. Reprinted from Sorensen *et al* (2019), Copyright (2019), with permission from Elsevier.

The prediction of equations (7.42)–(7.46) is compared to the exact Mie prediction as a ratio in figure 7.28. The prediction is shown to work well for $kR > 10$ and $\kappa > 3$ and becomes equal to the Mie prediction as both of these quantities grow larger. There is yet another 'interference' hump in the ratio near $kR \simeq 1$. Notice that the prediction continues to work well when $kR < 1$, as long as $\kappa kR \gtrsim 10$. This is remarkable because the concept of geometric, specular reflection at the foundation of our prediction is surprising for subwavelength-size particles with $kR < 1$!

Equations (7.42)–(7.46) can be simplified. Assume that the incident side of the sphere is flat with area $G = \pi R^2$ perpendicular to the incident light. Then we apply Fresnel reflection at normal incidence, where the reflectivity is

$$\boldsymbol{R} = \frac{(n-1)^2 + \kappa^2}{(n+1)^2 + \kappa^2}. \tag{7.47}$$

Now convert equation (7.42) to

$$C_{\mathrm{abs}} = G(1 - \boldsymbol{R}) \tag{7.48}$$

Substitute (7.47) to obtain the 'normal reflectance' absorption cross section

$$C_{\mathrm{abs\,norm\,refl}} \simeq G\frac{4n}{(n+1)^2 + \kappa^2}. \tag{7.49}$$

In the limit of large κ such that $\kappa \gg n$,

$$C_{\mathrm{abs\,norm\,refl}} \simeq 4nG/\kappa^2. \tag{7.50}$$

The result in equation (7.50) is compared to the exact Mie results in figure 7.23. The comparison is good in magnitude and exact in the G and κ^{-2} dependencies when $\kappa > 3$. These dependencies are hidden in the more exact equations (7.42)–(7.46). One can show that in the limit of $\kappa \rightarrow \infty$, equation (7.50) under-predicts the reflectance absorption cross-section result from equations (7.42)–(7.46) by about 33% for all kR, and hence misses the Mie prediction by the same amount.

It is rather remarkable that Fresnel reflection is relevant for sub-wavelength-size spheres. This is true for the functionality on the physical properties of the sphere and, to a good approximation, for the magnitude. Perhaps this is not so surprising when one considers the fact that both the Mie scattering and Fresnel equations are based on the electromagnetic boundary conditions, the only difference being the different symmetries, spherical and planar, respectively.

It is useful to examine the spherical particle absorption cross section as a function of the radius in analogy to figure 7.22 that shows the scattering cross section as a function of the radius. This is shown in figure 7.29. When $\kappa \leqslant 1$, volume dependent R^3 and surface area dependent πR^2 regimes are seen to crossover near $\kappa kR \sim 1$. When $\kappa = 10$, the absorption cross section is much smaller due to reflection. Also, note the ripples for $\kappa = 0.01$ and $0.1 < R < 1$ microns, which corresponds to $0.012 < \kappa kR < 0.12$. Small values, no ripple suppression.

Our discussion of absorption ends with figure 7.30 that shows the Mie calculated absorption cross section for a sphere with index of refraction $m = 1.5 + i0.0001$. The simple approximation of equation (7.37) is plotted as well and seen to do a good job of representing the trend in the absorption cross section as a function of sphere radius. Superimposed upon this trend in the Mie result is the ripple structure. It is not suppressed because $\kappa kR < 0.02$ in this graph. A physical interpretation of the ripple is that it is due to internal surface waves in resonance around the inner

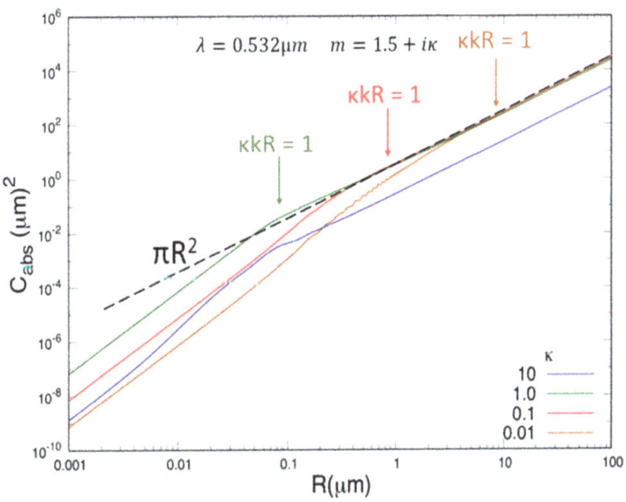

Figure 7.29. The absorption cross section for spherical particles versus radius for a systematic variety of imaginary parts of the refractive index κ. The light wavelength is $\lambda = 532$ nm.

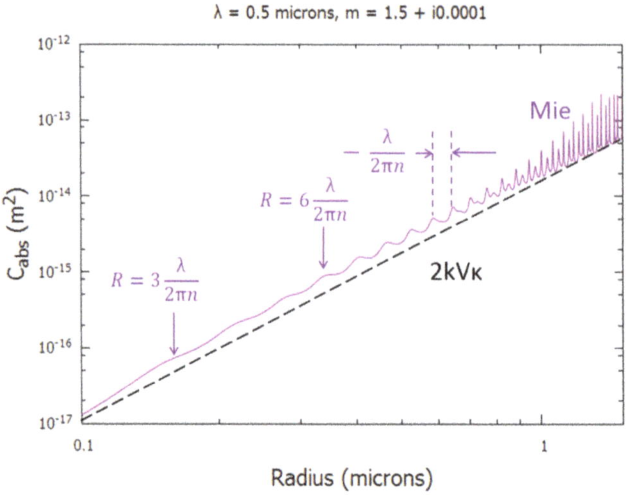

Figure 7.30. The Mie calculated absorption cross section versus sphere radius for light with a wavelength of 0.5 micron and a sphere with index of refraction $m = n + i\kappa = 1.5 + i0.0001$. Also plotted is the dashed trend line from equation (7.37). Resonances predicted by equation (7.52) are also indicated for $N = 3$ and 6.

circumference. This has been discussed above in section 7.2.3 and in the discussion regarding figure 7.16. The resonance condition is the inner path length is an integer number of wavelengths. The sphere largest inner circle is a circle of radius R. Thus, we recall equation (7.25) as

$$N\lambda_{\text{vac}}/n = 2\pi R \tag{7.51}$$

where n is the real part of the refractive index. From whence one obtains

$$R = N\lambda_{\text{vac}}/2\pi n \quad N = 1, 2, 3, \ldots \tag{7.52}$$

Equation (7.52) indicates radii at which resonances occur under our assumption. A few of the resonances calculated with equation (7.52) are marked in figure 7.30 and are seen to reside near local peaks in the absorption cross section. This reinforces the physical picture of the source of the ripple. Note, however, the situation gets more complex as the radius increases.

7.3 Effects of absorption on scattering

Figure 7.31 displays the effects of absorption as caused by a non-zero imaginary refractive index, $\kappa > 0$, on scattering. Scattering, as calculated using the Mie equations, by two spheres is shown; a small one, roughly a factor of 10 larger than the Rayleigh limit, and another 30 times larger than the first. The refractive index is $m = 1.5 + i\kappa$ for each and κ is adjusted so that the relative skin depth parameter, κkR, is either 0, which leads to no effect of absorption, or 3, which is quite significant.

Figure 7.31 shows that $\kappa kR = 3$ quenches, for the most part, the ripple structure. The remaining ripple for the smaller sphere has a spacing of ~18°, which is equal to the diffraction angle $\lambda/2$R. The larger sphere lost its ripple structure when $\kappa kR = 3$ as

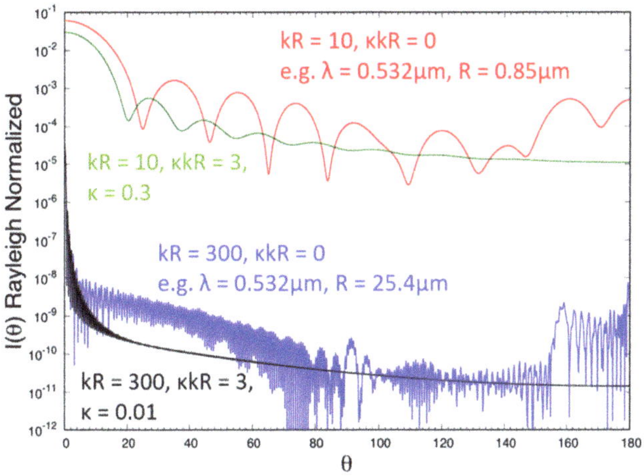

Figure 7.31. Rayleigh normalized scattering versus scattering angle for two spheres with a refractive index of $m = 1.5 + i\kappa$. Two different, and rather extreme, values of the skin depth parameter, κkR, are used to show the effects of absorption on scattering. Note that the black plot covers the blue plot when $\theta < 4°$.

well as other features for $\theta >$ ~20°, but retains the ripple as diffraction fringes below ~20°. Note that in each case the absorption did not significantly affect the scattering below $\theta \simeq \lambda/2R$. Now recall figures 7.24 and 7.25 to realize that $\kappa kR = 3$ quenched the interior fields as well. With all this, the conclusion is that when $\kappa kR = 3$, the banished interior field reduces the sphere to a simple diffracting object.

Figure 7.32 shows the forward normalized intensity of light scattered by four large spheres. The incident light is un-polarized light with a wavelength $\lambda = 532$ nm. The scattered intensity was calculated using the Mie equations. The ripple pattern was first eliminated by averaging over a zero-order lognormal size distribution with a geometric width of $\sigma = 1.2$. The refractive index is $m = 1.33 + i\kappa$. The imaginary part of the refractive index is varied for each radius, such that the relative skin depth parameter, κkR, ranges geometrically from 0 to 10.

Figure 7.32 displays a number of features. First, at very small angles such that $\theta < \lambda/2R$, there are sharp spikes in the intensity, which are due to diffraction from the particle. Next, at larger angles, follow three refraction effects. For $\lambda/2R < \theta < 90°$ there is a broad hump. We will often refer to this as the refraction hump because, as we will see in section 8.2.7, it is due to refraction of light into and out of the sphere. At larger angles, the secondary and primary rainbows appear near 130° and 137.5°, respectively. Between these rainbows is a dark region known as 'Alexander's band'. Finally, in the back direction as θ approaches 180° there is an increase in the intensity with the glory.

The black curves in figure 7.32 are an asymptotic solution reached by disregarding any refraction into or out of the particle and thus leaving only the sum of diffraction and reflection (Yang *et al* 2001). One might call this 'Mie minus refraction'.

Figure 7.32 shows that as the relative skin depth parameter κkR increases, the refraction effects systematically decrease until when $\kappa kR \gtrsim 3$, where they are essentially

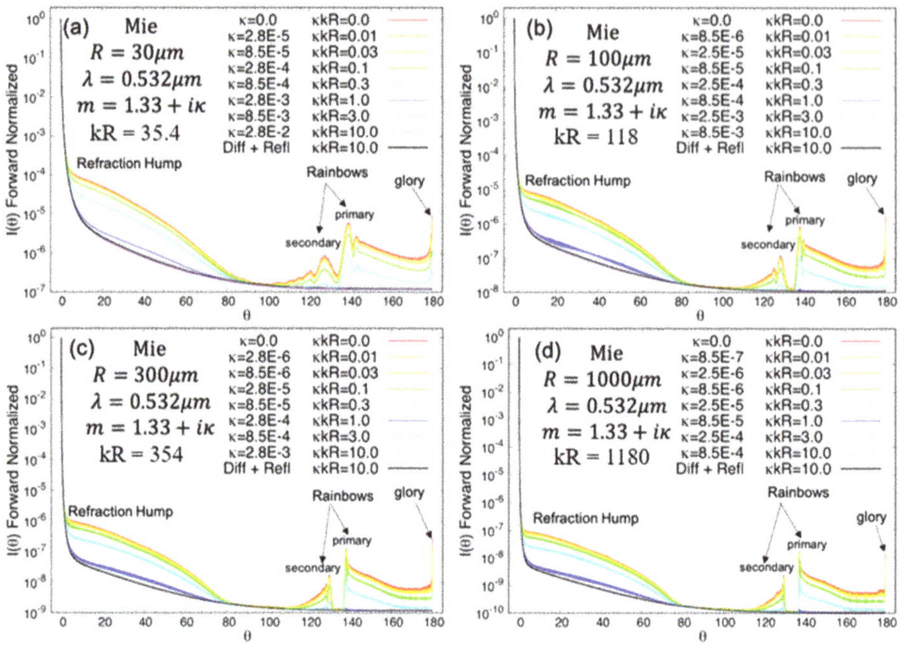

Figure 7.32. Forward normalized scattered intensity versus the scattering angle θ for four different size spheres calculated with the Mie equations. The incident light has a wavelength of λ = 532 nm and is unpolarized. The spheres have a refractive index of $m = 1.33 + i\kappa$, and κ is varied widely to achieve the same range of κkR values for all the spheres. The black curves are a calculated asymptotic limit for the spheres, which contains no refraction (Yang *et al* 2001). Reproduced from Maughan and Sorensen (2020). © The Optical Society.

gone. Then the Mie calculation is identical to the asymptotic black curve, which is devoid of refraction. We conclude that when $\kappa kR \geqslant 3$, all the refraction effects are gone leaving only diffraction and reflection to contribute to the scattering by the spheres. This conclusion is consistent with that reached with figure 7.31 above.

Figure 7.32 also demonstrates the universal nature of the skin depth parameter, κkR. The evolution of the loss of refraction effects is characterized by κkR, not κ.

The efficacy of the relative skin depth parameter, κkR, has now been well established for description of the effects of the imaginary refractive index, κ, on absorption and scattering. Naked κ is insufficient. The literature abounds with considerations of scattering and absorption by real particles with complex indices of refraction ignorant of this fact.

7.4 Efficiencies

An *efficiency* or *efficiency factor* Q is defined as the ratio of an optical cross section to the geometric projected area of the particle onto a plane perpendicular to the incident direction, a quantity we have been designating as G. Thus, one has the dimensionless quantity (see equation (2.11))

$$Q = C/G \tag{7.53}$$

Because there are scattering, absorption and extinction cross sections, there are also the corresponding efficiencies. The aptly named efficiency describes how efficient the geometric cross section is at interacting with the light. It compares what really happens with scattering, absorption or extinction compared to what would happen if one naively viewed light without its wave nature, i.e. if light was solely a flux of classical particles. Said differently, it is the ratio of the wave cross section to the geometric optics cross section. Efficiencies apply to all shapes. As simple as the concept of an efficiency is, efficiencies uncover some facts about scattering that will require close attention and perhaps adjust our view of how waves work.

The upper part of figure 7.33 shows the scattering efficiency for spheres with real refractive indices of $m = n = 1.2$, 1.5 and 2.0 (real m hence $Q_{\mathrm{abs}} = 0$ and $Q_{\mathrm{sca}} = Q_{\mathrm{ext}}$)

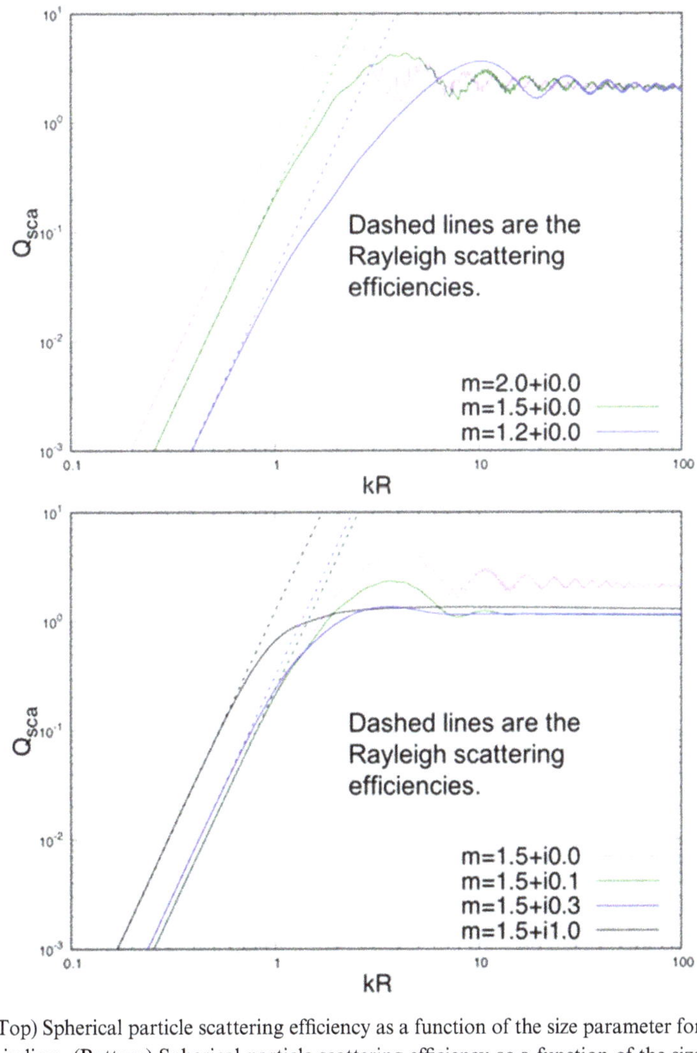

Figure 7.33. (Top) Spherical particle scattering efficiency as a function of the size parameter for three different real refractive indices. (Bottom) Spherical particle scattering efficiency as a function of the size parameter for four different complex refractive indices with various imaginary parts.

as a function of size parameter, kR, calculated with the Mie equations. One sees that the scattering efficiency rises rapidly from zero out of the Rayleigh limit where $kR < 1$ with a $(kR)^4$ functionality to maximum values in the range of 3.7 to 5.6 depending on the refractive index as kR increases. After that, it goes through a series of decaying oscillations, the interference structure, with increasing kR. Superimposed on these long period interference structure oscillations is a shorter period ripple structure. Although the ripple has smaller amplitude than the oscillations, its amplitude increases with the real refractive index.

With these results, it is interesting to recognize that it is only when the scattering efficiency is equal to one that the light is detecting the true size of the particle. When very small (compared to the wavelength of light), the particles scatter much less light than their geometric cross section could intercept; hence the light does not see them very well; it 'under-detects' them. On the other hand, notice that the limit $kR \to \infty$, a limit which brings us to everyday objects like baseballs, figure 7.33 shows that $Q_{sca} \to 2$, which means that the light is scattering twice as much light as our senses tell us it should be by casual visual observation. This alarming observation is the *extinction paradox*, which was mentioned above in reference to figure 7.22 and will be addressed in section 8.4. Finally, a third feature is the crossover between these small and large regimes in the vicinity of $kR \simeq 1$ to 3, which corresponds to when the diameter is about equal to the wavelength. Remarkably, in the crossover region the efficiency can be about four or larger, very efficient! Such are the enigmas of waves.

The lower part of figure 7.33 holds the real part of the refractive index constant at $n = 1.5$ but varies the imaginary part, κ. One can see the effect of increasing κ is to decrease the amplitudes of both the interference structure and the ripple and cause the large kR limit to fall from ca. 2 to 1.

Figure 7.33 shows that the real part of the refractive index has two effects: with increasing m, the period of the interference structure in Q_{ext} shortens with respect to kR, and, also with increasing m, the higher frequency ripple increases in amplitude. The interference oscillatory structure can be tamed by plotting the efficiency versus the phase shift parameter (Kerker 1969)

$$\rho = 2kR(n - 1) \tag{7.54}$$

Such a plot is shown in figure 7.34 (left), along with plotting versus the internal coupling parameter ρ' (right), for comparison. Figure 7.34 uses some of the same data found in figure 7.33.

Figure 7.34 shows the scattering efficiencies for a sphere with real refractive indices of $m = n = 1.2$, 1.5 and 2.0 (real m hence $Q_{sca} = Q_{ext}$) as a function of either the phase shift parameter ρ or the internal coupling parameter ρ'. For a given m, kR was varied to vary ρ or ρ'. When plotting versus ρ, one sees that all the interference structure has a periodicity $\Delta\rho \simeq 2\pi$. Thus, the phase shift parameter provides a quasi-universal description of the extinction and scattering efficiencies. When plotting versus ρ', the interference structure has a periodicity dependent upon the refractive index. We conclude that the phase shift parameter ρ provides the better

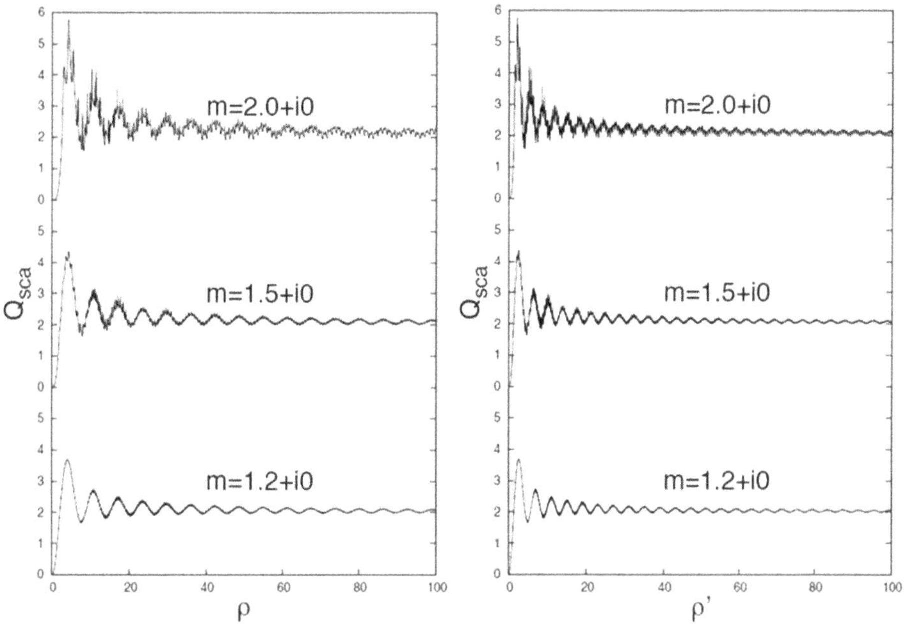

Figure 7.34. Spherical particle scattering efficiency for spheres with real refractive indices of m = 1.2, 1.5 and 2.0 as a function of the phase shift parameter ρ (left), and the internal coupling parameter ρ' (right). These are the same calculated data plotted in figure 7.33. The oscillation spacing of the interference structure with ρ is $\Delta\rho \simeq 2\pi$, with ρ' is $\Delta\rho' \simeq 4.0$, 3.6 and 3.2 for m = 1.2, 1.5 and 2.0, respectively.

description of the periodicity of the interference structure of the scattering (and extinction) efficiency. Now recall the observations made in regard to forward scattering, figure 7.6, that indicated that when wave phase is involved, the phase shift parameter ρ provides the best description, but when scattered intensities (field amplitudes squared) are involved, the internal coupling parameter ρ' is the best descriptor. We hope future research will continue to refine our understanding of these two parameters.

It is an empirical fact that the periodicity of the interference structure in the scattering efficiency is 2π when plotted versus ρ. This interference structure has been explained as a result of light passing through the particle and light passing past the particle and then these mixing in the far field (Chylek and Zhan 1989, Lock and Yang 1991). The phase change of light passing through the particle is $2nkR$ while that for light passing the particle is $2kR$ to yield a phase difference of $\rho = 2kR(n - 1)$, the phase shift parameter. Hence the term 'interference structure'. This is an attractive explanation with some merit, but as our discussion of the extinction paradox will show, the concept of diffraction around an edge has faults (see section 8.4).

The ripple structure increases with real refractive index n and becomes quite severe for $n > 2$ (not shown). The physical explanation for this structure is that it is due to internal resonance modes involving waves propagating around the inner circumference of the sphere (van de Hulst 1957, Chylek and Zhan 1989, Lock and Yang 1991). These internal modes were discussed above in our discussion of ripples

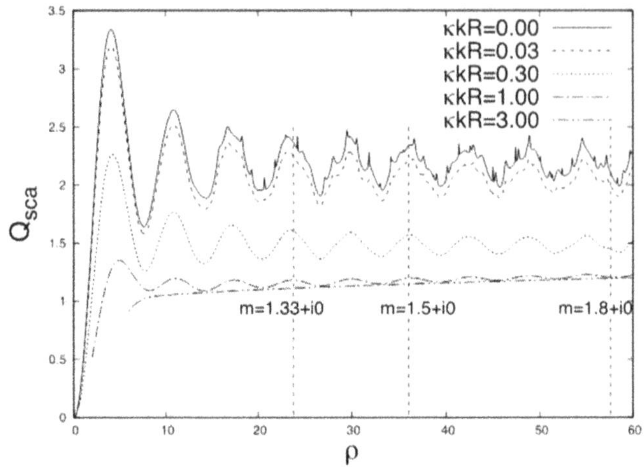

Figure 7.35. The scattering efficiency versus the phase shift parameter for different values of κkR. The size parameter is $kR = 37$. The phase shift parameter was varied by changing the real part of the refractive index, n, and some values of $m = n + i\kappa$ are marked with vertical dashed lines.

in the glory, see figure 7.16, and in the absorption, see figure 7.30. The internal reflections grow in strength with increasing refractive index real part n.

Returning to the lower part of figure 7.33, the interference structure and ripple damp out with significant imaginary part κ of the refractive index. We expect from our study of the absorption cross section that the significance is measured universally by κkR. Figure 7.35 further explores the effect of the imaginary part of the refractive index κ on the scattering efficiency. Plotted there (linear axes, not logarithmic) is the scattering (not extinction) efficiency Q_{sca} versus the phase shift parameter ρ for a $kR = 36$ sphere for a variety of κkR values. This size parameter corresponds to a sphere with diameter $2R \simeq 7.1$ microns when the wavelength is $\lambda = 532$ nm; ρ was varied by changing the real part of the refractive index. Note that when $\kappa kR = 0$, $Q_{sca} = Q_{ext}$. As κkR increases from zero, two effects occur:

1. the scattering efficiency begins to decrease for all ρ, and
2. the amplitudes of both the interference structure and the ripple damp out.

Both these effects are related to the fact that the relative skin depth parameter, κkR, is the ratio of the spherical particle's radius to the optical penetration depth as shown in figure 7.24. For the first effect as κkR increases, the sphere is less illuminated hence scatters less light. Another explanation (with the same physical basis) is that in the large ρ limit, $Q_{ext} = 2$, independent of all parameters. Since $Q_{ext} = Q_{sca} + Q_{abs}$, if Q_{abs} increases, which it does with increasing κkR, Q_{sca} must decrease. Figure 7.35 shows that when $\kappa kR \gtrsim 1$ $Q_{sca} \simeq 1$, except when $\rho \to 0$. For the second effect with large κkR, the diameter beam is absorbed before it can get through the particle. Then there can be no interference with the edge beam; hence no interference structure. Furthermore, internal modes cannot survive when $\kappa kR \gtrsim 1$; hence no ripple structure.

7.5 The single scattering albedo

The single scattering albedo (SSA), ω, is the ratio of scattering and extinction cross sections as described in section 2.11. Recall that the extinction cross section is the sum of the scattering and absorption cross sections. Thus

$$\omega = \frac{C_{sca}}{C_{ext}} = \frac{C_{sca}}{C_{sca} + C_{abs}} \tag{7.55}$$

As such, it is bounded between zero and one, i.e. $0 \leqslant \omega \leqslant 1$. For a beam of light it is the fraction of energy lost from an incident beam due only to scattering (Kerker 1969), the rest being lost to absorption. For an object, it is the fraction of light diffusively reflected. Thus, a white object, like a sheet of paper, has a large albedo, perhaps 0.95 (its brightness), while a black object, like powdered fractal graphene, has a small albedo, perhaps 0.05. Recognize that for particles, when the refractive index imaginary part $\kappa = 0$, $C_{abs} = 0$, hence $\omega = 1.0$.

The albedo in the Rayleigh regime can be calculated with the cross sections equations (6.13) and (6.19) to find

$$\omega_{Ray} = \left(1 + \xi_{Ray}^{-3}\right)^{-1}, \tag{7.56}$$

where

$$\xi_{Ray}^{3} = \frac{2}{3}\frac{F(m)}{E(m)}(kR)^{3}, \tag{7.57}$$

and

$$F(m)/E(m) = \frac{(n^2 - \kappa^2 - 1)^2 + 4n^2\kappa^2}{6n\kappa}. \tag{7.58}$$

We call ξ_{Ray} the *generalized Rayleigh size parameter* (Moosmuller and Sorensen 2018) combining particle size, refractive index and optical wavelength.

In the Rayleigh regime $kR < 1$ and $(m)kR < 1$ so $\xi_{Ray} < 1$ then equation (7.56) leads to $\omega_{Ray} \simeq \xi_{Ray}^{3}$, which should provide a single, unified description of the albedo within the Rayleigh regime. Figure 7.36 demonstrates this by plotting the Mie calculations of the albedo as function of the generalized Rayleigh size parameter resulting in a single curve in the Rayleigh regime. Three different regimes can be distinguished in this figure: the Rayleigh regime with a single functionality on ξ_{Ray}, a peak regime where the albedo gets large approaching unity, and finally, a geometric regime where $\omega \simeq 0.5$.

The peak regime signals the end of the Rayleigh regime where the particles are dark but brighten up with increasing size. The peak can also be viewed as a remnant of when $\kappa = 0$, $\omega = 1.0$. However, since all the plots in figure 7.36 are for $\kappa > 0$, the albedo never gets to 1.0. Note that the relative penetration depth parameter, κkR, ends the peak regime as it ascends towards unity, not the value of κ itself. The geometric regime occurs for $\kappa kR > 0.5$. Large particles at large κkR, have both scattering and absorption cross sections approximately equal and equal to the

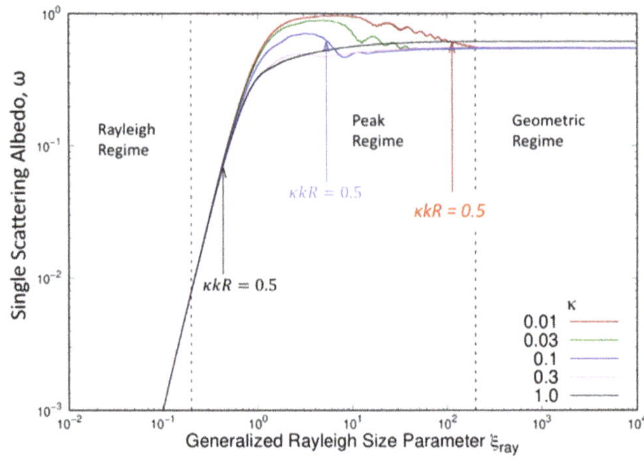

Figure 7.36. Mie calculations of the single scattering albedo ω as function of the generalized Rayleigh size parameter, equation (7.57), for a refractive index $m = 1.5 + i\kappa$. Vertical dashed lines indicate approximate boundaries between Rayleigh, peak, and geometric regimes. Also, note that $\kappa kR \simeq 0.5$ describes the end of the peak regime.

geometric cross section, $G = \pi R^2$. Hence $\omega \simeq 0.5$. Note that when $\kappa = 1.0$, ω is getting larger. This is due to the onset of reflection, which will be described more thoroughly below. As we have seen in our discussion of absorption, section 7.2, reflection is more directly controlled by κ, not κkR.

It is instructive to plot the albedo versus the refractive index imaginary part, κ. This is done in figure 7.37 for a sphere with a size parameter $kR = 0.1$, hence in the Rayleigh regime, for a series of refractive index real parts. Figure 7.7 demonstrates that for small particles, the albedo is large and scattering dominates until κ becomes large enough. For a given size parameter 'large enough' is determined by the ratio $F(m)/E(m)$, equation (7.58). The albedo is equal to one-half when the scattering and absorption cross sections, equations (6.13) and (6.19), are equal. Setting these equal for a given refractive index real part n and under the condition $\kappa \ll 1$ and $n > 1.1$, the value of κ when $\omega = 0.5$ is

$$\kappa_{1/2} \simeq \frac{1}{9}\frac{(n^2 - 1)^2}{n}(kR)^{-3} \quad kR \leqslant 1 \tag{7.59}$$

The predictions of equation (7.59) are included in figure 7.37.

The story that figure 7.37 tells us for this Rayleigh sized particle is that when κ is very small, small compared to $\kappa_{1/2}$, the scattering dominates, hence the albedo is large and the particle is bright. When $\kappa > \kappa_{1/2}$, absorption dominates so that the albedo plunges to zero and the particle is very dark. This is the regime in which the classic statement that 'absorbing Rayleigh particles are black' holds. However, when $\kappa > 3$, the albedo rises from zero and the particle becomes bright again. From our studies of absorption above, we know that this increase occurs when κ ceases causing absorption and instead causes reflection. The reflection is mirror-like, metallic and

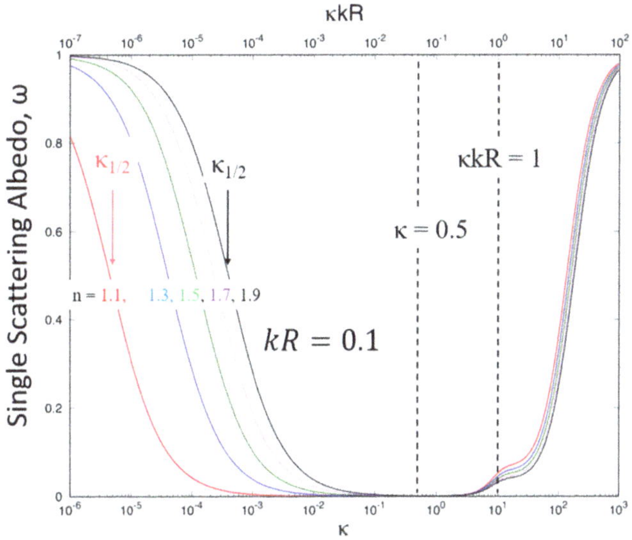

Figure 7.37. Mie calculations of the single scattering albedo ω of a sphere with size parameter $kR = 0.1$ versus the imaginary part of the refractive index κ, bottom scale, and the relative skin depth parameter κkR, top scale. Plots for different refractive index real part, n, are as indicated. Arrows mark the values of $\kappa_{1/2}$ calculated with equation (7.59). Also indicated are dashed lines where $\kappa = 0.5$ and $\kappa kR = 1$.

bright. Note that the inflection in the albedo in the region $0.3 < \kappa kR < 3$ is due to the absorption cross section crossover regime shown in figure 7.23, above.

Whereas small spheres (and particles of any shape) can be very dark, this ability is lost as the size increases. This is illustrated in figures 7.38 and 7.39 where the albedo for spheres of size parameters $kR = 1$ and 100, respectively, is displayed as a function of κ and hence κkR, similar to figure 7.37. When $kR = 1$, the parameter $\kappa_{1/2}$ of equation (7.59) still works well for describing when absorption starts to effect the albedo to darken the sphere.

On the other hand, for larger particles with $kR > 1$ equation (7.59) no longer applies, and figure 7.39 shows that $\kappa kR = 1$ indicates when $\omega \simeq 1/2$. Thus, equation (7.59) yields to

$$\kappa_{1/2} \simeq 1/kR \quad kR > 1 \tag{7.60}$$

Using the Mie equations, we find empirically that equation (7.59) gives over to equation (7.60) with increasing size. However, the physics behind these two equations is different. Equation (7.59) was derived under the conditions of the Rayleigh regime with the incident light energy interacting with the entire sphere (or particle). Then the equations for the scattering and absorption cross sections were set equal and the result followed. Equation (7.60) occurs because κkR becomes large enough to eliminate the interaction of the light with much of the volume of the sphere, allowing only access to the surface.

Figure 7.39 also shows that for large spheres the albedo develops a central region in the range $\kappa kR > 1$ but $\kappa < 0.5$. This range is marked in figure 7.39 with dashed

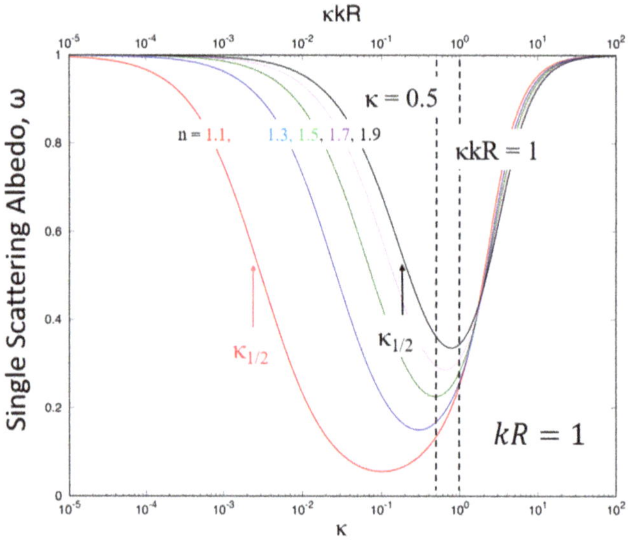

Figure 7.38. Mie calculations of the single scattering albedo ω of a sphere with size parameter $kR = 1$ versus the imaginary part of the refractive index κ, bottom scale, and the relative skin depth parameter κkR, top scale. Plots for different refractive index real part, n, are as indicated. Arrows mark the values of $\kappa_{1/2}$ calculated with equation (7.59). Also indicated are dashed lines where $\kappa = 0.5$ and $\kappa kR = 1$.

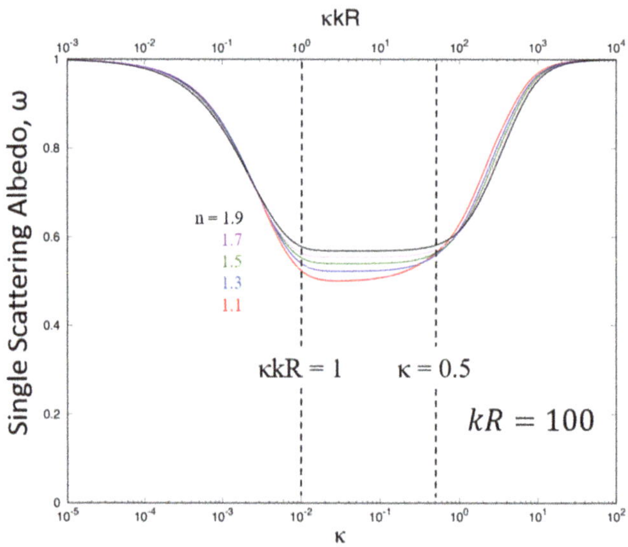

Figure 7.39. Mie calculations of the single scattering albedo ω of a sphere with size parameter $kR = 100$ versus the imaginary part of the refractive index κ, bottom scale, and the relative skin depth parameter κkR, top scale. Plots for different refractive index real part, n, are as indicated. Also indicated are dashed lines where $\kappa = 0.5$ and $\kappa kR = 1$. Note that for this large size parameter $\omega \simeq 0.5$ begins when $\kappa kR \simeq 1$.

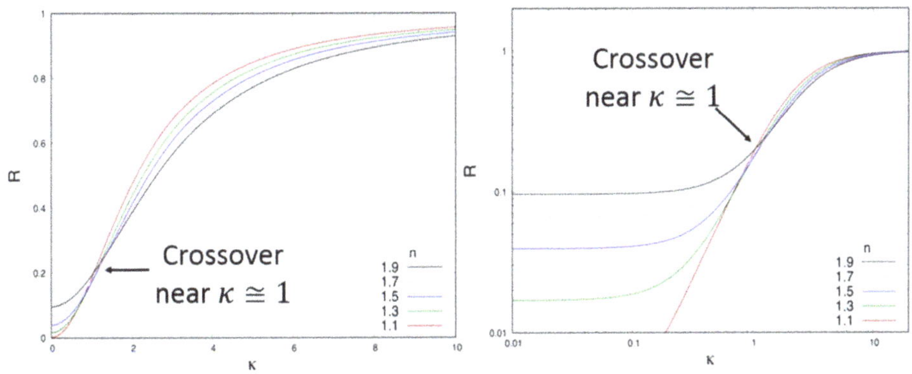

Figure 7.40. Normal incidence reflectance ***R*** from a planar interface from vacuum to a medium with refractive index $m = n + i\kappa$ as a function of κ for a variety of n values. (Left) ***R*** plotted versus κ, linear-linear; (right) ***R*** plotted versus κ, log–log.

vertical lines. In this central region, the albedo lies in the range $0.5 \lesssim \omega \lesssim 0.6$, independent of κ for very large spheres but with minor dependence upon the refractive index real part, n. The central region ends when $\kappa > 0.5$ where, as we have seen in figure 7.23, κ stops causing absorption and starts causing reflection, a form of scattering. Ultimately, for all sphere sizes the albedo approaches one as $\kappa \to \infty$, the highly reflective limit.

The reflective nature of large spheres when $\kappa kR > 1$, i.e. $\kappa_{1/2} > 1/kR$ is nicely illustrated by looking at the reflectivity of a planar interface as calculated with the Fresnel equations. At normal incidence, the Fresnel reflectance ***R*** is given above in equation (7.47) and plotted versus κ in figure 7.40. In particular, compare the right hand side of figure 7.40, below, to the bottom plot of figure 7.39, which is the albedo for a sphere of size parameter $kR - 100$. In the albedo's central region, the scattering cross section for such a large sphere is approximately equal to the geometric cross section, G. The physical basis of this cross section is the forward diffraction within an angle of $\theta \simeq \lambda/2R$, which accounts for nearly all of the cross section and a minor amount of surface reflection when $\kappa \lesssim 0.5$. Note how the central region albedo increases in semi-quantitative agreement with the reflectance in figure 7.39 with increasing refractive index real part, n. When $\kappa \gtrsim 0.5$, both the albedo and the reflectivity increase in a very similar manner including a crossover near $\kappa = 1.0$.

References

Abramowitz M and Stegun I A 1972 *Handbook of Mathematical Functions, with Formulas, Graphs, and Mathematical Tables* (New York: Dover)

Adam J A 2002 The mathematical physics of rainbows and glories *Phys. Rep. Rev. Sect. Phys. Lett.* **356** 229–365

Berg M J, Sorensen C M and Chakrabarti A 2005 Patterns in Mie scattering: evolution when normalized by the Rayleigh cross section *Appl. Opt.* **44** 7487–93

Bohren C F and Huffman D R 1983 *Absorption and Scattering of Light by Small Particles* (New York: Wiley)

Chylek P and Zhan J Y 1989 Interference structure of the Mie extinction cross-section *J. Opt. Soc. Am.* A **6** 1846–51
Greenler R 2020 *Rainbows, Halos, and Glories* (Bellingham, WA: SPIE)
Guinier A and Fournet G 1955 *Small-Angle Scattering of X-rays* (New York: Wiley)
Hecht E 2002 *Optics* (Reading, MA: Addison-Wesley)
Heffernan B M, Heinson Y W, Maughan J B, Chakrabarti A and Sorensen C M 2016 Backscattering measurements of micron-sized spherical particles *Appl. Opt.* **55** 3214–8
Heinson W R, Chakrabarti A and Sorensen C M 2015 A new parameter to describe light scattering by an arbitrary sphere *Opt. Commun.* **356** 612–5
Kerker M 1969 *The Scattering of Light and Other Electromagnetic Radiation* (New York: Academic)
Kerker M, Smith L B, Matijevic E and Farone W A 1964 Determination of particle size by minima and maxima in angular dependence of scattered light. Range of validity of method *J. Colloid Sci.* **19** 193–200
Laven P 2003 Simulation of rainbows, coronas, and glories by use of Mie theory *Appl. Opt.* **42** 436–44
Lock J A and Yang L 1991 Interference between diffraction and transmission in the Mie extinction efficiency *J. Opt. Soc. Am.* A **8** 1132–4
Maughan J B and Sorensen C M 2020 Universal parameter to describe the reduction of refraction effects in the scattering of absorbing spheres *J. Opt. Soc. Am.* A **37** 1456–64
Mie G 1908 Beitrage zur optik truber medien speziell kolloidaler metallosungen *Ann. Phys.* **25** 377–445
Minnaert M 1993 *Light and Color in the Outdoors* (New York: Springer)
Mishchenko M I, Travis L D and Lacis A A 2002 *Scattering, Absorption, and Emission of Light By Small Particles* (Cambridge: Cambridge University Press)
Moosmüller H and Arnott W P 2003 Angular truncation errors in integrating nephelometry *Rev. Sci. Instrum.* **74** 3492–501
Moosmüller H and Arnott W P 2019 Erratum: angular truncation errors in integrating nephelometry *Rev. Sci. Instrum.* **74** 3492
Moosmüller H and Arnott W P 2003 *Rev. Sci. Instrum.* **90** 019901
Moosmuller H and Sorensen C M 2018 Small and large particle limits of single scattering albedo for homogeneous, spherical particles *J. Quant. Spectrosc. Radiat. Transf.* **204** 250–5
Nussenzveig H M 1977 Theory of rainbow *Sci. Am.* **236** 116–27
Nussenzveig H M 1979 Complex angular-momentum theory of the rainbow and the glory *J. Opt. Soc. Am.* **69** 1068–79
Nussenzveig H M 2002 Does the glory have a simple explanation? *Opt. Lett.* **27** 1379–81
Sorensen C M, Maughan J B and Moosmuller H 2019 Spherical particle absorption over a broad range of imaginary refractive index *J. Quant. Spectrosc. Radiat. Transf.* **226** 81–6
Sorensen C M and Shi D 2000 Guinier analysis for homogeneous dielectric spheres of arbitrary size *Opt. Commun.* **178** 31–6
Sorensen C M and Shi D 2002 Patterns in the ripple structure of mie scattering *J. Opt. Soc. Am.* A **19** 122–5
van de Hulst H C 1957 *Light Scattering by Small Particles* (New York: Wiley)
van de Hulst H C 1947 A theory of the anti-coronae *J. Opt. Soc. Am.* **37** 16–22
Waez M S, Eckels S J and Sorensen C M 2018 A refractive-index and position-independent single-particle detector for large, nonabsorbing, spherical particles *Aerosol Sci. Technol.* **52** 1429–36

Waez M S, Eckels S J and Sorensen C M 2021 Determination of size and complex index of refraction of single particles with elastic light scattering *Appl. Opt.* **60** 600–5

Wang G, Chakrabarti A and Sorensen C M 2015 Effect of the imaginary part of the refractive index on light scattering by spheres *J. Opt. Soc. Am.* A **32** 1231–5

Yang P, Gao B C, Baum B A, Hu Y X, Wiscombe W J, Mishchenko M I, Winker D M and Nasiri S L 2001 Asymptotic solutions for optical properties of large particles with strong absorption *Appl. Opt.* **40** 1532–47

Chapter 8

Q-space analysis of light scattering by spherical particles

The previous chapter studied the scattering and absorption of light by spherical particles. Descriptions of the scattering as function of scattering angle involved plotting the differential cross section, or the proportional scattered intensity, versus the scattering angle along a linear axis.

In this chapter, we embrace the clues given by diffraction theory to gain a fundamental understanding of light scattering. The chapters of this book have laid the foundation for us to interpret these clues. The first clue is that light is first and foremost a wave, its electromagnetic character notwithstanding. The second clue is that diffraction from macroscopic objects involves the addition of waves and this addition shows itself as the Fourier transform of the object. Note that diffraction from macroscopic objects uses only light's wave nature ignoring the electromagnetic character. The Fourier transform brings with it the third clue, the transform variable q with units of inverse length. We have seen how q^{-1} is a length scale that senses the length scales of the object, and q itself is determined by the wavelength and the scattering angle. The fourth clue comes from experience: nature reveals simplicity when it is viewed without scale of any kind and to gain such a view one can use dimensionless variables. Finally, again from experience, we have learned that this is, by and large, a geometric world, not an arithmetic one.

Given all this, we propose that the angular light scattering functionality is best viewed as a function of the scattering wave vector, q, and the plotting should be logarithmic along both axes. Such a plot is at the foundation of what I call 'Q-space analysis'.

8.1 Motivation for Q-space analysis

Clues to the nature of light scattering are taken well from the Rayleigh–Debye–Gans (RDG) formulation for diffractive scattering by a 'soft' sphere, discussed in section 7.1.2. The RDG formulation assumes a relatively un-refractive sphere, such that

doi:10.1088/978-0-7503-5157-7ch8

$m \to 1$, a condition consistent with the electromagnetic aspect of light being ignored, leaving only its wave nature. Thus, the scattered wave amplitude is the Fourier transform of the solid spherical shape; the scattered intensity is the square of this. The result is the structure factor for the sphere, equation (4.49)

$$S(q) = \left[\frac{3}{u^3}(\sin u - u\cos u)\right]^2 \tag{8.1}$$

where

$$u = qR \tag{8.2}$$

and

$$q = (4\pi/\lambda)\sin(\theta/2) \tag{8.3}$$

Recall that q is both the Fourier transform variable and the magnitude of the scattering wave vector. Also recall that $S(q)$ is known as both the Fraunhofer diffraction pattern and the RDG scattering pattern of the 3d sphere. The diffracted or scattered intensity is proportional to $S(q)$. According to equation (7.1), the Rayleigh–Debye–Gans (RDG) formulation for the scattering is to multiply $S(q)$ by the Rayleigh scattering differential cross section of the sphere. The result of this is given in equation (7.7). The scattered intensity is proportional to the differential cross section, and if we normalize by the forward scattering at $q = 0$, and we will, one obtains $S(q)$ again. This forward normalized scattering for a soft sphere was plotted in figure 7.1, and is replotted here for convenience as figure 8.1.

Figure 8.1 shows the RDG scattering pattern (or, equivalently for this case, diffraction pattern) for a sphere of radius R when the incident polarization is perpendicular to the scattering plane. On the left, figure 8.1 shows the scattered intensity normalized by the scattering at zero angle versus the scattering angle θ for a

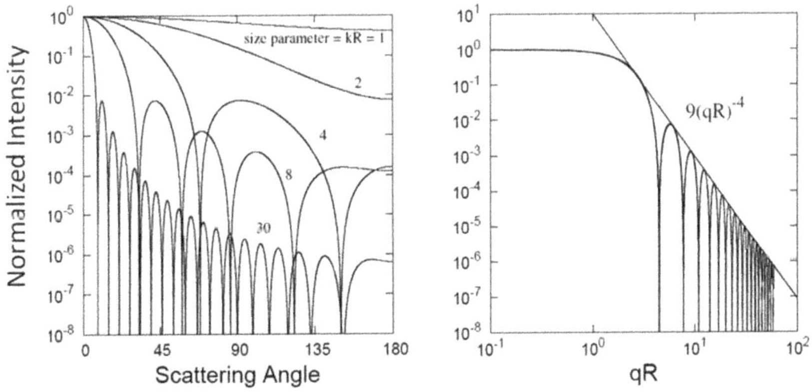

Figure 8.1. Plots of the Rayleigh–Debye–Gans scattering for uniform spheres with size parameters kR. The incident light polarization is perpendicular to the scattering plane, which contains the scattering angle. Left: forward normalized scattering versus scattering angle θ, right: forward normalized scattering versus qR. In comparing, the left and right sides note that the maximum qR for a given size is $2kR$.

variety of size parameters kR. Curves with interference wiggles are seen with an identifiable trend to enhanced forward scattering as the size parameter kR increases. In contrast, on the right, figure 8.1 shows the same results plotted versus the dimensionless qR and plotted double logarithmically. Figure 8.1 has managed to condense the plethora of plots on the left into a single plot, a virtue attributable to the intensity normalization and the use of the dimensionless variable qR. Furthermore, beyond this simple virtue, the plot on the right uncovers an important fact that the left plot cannot disclose: a power law envelop appears when $qR > 1$ (a geometric world indeed!). This power law was uncovered by plotting log–log.

Given this comparison, the question naturally arises, what if the Mie results were plotted, not as a function of scattering angle, as has been done since 1908 (I assume), but as a function of the magnitude of the scattering wave vector q or the dimensionless qR? Certainly, we expect the Mie curves to evolve from the RDG result with increasing refractivity. In 2000, we (Sorensen and Fischbach 2000) began the exploration of plotting Mie scattering as a function of either q or qR on a log–log plot, a method we have come to call 'Q-space analysis'. When this is done, unforeseen patterns in the scattering are uncovered that provide useful new descriptions of the scattering. Figures 8.2 and 8.3 show replications of our initial results.

Figure 8.2 (top), a replication of figure 7.3 from chapter 7, shows an example of Mie scattering of light with an incident polarization perpendicular to the scattering plane for a sphere with an index of refraction $m = 1.05$ and a variety of sizes expressed as the size parameter kR. We refer to the Mie differential cross section as simply the Mie scattered intensity (they are directly proportional). The intensity normalized by its value at zero scattering angle $I(\theta)/I(0)$ is plotted versus the scattering angle θ. A series of bumps and wiggles are seen with some periodicities, but with no particularly coherent pattern. The scattering angle θ, although conveniently measured in the laboratory, uncovers no patterns. Figure 8.2 (bottom) shows the same data but plotted versus the dimensionless qR double logarithmically. Now the plots for the different size parameters appear to be related in a coherent manner. A common forward scattering lobe and Guinier regime (near $qR \simeq 1$) occur and power laws describing the envelopes of the plots at large qR are apparent.

Figure 8.3 (top), a replication of figure 7.4 from chapter 7, shows the same comparison as seen in figure 8.2 only for an index of refraction of 1.50. Figure 8.3 (top) shows the conventional plot versus scattering angle and again chaos reigns. Figure 8.3 (bottom) shows the same data plotted in q-space and again a systematic description with power laws appear.

Figures 8.2(b) and 8.3(b) *are examples of Q-space analysis, the double logarithmic plotting of the scattered intensity versus either q or qR.* Q-space analysis uncovers quantitative patterns unseen or unemphasized previously.

Figures 8.2 and 8.3 are suggestive but need refinement. Refinement and a complete description of the Q-space patterns will come with inclusion of the internal coupling parameter, ρ', and grounding the scattering with the Rayleigh limit.

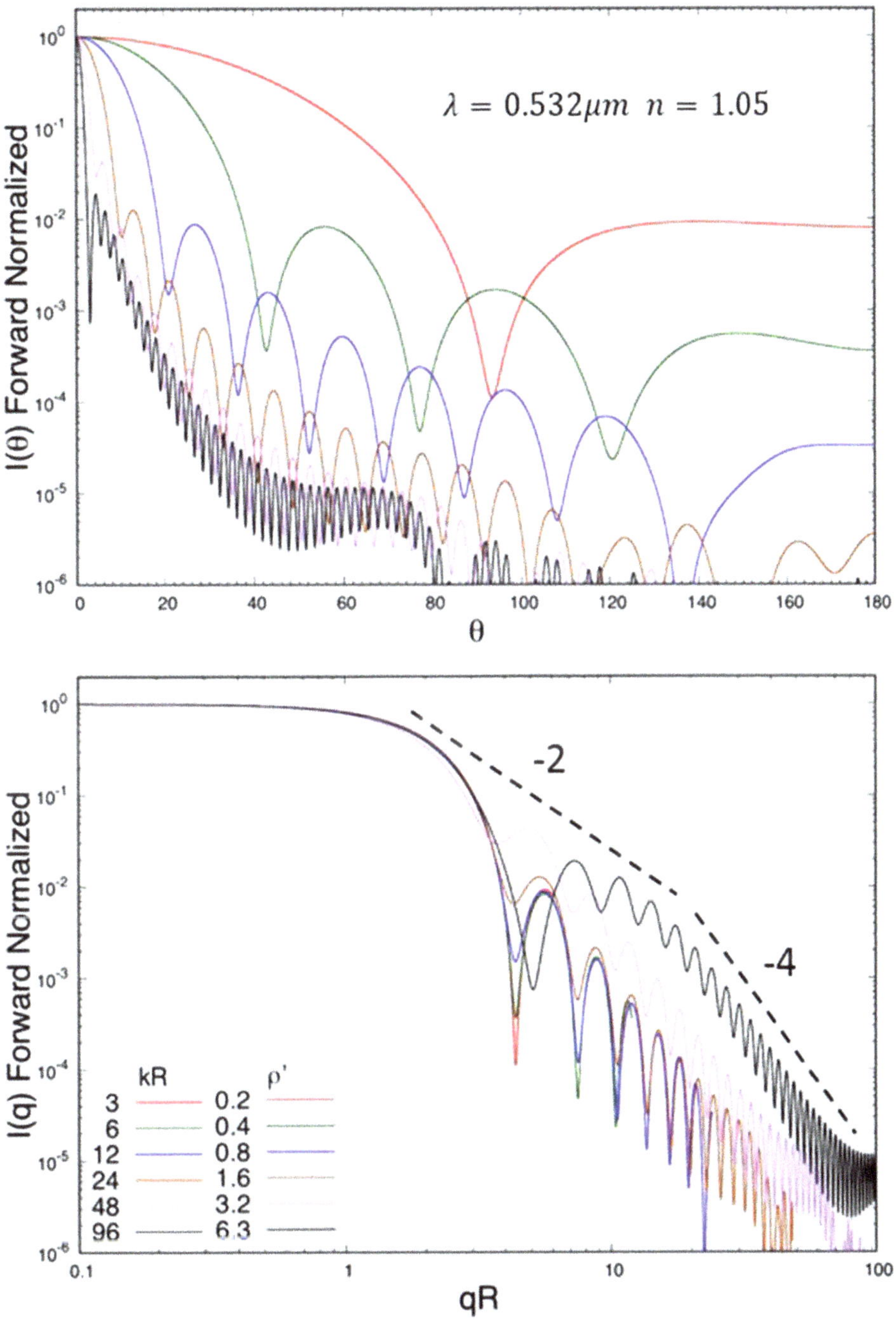

Figure 8.2. (Top) Forward normalized Mie scattered intensities as a function of scattering angle for spheres of refractive index $m = 1.05$ and a variety of size parameters kR and concomitant internal coupling parameters ρ'; and (bottom) the same as above but plotted logarithmically versus qR, the method of Q-space analysis. For both the incident light is vertically polarized perpendicular to the scattering plane containing the scattering angle. Lines with slope −2 and −4 are shown. Reprinted from Sorensen and Fischbach (2000), Copyright (2000), with permission from Elsevier.

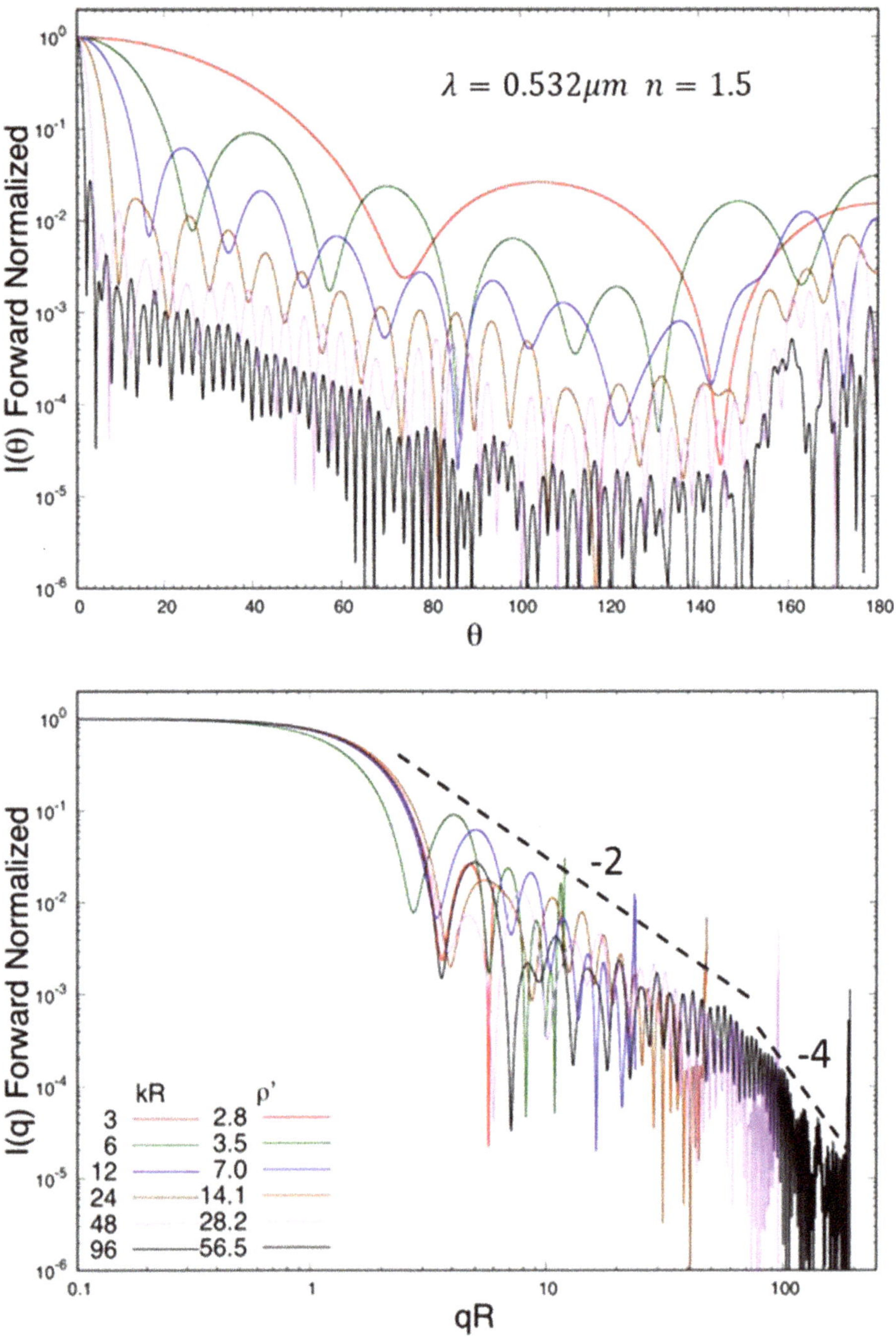

Figure 8.3. (Top) Forward normalized Mie scattered intensity as a function of scattering angle for spheres of refractive index $m = 1.50$ and a variety of size parameters and concomitant internal coupling parameters ρ'; and (bottom) the same as above but plotted logarithmically versus qR, the method of Q-space analysis. For both the incident light is vertically polarized perpendicular to the scattering plane containing the scattering angle. Lines with slope −2 and −4 are shown. Reprinted from Sorensen and Fischbach (2000), Copyright (2000), with permission from Elsevier.

8.2 Q-space analysis of scattering by an arbitrary sphere

8.2.1 Scattering at any angle (any q)

Normalization by $I(0)$ was useful to motivate and uncover the Q-space patterns seen in figures 8.2 and 8.3, above, but it negates the overall magnitude of the scattering. However, we have also learned with figure 7.6 that the forward scattering for a sphere is based on Rayleigh scattering controlled by the internal coupling parameter. This suggests that it would be useful to apply Q-space analysis to the Rayleigh normalized scattering.

Figures 8.4 and 8.5 show Q-space analysis of Mie scattering both with and without normalization by the forward scattering. The Mie scattering calculation was done with a wavelength of $\lambda = 200\ \pi$ nm = 628 nm. Values for the internal coupling parameter are calculated with equation (7.17), which is

$$\rho' = 2kR\ |(m^2 - 1)/(m^2 + 2)| \tag{8.4}$$

Tabulated values are in the figures. The scattering situation is the usual lab one with light propagating in the positive z direction and the xy plane is the scattering plane. The incident light is polarized in the x direction, perpendicular to the scattering plane. The x polarized scattered intensity is proportional to the differential cross section. The results for y polarized incident light are essentially the same except that when ρ' is small a dip near $qR \simeq 2kR$, which corresponds to a scattering angle of $\theta = 90°$, appears. To eliminate the ripple structure and leave only the envelopes, these plots include a small degree of size polydispersity. This was done by integrating the Mie solution over a log-normal size distribution for the radius. The ripples were sufficiently washed out when the geometric standard deviation was 1.22. The radius specifying the size is the most probable size of the distribution.

Figures 8.4(a) and 8.5(a) show the Mie results for the differential scattering cross section without any normalization. Following Berg *et al* (2005) several notable features appear in these figures:

(i) The overall scattering increases rapidly with the size parameter kR.
(ii) For $qR < 1$, a flat, forward scattering lobe is seen with no functionality on qR, hence no functionality on q nor the scattering angle θ. This feature was seen above in figures 8.1–8.3. For large kR the forward scattering lobe is independent of m.
(iii) There is a Guinier crossover regime near $qR \simeq 1$
(iv) For $qR > 1$, the plots fall off quickly, the functionality of which will be described below.
(v) A complex backscattering, including 'rainbows' and the glory, near $qR = 2kR$. For $m = 1.05$ this develops when $\rho' \gtrsim 8.7$; for $m = 1.50$ this develops when $\rho' \gtrsim 1.2$.

Figures 8.4(b) and 8.5(b) show the Rayleigh-normalized Mie scattering intensity, i.e. the scattered intensity divided by $R^6k^4F(m)$. To realize this normalization the Mie calculation and the Rayleigh calculation must be performed with the same wavelength and refractive index. Now patterns appear with several notable features:

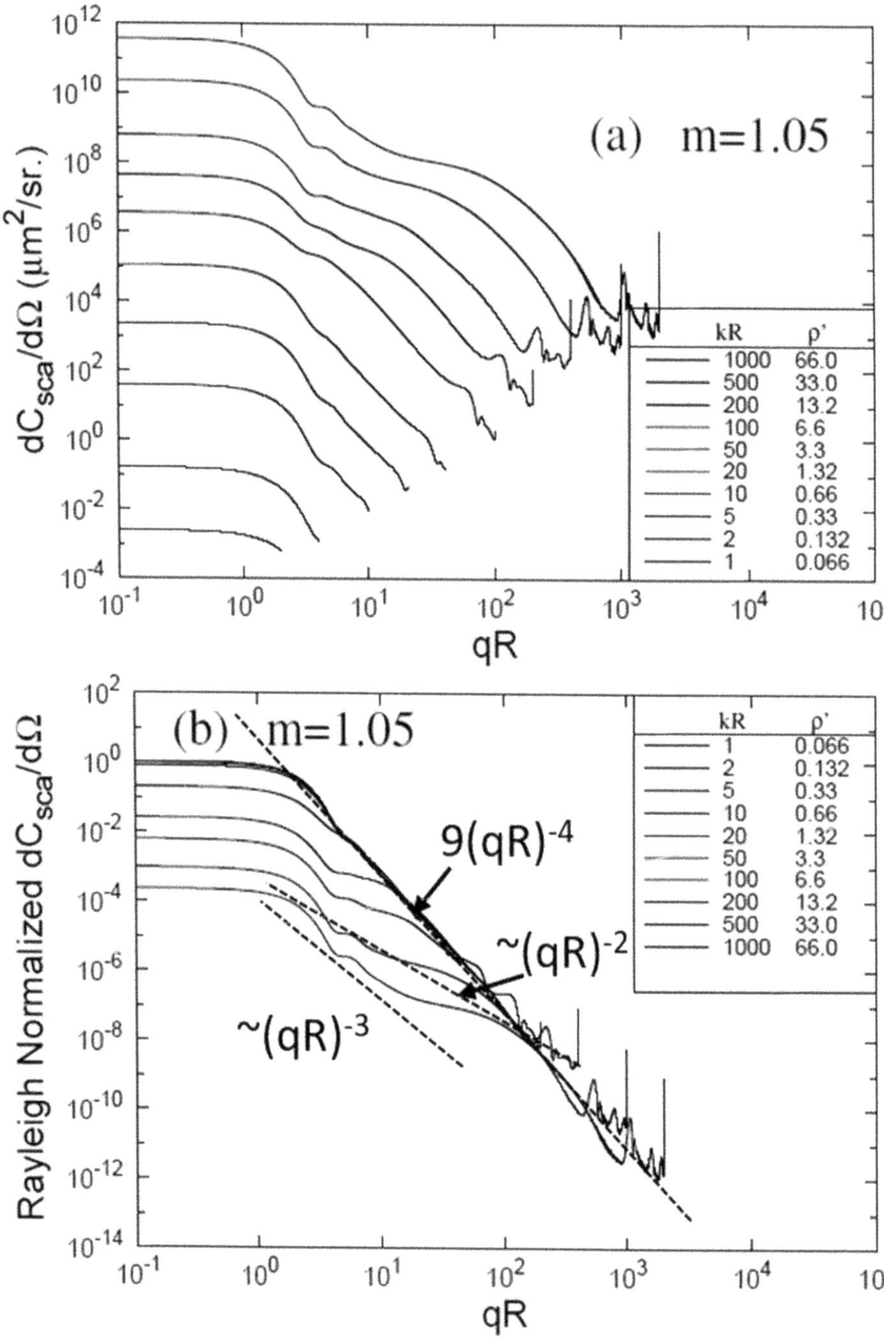

Figure 8.4. (a) Q-space analysis of the Mie scattered intensity (with units of differential cross section) plotted as a function of qR for spheres with $m = 1.05$ and various size parameters kR, hence various internal coupling parameters ρ'. The incident light is polarized perpendicular to the scattering plane. (b) Mie scattered intensity normalized by the Rayleigh differential cross section for the same spheres as in (a). The RDG envelope, $9(qR)^{-4}$, is shown as a black dashed line, and the $(qR)^{-2}$ and $(qR)^{-3}$ power laws are also shown for comparison. For clarity, the legends in both figures show the data sets in the same order as they appear in the figure. A small size distribution was included to eliminate the ripple structure.

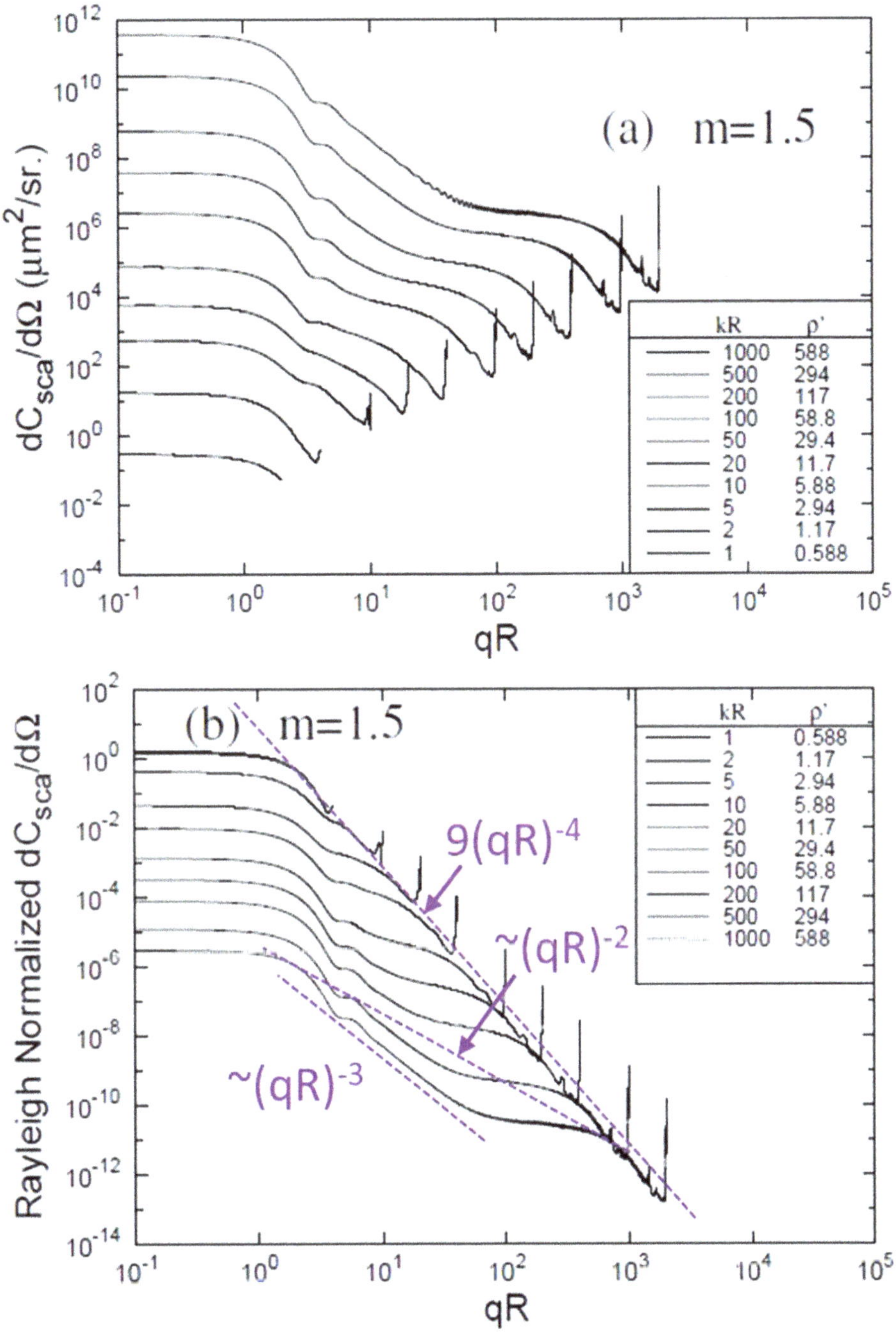

Figure 8.5. (a) Q-space analysis of the Mie scattered intensity (with units of differential cross section) plotted as a function of qR for spheres with $m = 1.50$ and various size parameters kR; hence various internal coupling parameters ρ'. The incident light is polarized perpendicular to the scattering plane. (b) Mie scattered intensity normalized by the Rayleigh differential cross section for the same spheres as in (a). The RDG envelope, $9(qR)^{-4}$, is shown as a black dashed line, and the $(qR)^{-2}$ and $(qR)^{-3}$ power laws are also shown for comparison. For clarity, the legends in both figures show the data sets in the same order as they appear in the figure. A small size distribution was included to eliminate the ripple structure.

(i) For $\rho' < 1$ the curves fall on top of each other with $(qR)^0$ and $(qR)^{-4}$ regimes. This is expected because this is the RDG limit where diffraction determines the shape of the curves. For all kR at large qR the $(qR)^{-4}$ regime is consistent, but not identical to (see below), the RDG limit.

(ii) As ρ' increases above 1, the curves begin to separate and decrease from the $\rho' < 1$ curves for $qR \lesssim 2\rho'$. Hence Mie scattering is smaller than the Rayleigh prediction when $\rho' > 1$. Recall that this is well displayed in figure 7.6. Also, for $\rho' > 1$ an approximate $(qR)^{-2}$ regime develops between $qR \simeq 1$ and $qR \simeq 1.5\rho'$. In general, the curves sag below $(qR)^{-2}$ between these limits, more so with increasing ρ', but the limits are well connected by an equal tangent power law of $(qR)^{-2}$ spanning between the Guinier regime and a hump structure. Finally, for $\rho' > 30$ this sagging region begins to develop another power law of $(qR)^{-3}$. This is the onset of 2d diffraction and will be described in more detail below.

(iii) For $\rho' \gtrsim 30$ a broad 'hump' appears centered near $qR \simeq \rho'$.

(iv) For $qR \gtrsim 2\rho'$ all curves approximately resume the same $9(qR)^{-4}$ power law of the RDG limit (if the enhanced backscattering is ignored).

The figures above demonstrate the power of the Q-space analysis. The simple change of plotting scattered intensity as a function of q or qR rather than the scattering angle θ and making the plots with both axes logarithmic uncovers patterns in the Mie scattering heretofore unseen. We now study the Mie patterns individually in detail.

8.2.2 The intermediate, −2 power law regime, $1 < qR < 1.2\rho'$

As the internal coupling parameter ρ' increases through unity, a rough power law $(qR)^{-2}$ regime appears near $1 < qR < 1.2\rho'$. As figures 8.4(b) and 8.5(b) show, an equal tangents line of slope −2 can be placed touching the Mie curves at $qR \simeq 2$ (near the Guinier regime) and a hump-like feature ('The Hump', see below) at approximately $qR \simeq 1.2\rho'$. Note that the value of qR of the latter touch point tends to decrease with increasing m. Under this power law, the Mie curves sag significantly. For small ρ' this sag appears to be related to the first minimum in the interference ripple structure present in the sphere's structure factor near $qR \simeq 4$, see figure 8.1. However, for very large ρ' the region of the sag develops a power law of its own, $\sim(qR)^{-3}$, as best shown in figure 8.5(b). Notice that the −2 envelope seen in figures 8.2 and 8.3 obscure this sag with their overlapping multiplicity of size parameters.

This sagging region in the $(qR)^{-2}$ regime is related to the crossover of the Mie curves for spherical particles to the 2d diffraction by a circular aperture, which is the geometric limit of a sphere when $R \gg \lambda$, to some degree, but more precisely when $\rho' \to \infty$. A refractive sphere very large compared to the wavelength of light acts as a two-dimensional, circular obstacle for the light. As such, its refractive index is not relevant, unless, of course, it is so small that ρ' is small as well. By Babinet's principle (section 1.5.2), an obstacle has the same diffraction pattern as an aperture of the same size and shape in the far field, Fraunhofer limit. Circular aperture Fraunhofer diffraction is well known (Hecht 2002) and is described by the first Bessel function of the first kind, equations (7.11) and (8.5). The diffracted intensity has zeros out to $u = qR \to \infty$, but application of the asymptotic expansion for the large u form of the Bessel function yields a diffracted intensity for large u of (Abramowitz and Stegun 1972, equation (9.2.1))

$$I(u) = (k^2R^4/4)\left[\frac{2J_1(u)}{u}\right]^2 = (k^2R^4/4)(8/\pi)u^{-3}\cos^2(u - 3\pi/4),\ u = qR \quad (8.5)$$

Equation (8.5) supports the claim that the −3 power law in the sag region is diffraction by the sphere acting as a circular aperture. To finish this contention, recognize that the coefficient of the power law in equation (8.5) is $(k^2R^4/4)(8/\pi)$. Figure 8.6 shows Mie scattering in two extreme limits of $\rho' = 0.4$ and $\rho' = 1200$. When $\rho' = 0.4$, the RDG limit is seen to hold with an envelope at $qR > 1$ of $9(qR)^{-4}$.

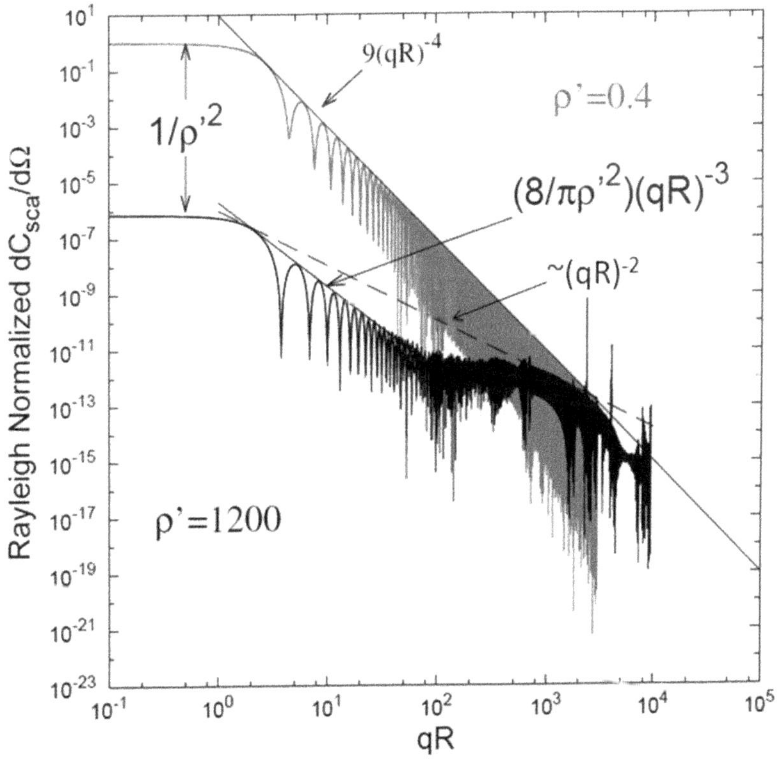

Figure 8.6. Rayleigh normalized Mie differential cross section for spheres versus qR for $\rho' = 0.4$ and $\rho' = 1200$. The incident light is polarized perpendicular to the scattering plane. These are composites of scattering from a range of size parameters kR and refractive indices m to achieve either ρ'. The envelope for $\rho' = 0.4$ is $9(qR)^{-4}$, which is equivalent to 3d Fraunhofer diffraction from a sphere, the RDG limit. For $\rho' = 1200$ a major $(8/\pi\rho'^2)(qR)^{-3}$ envelope develops, which is equivalent to 2d Fraunhofer diffraction from a circular aperture. Reproduced from Heinson *et al* (2014). © The Optical Society.

When $\rho' = 1200$, the regime where $1 \lesssim qR \lesssim 100$ has a power law envelope with -3 exponent. Because the plot in figure 8.6 is normalized by the Rayleigh differential scattering cross section, the coefficient of this envelope has also been so normalized for the plot. From equation (8.5) the normalization yields a coefficient of $8/\pi\rho'^2$ which is included in figure 8.6 and gives excellent agreement with the Mie calculation. We have also made a direct comparison of the 2d diffraction equation (8.5) to the Mie solution, including the ripples at large ρ' to find excellent agreement (Heinson *et al* 2014).

Before we leave figure 8.6, notice that although the $\rho' = 1200$ situation is dominated by 2d diffraction, the hump ultimately appears, brushing against the $9(qR)^{-4}$ envelop of the 3d diffraction, as well as the $(qR)^{-2}$ equal tangent with the Guinier regime.

8.2.3 The large qR, −4 power law regime, $qR > 1.2\rho'$

Figures 8.4, 8.5 and 8.6 show that the Mie scattering curves approach the RDG limit envelope of $9(qR)^{-4}$ at large qR regardless of ρ'. This is a surprising result given that RDG is essentially wave diffraction and large ρ' contains all the complexity of Maxwell's equations applied to spherical boundary conditions. However, let's not forget that here the RDG diffraction was normalized by Rayleigh scattering, which contains the refractive index.

This is a good point in which to stress a weakness of the Q-space analysis and that is the large scattering angles are squashed into the very end of the logarithmic qR axis. This happens both due to the logarithmic plotting and the $\sin(\theta/2)$ functionality, e.g. $\sin(90°/2) = 0.707$ and $\sin(180°/2) = 1.0$. Thus, features like rainbows and glories are compressed beyond recognition. On the other hand, we will soon show many very favorable features of q-space. Moreover, we will show that approximately half the scattered light resides in the forward scattering lobe and another approximate half lies in the hump. As the book progresses, we will stress that a combined Q-space/θ-space analysis of data or computational results will yield the most comprehensive description of the scattering.

8.2.4 Quasi-universality with the internal coupling parameter ρ'

Figure 8.7 demonstrates a quasi-universality of Mie scattering on the internal coupling parameter ρ'. For each internal coupling parameter value shown, the values of the refractive index m and the size parameter kR vary widely. Despite this variation, the curves for the same ρ' lie with each other (on this low resolution, logarithmic graph) to imply a universality with ρ'. This universality is not perfect, with variations of as much as a factor of 3 for the same ρ', so we will use the term 'quasi-universal' to describe it.

8.2.5 Summary of the Q-space analysis of scattering by a sphere: a new, general description of spherical particle light scattering

A summary of the results above leads to a new general description of angular light scattering by a sphere of arbitrary size and real refractive index, i.e. Mie scattering

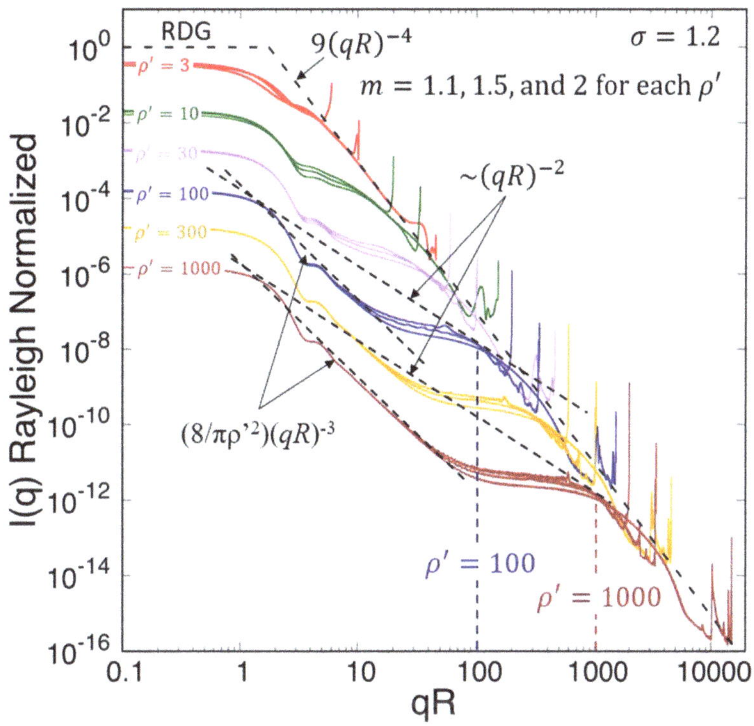

Figure 8.7. A demonstration of the quasi-universal parameterization given by the internal coupling parameter, ρ'. Rayleigh-normalized Mie scattered intensity (equal to the Rayleigh normalized differential cross section) versus qR plotted in the mode of Q-space analysis. The incident light is polarized in the x-direction, perpendicular to the scattering plane. For a given internal coupling parameter ρ' and refractive index m, a size parameter kR was determined for the calculation. These size parameters ranged from 3 to 7745. Even though the refractive indices and size parameters vary widely for a given ρ', plots with the same ρ' approximately fall together except for the backscattering regime. Also shown are dashed lines indicating the RDG limit and -2 and -3 power laws. Stronger vertical dashed lines indicate the position of the hump at $\rho' = 100$ and 1000. A small size distribution with $\sigma = 1.2$ was included to eliminate the ripple structure.

by spheres. This description is for scattering in the scattering plane defined by the incident and scattering directions. It holds rigorously for incident polarization perpendicular to the scattering plane, vertical polarization, and also for when the incident polarization is parallel to the scattering plane, horizontal polarization, when ρ' is large.

To repeat previous reminders, our notation uses the scattered intensity, but the formulae given here are for the differential cross section as well, i.e. we use the proportionality $I \sim dC_{sca}/d\Omega$. Other synonymous terms are the 'phase function' (see section 2.3.1) and the Mueller matrix combination $S_{11} - S_{12}$. There are five major regimes/features for the scattering. These are:

1. *The forward scattering lobe.* The forward scattering lobe lies in the region $qR < 1$, it becomes isotropic therein as $qR \ll 1$ and in that manner is Rayleigh-like. It is described by

$$I(qR < 1) = k^4R^6F(m) \qquad \text{when } \rho' < 1 \tag{8.6a}$$

$$I(qR < 1) \simeq \frac{k^4R^6F(m)}{\rho'^2} \qquad \text{when } \rho' \gtrsim 5 \tag{8.6b}$$

$$I(qR < 1) = k^4R^6F(m)/\rho'^2 = \frac{k^2R^4}{4} \quad \text{when } \rho' \gtrsim 100 \tag{8.6c}$$

where $F(m) = [(m^2 - 1)/(m^2 + 2)]^2$ is the square of the Lorentz–Lorenz factor. Note that $qR < 1$ corresponds to $\theta < \lambda/2\pi R$. Also note that when $\rho' < 1$, Rayleigh scattering occurs in the forward lobe for any size particle. These results repeat equations (7.18).

2. *The Guinier regime.* When $qR \simeq 1$,

$$I(qR \lesssim 1) \simeq 1 - \left(qR_{g,\,G}\right)^2/3 \tag{8.7}$$

where $R_{g,G}$ is related to the actual radius of gyration of the scattering object via curves as shown in figure 7.7, which indicates $R_{g,G}$ can be as much as 50% larger.

3. *The power law regime.* The power law regime is complex because it evolves with increasing ρ'. When $\rho' < 1$, the RDG limit holds with its $(qR)^{-4}$ power law when $qR > 1$. As ρ' increase above unity, the forward scattering falls relative to the Rayleigh limit and then three rough power laws can ensue. For $3 < \rho' < 30$ there are three power law regimes (including the forward scattering lobe) that are quasi-universal with ρ'

$$\text{When } 3 < \rho' < 30 \qquad I \sim (qR)^0 = \text{constant when } qR < 1 \tag{8.8a}$$

$$I \sim (qR)^{-2} \qquad \text{when } 1 < qR < \rho' \tag{8.8b}$$

$$I \sim (qR)^{-4} \qquad \text{when } qR > \rho' \tag{8.8c}$$

One can show that the extent of (8.8c) in qR is equal to the inverse of the Lorentz–Lorenz factor and thus requires a small refractive index.

As ρ' increases above ca. 30, the −2 power law begins to sag significantly below the power law line. By $\rho' \approx 100$, a fourth power law is quite apparent with −3 exponent. This −3 power law is the Fraunhofer diffraction from the spherical particle acting as a two-dimensional, circular obstacle. The slope of an equal tangent line between the Guinier regime at the end of forward scattering lobe and a hump that becomes very apparent with increasing ρ' remains −2. Thus,

$$\text{When } \rho' > 100,\ I \sim (qR)^{-3} \quad \text{when } 1 < qR < \rho'/10 \tag{8.8d}$$

In anticipation of patterns to be described below, one could say the $(qR)^{-2}$ power law spans from the Guinier hump to another hump ('The Hump', section 8.2.1) centered near $qR \simeq \rho'$.

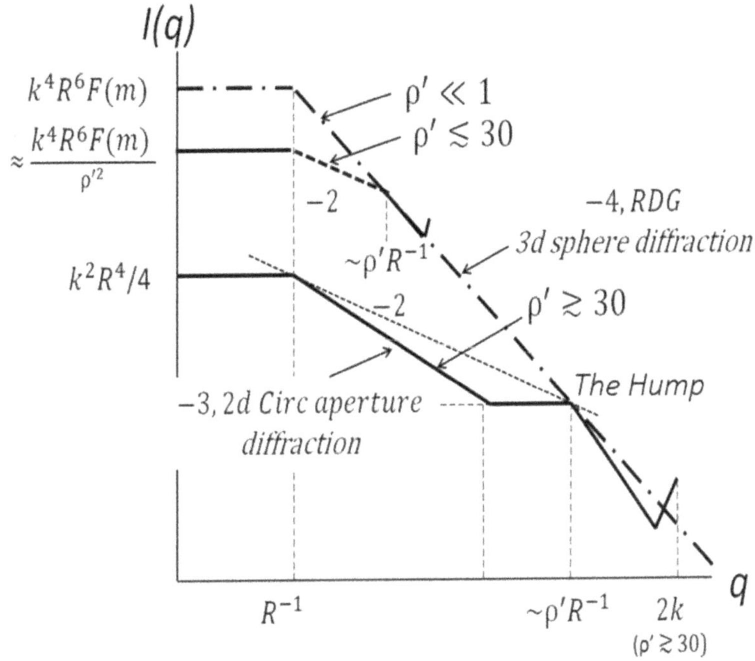

Figure 8.8. General Q-space schematic diagram for the scattering envelopes produced by an arbitrary sphere with a real refractive index, Mie scattering. The unnormalized scattering is plotted versus q (not qR). An interference ripple structure, not shown, following equations (8.9) lies under these envelopes. Shown are three stages of the scattering as its functionality evolves from the 3d diffraction limit (RDG), when $\rho' \ll 1$, with increasing ρ'.

3a. The ripples. Ripples appear beyond the forward scattering lobe, $qR > 1$. Within sometimes significant deviations and typically for $\theta \lesssim 90°$, the ripple spacing is

$$\delta(qR) = \pi \quad \text{when } \rho' < 3 \tag{8.9a}$$

$$\delta\theta = \lambda/2R \quad \text{when } \rho' > 3 \tag{8.9b}$$

4. Enhanced backscattering appears roughly when $\rho' > 5$.

This general description is portrayed schematically (and perhaps more effectively) in figure 8.8, which shows the unnormalized envelopes of the scattering curves, ignoring the ripples, plotted versus q (not qR).

8.2.6 The story of light scattering

More than a summary, figure 8.8 shows how Mie scattering patterns evolve with increasing internal coupling as parameterized by ρ'. The story of light scattering by a dielectric sphere with a real refractive index starts at $\rho' = 0$ with the Rayleigh–Debye–Gans limit. This is the diffraction limit. In this limit, each sub-volume of the scattering object sees only the incident wave and not the scattering from other parts of the object for this diffraction limit. For $q < R^{-1}$ the forward scattering lobe of the RDG curve is flat, i.e. q^0, and the magnitude of the scattering is equivalent to the Rayleigh scattering prediction, equation (8.6a). There is a change in slope near $q \simeq R^{-1}$, the Guinier regime, indicating the length scale of the sphere, its radius. For $q > R^{-1}$ the envelop functionality falls off with a negative four power law, the Porod limit, with an envelope magnitude $9(qR)^{-4}$ times the Rayleigh scattering cross section. This description includes Rayleigh scattering in the limit $\lambda \gg R$ because then $q < R^{-1}$ at all angles and only the flat (isotropic) part of the scattering function is obtained.

When $\rho' > 1$, scattering in the forward lobe regime *decreases relative to* the Rayleigh scattering value. The *relative* decrease (remember, the unnormalized scattering increases with ρ') is proportional to a factor of ρ'^2 as depicted in figure 8.8. This decrease indicates that the internal coupling is now significant. One can rephrase and say that 'internal multiple scattering' is occurring, which to some extent darkens, but more so dephases, the fields in the interior of the particle. Near $q \simeq R^{-1}$, a Guinier regime indicates the particle size. For $q > R^{-1}$ and $\rho' \lesssim 30$ the scattering now falls off roughly as q^{-2} until this functionality tags up with the RDG curve near $q \simeq \rho' R^{-1}$. For $q > \rho' R^{-1}$ the scattering comes close to RDG scattering, falling off approximately as $9(qR)^{-4}$ times the Rayleigh cross section for all R and m as long as $q > \rho' R^{-1}$. The q^{-2} functionality between $q \simeq R^{-1}$ and ca. $\rho' R^{-1}$ is exact only at those limits; the average Mie curves sag below the q^{-2} line. This sag begins as the first interference minimum present in all Mie curves and is the strongest of all the minima. Its position is near $q \sim 4R^{-1}$. By the point where $\rho' \gtrsim 30$, the small q part of the sag becomes the Fraunhofer diffraction pattern for a 2d circular aperture of the same radius as the sphere with a q^{-3} power law. This power law ends in a bit of a plateau which grows to the 'hump' relative to the overall decrease in normalized scattering with increasing ρ'. This plateau ends near $q \simeq \rho' R^{-1}$ where the hump is centered, after which it regains a strong but inexact similarity to the RDG, q^{-4} Porod regime of 3d spherical particle diffraction. In all cases, once ρ' is significant, an enhanced backscattering occurs including rainbows and a glory near $q \simeq 2k$, which is equivalent to a scattering angle of 180°.

We see that scattering begins at $\rho' \to 0$ in the 3d, diffraction limit, the mathematics of which is a 3d Fourier transform. As the internal coupling increases, as indicated by the unifying internal coupling parameter ρ', the functionalities of this limit are severely altered. However, in the limit $\rho' \to \infty$ the sphere becomes an obstructing object for the light. Then the 2d, circular aperture, diffraction limit describes the scattering (via Babinet's principle), the mathematics of which is a 2d Fourier transform.

In fact, light scattering for all shapes is bounded by 3d diffraction for no internal coupling to 2d diffraction with extreme internal coupling as quantified by ρ'.

8.2.7 The effects of the imaginary part of the refractive index on scattering

Our study in chapter 7 of the effects of the imaginary part of the refractive index, κ, on absorption and scattering by spheres indicated these phenomena were governed, to a large extent, by the relative skin depth parameter κkR, section 7.2.1. Here we study the effects of κ on scattering from the vantage of Q-space (Wang *et al* 2015).

Figure 8.9 shows plots of the Rayleigh normalized differential scattering cross section versus qR for three spherical particles with size parameters $kR = 5$, 50 and 500 with refractive index $m = 1.5 + i\kappa$ scattering $\lambda = 532$ nm green light. These plots include a distribution of particle sizes so the interference ripples are eliminated. The imaginary part of the refractive index κ varies for each kR to yield the same κkR values for each. universality on κkR is readily apparent for the two larger kR particles. In particular, figure 8.9 demonstrates the loss of the hump and back scattering features such as the nascent rainbow and the glory. One sees that $\kappa kR = 0.3$ removes approximately half of the hump and these features, and with $\kappa kR = 1.0$, the effect is nearly saturated as for $\kappa kR = 3.0$. It is important to stress, as we did in section 7.2 for absorption, that it is κkR that is the important variable, not κ by itself. For large kR, κ can be quite small and still have an effect. For example, when $kR = 500$, $\kappa kR = 1$ implies $\kappa = 0.002$.

Figure 8.10 is almost a repeat of figure 8.9 but without a small size distribution to reduce the ripples, which was necessary in 8.10 to see the evolution of the hump with κkR. Plots are for extreme values of $\kappa kR = 0$, no effect, and $\kappa kR = 3$ when the effect becomes saturated. Because the particles are monodisperse, ripples occur for $\kappa kR = 0$. When $\kappa kR = 3$, the ripples disappear in the hump region along with the hump, but remain unaffected in the 2d diffraction regime.

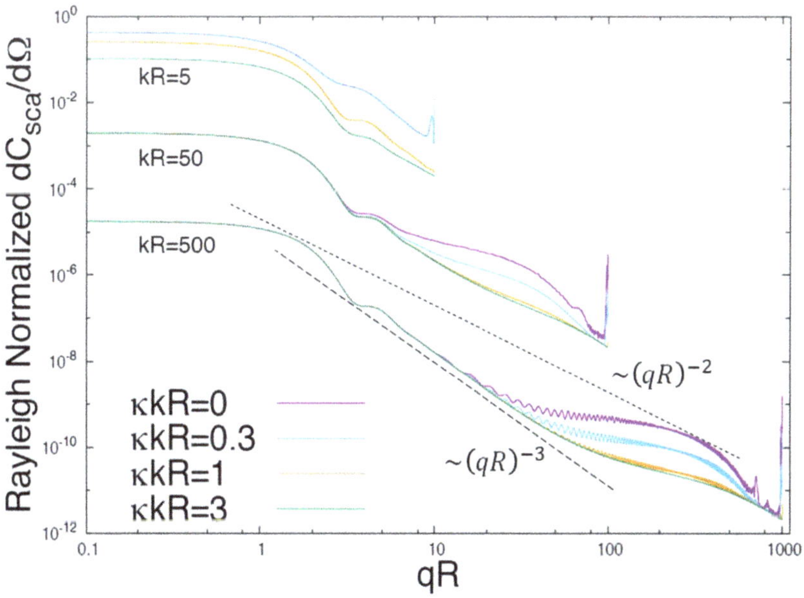

Figure 8.9. Rayleigh normalized differential scattering cross section versus qR for spheres with mean size parameters $kR = 5$, 50 and 500 with refractive index $m = 1.5 + i\kappa$ scattering $\lambda = 532$ nm green light. The incident light is polarized perpendicular to the scattering plane. The corresponding diameters are ~0.85, 8.5 and 85 microns, respectively. These plots are for a small size distribution with geometric width 1.2 to remove much of the ripple structure. Plots are for a series of κkR values as indicated. Note: (1) that when $kR = 5$, the forward scattering is affected by the value of κ, and (2) the plots for $\kappa kR = 1$ and 3 overlap or nearly do so.

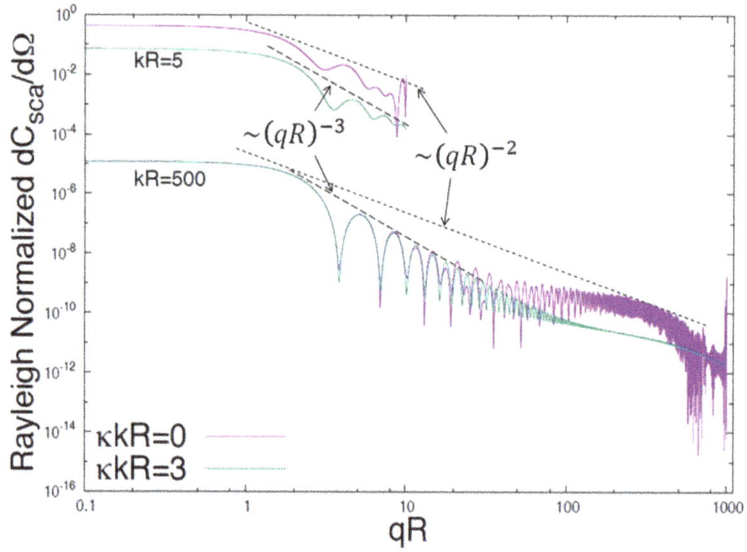

Figure 8.10. Rayleigh normalized differential scattering cross section versus qR for spheres with size parameters $kR = 5$ and 500 with refractive index $m = 1.5 + i\kappa$ scattering $\lambda = 532$ nm green light. The incident light is polarized perpendicular to the scattering plane. The corresponding diameters $2R$ are ~0.85 and 85 microns, respectively. Plots are for extreme values of $\kappa kR = 0$, no effect, and $\kappa kR = 3$, extreme effect ($\kappa = 0.6$ and 0.006, respectively). Note that when $kR = 5$, the forward scattering is affected by the value of κ (0.6) because it is significant compared to n.

The phenomena in figures 8.9 and 8.10 can be understood with the help of section 7.2.1. There, the physical significance of the relative skin depth parameter, κkR, is shown to be that it is the ratio of the radius of the particle to the optical skin depth, δ, equation (7.31). Figure 7.24 draws a sketch of what this might look like for $\kappa kR \simeq 3$, and figure 7.25 shows Mie solutions for the interior field of a large sphere. When $\kappa kR = 3$, the only interior light is located thinly at the side of the sphere facing the incident light. In section 8.2.8 below, we will show that the hump is largely due to waves or rays that refract into and back out of the sphere. For this to happen the rays must have a viable path length inside the sphere greater than the radius of the sphere. Such is not the case when κkR becomes large; hence, the hump disappears. This also explains the disappearance of the glory and rainbows with large κkR. Moreover, in section 8.3 we will show that roughly have of all the scattered light is in the hump. We conclude that the hump is due to refraction.

Finally, with light confined to the side of the sphere facing the incident light, the sphere diffracts like a 2d circular aperture, and this explains the 2d Fraunhofer diffraction pattern at small qR.

8.2.8 The Debye series expansion

The Debye series expansion (Debye 1908) allows for interpretation of scattering from spheres in terms ray optics. A ray encounters the surface of the sphere reflecting and refracting at the interface. The pth term in the series, where $p \geqslant 0$ is an integer, is interpreted as a ray that experiences $p - 1$ internal reflections before leaving the sphere via a second refraction through the surface; the $p = 0$ term represents mostly diffraction from the sphere. The series derives from an expansion of the Mie scattering coefficients into a series of Fresnel coefficients. Hence, like the Mie theory, it is an exact description of scattering of a plane wave by a homogeneous a sphere.

Figure 8.11 illustrates in Q-space the Debye series decomposition of scattering by a sphere with size parameter $kR = 125$ and refractive index $m = 1.33 + i0$. Such a sphere has a diameter of $D = 2R \simeq 20$ μm at visible wavelengths. Figure 8.11 shows the first four Debye terms, the sum of which very nearly equals the complete Mie result. The $p = 0$ ray is seen to be mostly the diffracted ray (wave!) due to the sphere being a circular obstruction. There is also a small reflection component at large qR. It dominates the scattering for $qR \lesssim 1$ (i.e. $\theta \lesssim \lambda/2\pi R$) for this large particle. Adding the $p = 1$ ray brings in the hump but the enhanced backscattering is missing. We conclude that the hump is largely the $p = 1$ ray, which involves refraction. Adding $p = 2$ brings in the enhanced backscattering but does little to the diffraction and hump regions. Adding $p = 3$ improves the backscattering fidelity to the complete Mie solution. Of course, these rays with $p \geqslant 1$, must propagate through the medium

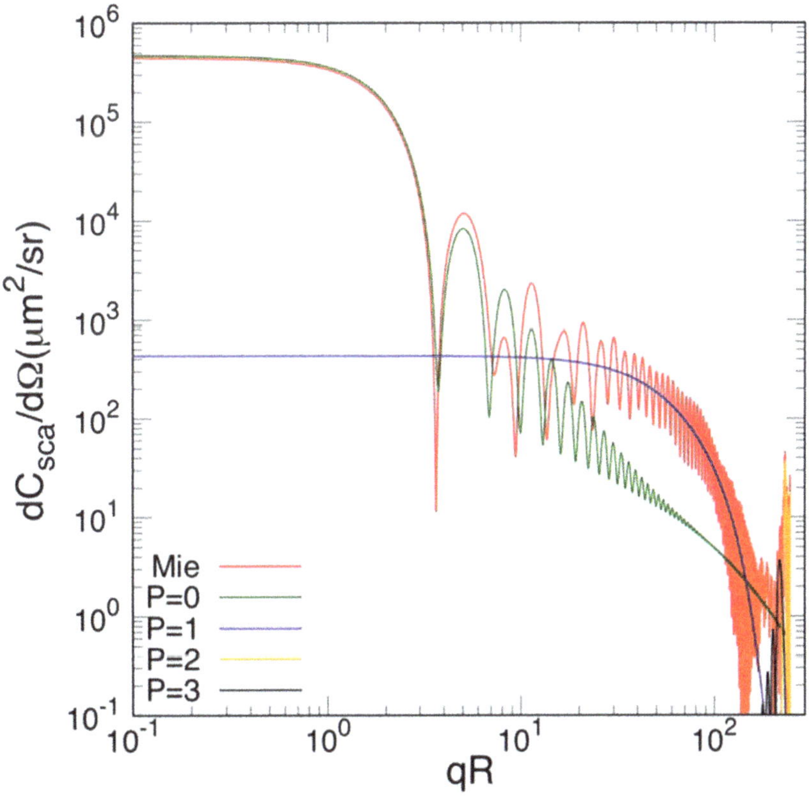

Figure 8.11. The first four Debye terms and the complete Mie theory for a sphere with size parameter $kR = 125$ and refractive index $m = 1.33 + i0$. The incident light is polarized perpendicular to the scattering plane.

of the sphere, hence they would be absorbed if the absorbance length was comparable to the size of the sphere, i.e. when $\kappa kR > 1$. This is an explanation of the universality of κkR. Adding higher order terms, not shown, is fine tuning, the first four terms yielding a good description of the scattering. It is apparent that the Debye series converges rapidly.

8.2.9 Diffraction and refraction

The Debye series results help differentiate the roles of diffraction and refraction in the total scattering process. Figure 8.18 shows that the $p = 0$ ray contributes the most to the forward scattering lobe and the Guinier regime. The $p = 1$ ray contributes the most to the hump. Rays $p \geqslant 2$ get involved in more complex refraction phenomena at large angles such as rainbows and the glory that I lump together with the term 'enhanced backscattering'. With these observations, Maughan (2020) has created an illustration of the various regions of scattering as displayed in figure 8.12.

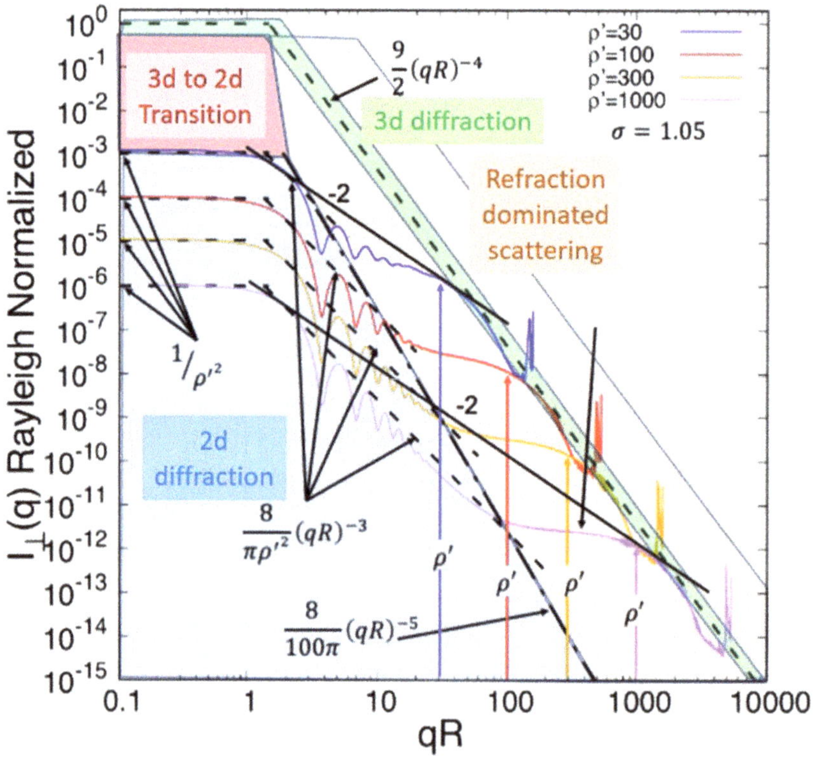

Figure 8.12. Rayleigh normalized Mie scattering curves for spheres with a relative refractive index of $m = 1.3 + i0.0$, and $\rho' = 30, 100, 300, 1000$ plotted versus q. The incident polarization is perpendicular to the scattering plane. The purpose of this figure is to indicate various regions where diffraction or refraction dominates the scattering. The scattering was calculated with a small log-normal size distribution with geometric width $\sigma = 1.05$ to eliminate the ripple structure. From Maughan (2020).

8.2.10 Comparison of Q-space and θ-space

It should be very apparent by now that Q-space provides a significantly different perspective on light scattering than plotting versus the scattering angle, i.e. plotting in θ-space. This is due to two reasons: (1) the magnitude of the scattering wave vector, q, has dimensions of inverse length, and hence can uncover length scales in the scattering object, whereas θ is dimensionless and thus has no such ability. (2) When plotting versus q, we have been plotting on a logarithmic scale, whereas when plotting versus θ, the scale is usually linear. In general, plotting logarithmically uncovers the geometric nature of the Universe, which usually prevails over arithmetic trends. This geometric nature (of Nature) is displayed as power laws. In fact, because the $\sin(\theta) \simeq \theta$ when $\theta \ll 1$, yields $q \simeq (2\pi/\lambda)\theta$ for small angles θ, many of the features uncovered by Q-space analysis would have appeared just as well had the plotting been versus $\log(\theta)$. Missing, of course, would have been the inverse length nature of q and the opportunity to plot versus the dimensionless qR which was a major factor in finding universal descriptions. Finally, plotting versus q emphasizes that the foundation of light scattering is diffraction. Figure 8.13 walks

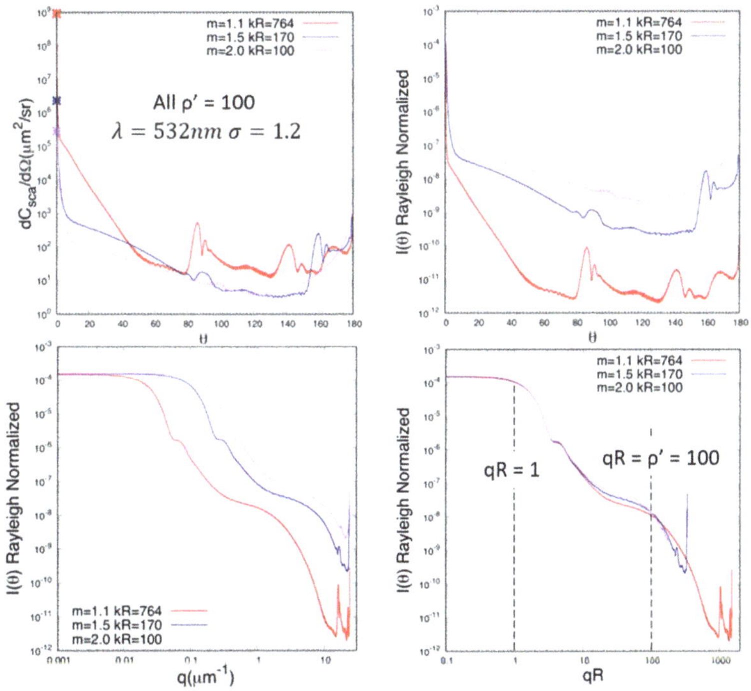

Figure 8.13. Evolution of the plots of the light scattering from three different spheres that have the same internal coupling parameter $\rho' = 100$. The incident light is polarized perpendicular to the scattering plane and the ripples have been removed by averaging over a small size distribution. Upper left shows the differential cross section versus linear θ, the scattering angle. The scattering increases sharply as $\theta \to 0°$, and the asterisks indicate the scattering at $\theta = 0$. Upper right uses the data in the upper left but normalizes the scattering intensity by the Rayleigh scattering for the sphere. Lower left takes another step by converting linear θ to logarithmic q, the scattering wave vector. We are now in Q-space. Lower right converts q to the dimensionless qR and hence completes the transformation of the scattering to dimensionless Q-space. universal features appear.

through four steps of the evolution of viewing scattering from the conventional θ-space to the Q-space perspective.

Nevertheless, as an experimental physicist in the lab with my apparatus, I speak θ in units of degrees, and this is equally true for my cohort. Therefore, it is valuable to make a direct comparison of scattering in both Q- and θ-space, as is done in figure 8.14.

Perhaps the first thing that becomes abundantly clear when viewing this comparison is that the Q-space analysis expands the small angle regime; e.g. note where $\theta = 5°$ and $90°$ are on the plot in the lower right for diameter $2R = 100$ microns. This is both a merit and a fault of Q-space analysis. In θ-space the constant forward scattering lobe, the Guinier regime and the onset of the 2d diffraction regime are not readily apparent and can be missed, especially for large particles. Moreover, below in section 8.3 when the concept of a partial scattering cross section is discussed, it will be shown that approximately half of the scattered energy is in the forward region where $\theta \lesssim \lambda/2R$. Thus, it is a region not to be ignored. However, by expanding the small angles, the large angle scattering gets squished, so that Q-space analysis misses detail past 90° such as the rainbows and the glory[1]. On the other

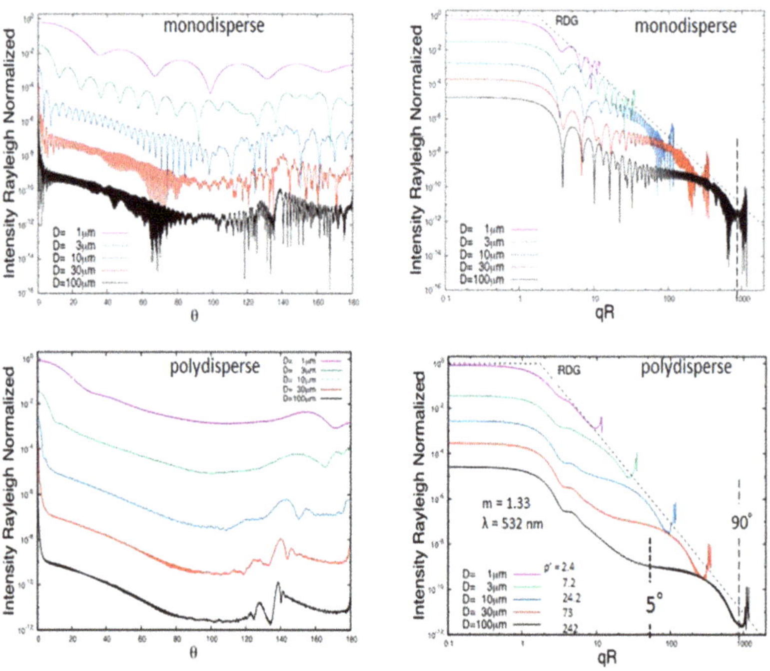

Figure 8.14. Comparison θ-space and Q-space. Plots of the Rayleigh normalized scattered intensity versus linear scattering angle θ (left) and logarithmic qR (right). The spherical particles have diameters $D = 2R$ and real refractive index $m = 1.33$. The light has a wavelength of 532 nm. Upper plots are for single sized particles, lower are with the incorporation of a small size distribution to eliminate the ripples. Lower right plot shows internal coupling parameter values and the positions of $\theta = 5°$ and $90°$ for $D = 100$ μm. The incident light is polarized perpendicular to the scattering plane.

[1] I hope the reader will forgive me for calling the rainbows and glory 'details'.

hand, the compression that Q-space provides at large θ accentuates the hump and demonstrates the ultimate approach to the RDG, 3d diffraction limit. I believe good intuition for scattering can be obtained best by musing over both manners of plotting as displayed in figure 8.14.

Figure 8.15 (top) reproduces figure 7.21, which demonstrated the stationary angle at $\theta \simeq 37°$ (stationary as the real refractive index changed) and replots the same data

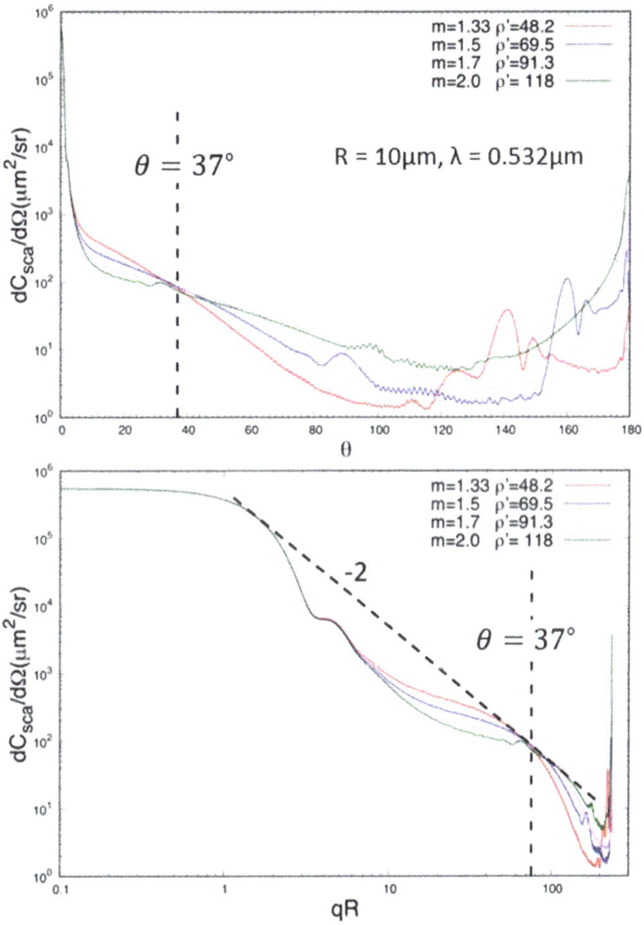

Figure 8.15. Comparison of the θ-space (top) and Q-space (bottom) plots of the differential scattering cross section for a sphere that displays a stationary angle at $\theta \simeq 37°$. Top plot reproduces figure 7.21. Ripples have been suppressed by averaging the scattering over a 10° range. Incident polarization was perpendicular to the scattering plane.

in Q-space (bottom). Q-space shows the hump sliding along the −2 power law as the real refractive index changes. Q-space also shows that the Guinier regime is stationary too. Finally, notice that the scattering as $qR \to 0$ doesn't change, consistent with equation (8.6c).

Figure 8.16 demonstrates the effect of a non-zero imaginary part of the refractive index, $\kappa \geqslant 0$ through the relative skin depth parameter, κkR in both the θ-space

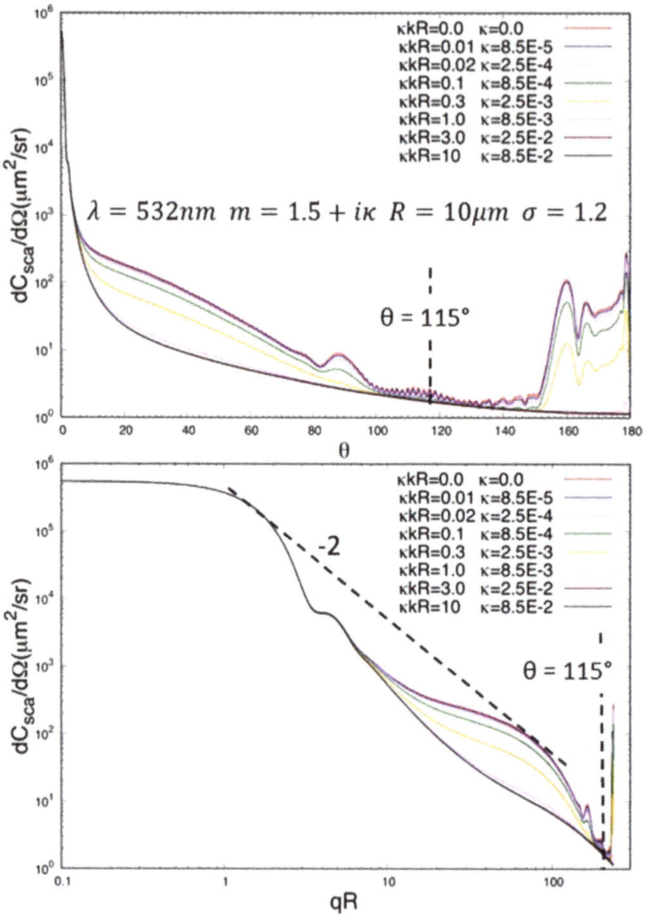

Figure 8.16. Comparison of the θ-space (top) and Q-space (bottom) plots of the differential scattering cross section for a series of different relative skin depth parameters, κkR for a sphere that displays a quasi-stationary angle at $\theta \simeq 115°$. Ripples have been suppressed by averaging over a small size distribution. Incident polarization was perpendicular to the scattering plane.

(upper) and Q-space (lower). This figure also demonstrates that the hump lies between approximately 5° and 90°.

Figure 8.17 (left) reproduces figure 8.7, which demonstrated the quasi-universality of Mie scattering on the internal coupling parameter ρ'. This master graph unifies the light scattering by spheres of all sizes and real refractive indices and demonstrates various salient regimes. These exact same Mie calculations are replotted on the graph to the right in linear θ-space to find no indication of universality on ρ', no power laws, no obvious regimes, and no indication of the evolution of scattering from the 3d to 2d Fraunhofer diffraction limits. Indeed, figure 8.17 represents one of the goals of this book.

The comparisons of Q-space and θ-space above make, I believe, a strong case for Q-space analysis. However, as has been stated elsewhere in this book, a problem with Q-space is that it does not display the large scattering angles well. Here, we give

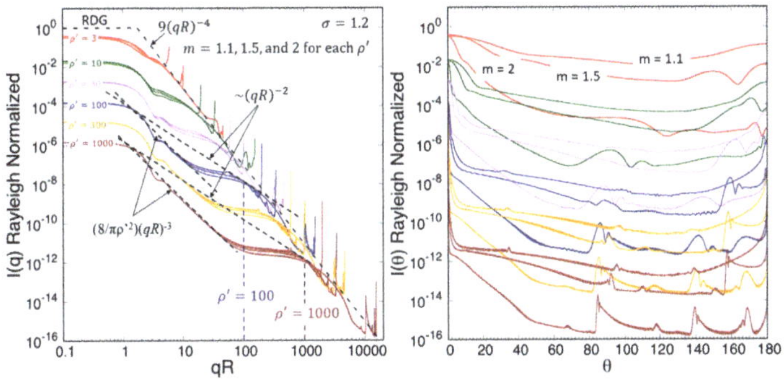

Figure 8.17. (Left) A reproduction of the master graph of figure 8.8, above, which demonstrated the quasi-universality of Mie scattering on the internal coupling parameter ρ'. These same Mie calculations are replotted on the graph to the right in linear θ-space to find no coherent patterns.

an example of this deficiency, the value of θ-space, and the need for a combined comprehensive analysis.

Figure 8.18 shows apparatus that we have recently built and published in (Gautam and Sorensen 2021). The angular scattering range is rather exceptional, from 0.32° to 177.6°. Figure 8.19 shows scattering data from a water drop aerosol that were obtained with this apparatus. The data are plotted in both Q-space and θ-space.

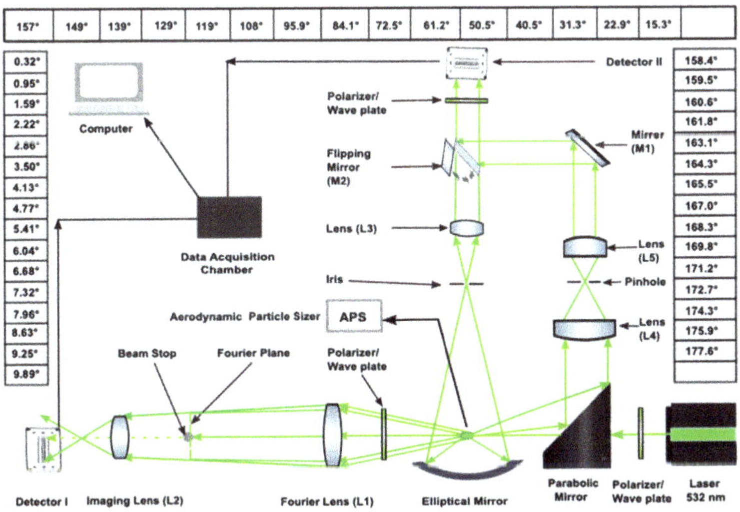

Figure 8.18. Schematic diagram of a multi-angle light scattering apparatus with an angular range of from 0.32° to 177.6°. There are two 16-channel detectors, but a flipping mirror allows access for one detector to detect scattering from either 15.3° to 157° or 158.4° to 177.6°. Obviously, a third detector could be added, and the flipping mirror mechanism eliminated so that all 46 angles could be detected simultaneously (funding limitations prevented us from doing so). The Aerodynamic Particle Sizer (APS) samples the aerosol after it passes through the scattering volume. Reprinted from Gautam and Sorensen (2021), with the permission of AIP Publishing.

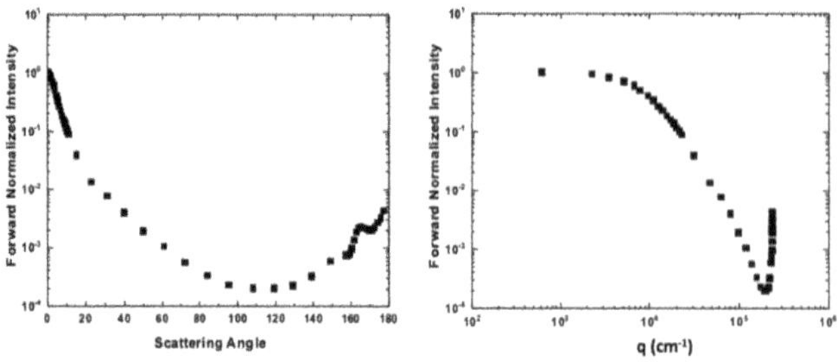

Figure 8.19. A demonstration of the complimentary viewpoints for scattering that θ-space and Q-space can provide. Scattered light intensity normalized to unity at $\theta = 0.32°$, the smallest forward scattering angle, for an aerosol of water droplets measured with the apparatus described in figure 8.18. (Left) Data plotted in θ-space, which shows detail in the back direction; (right) is the same data plotted in Q-space which uncovers a forward scattering regime, a Guinier regime and a short power law. Reprinted from Gautam and Sorensen (2021), with the permission of AIP Publishing.

As we concluded in our paper, the inclusion of small angles engenders the need for logarithmic plotting of the data versus the independent scattering variable. That independent variable could be θ, but the magnitude of the scattering wave vector q is better because it carries dimensions of inverse length. This method we have called 'Q-space analysis'. However, a weakness of Q-space analysis is the compression of the data at large scattering angles where useful data also reside. Thus, we advocate a combination of both viewpoints to gain a comprehensive description of the scattering as illustrated in figure 8.19.

8.3 The partial scattering cross section

Particles of any shape scatter light isotropically when they are significantly smaller than the wavelength of light (except for polarization effects) and anisotropically when they are larger than the wavelength, more so in the forward direction. This observation can be quantified with the asymmetry parameter, defined in equation (2.10)

$$g = \int_{4\pi} p(\theta, \varphi) \cos\theta \, d\Omega \tag{8.10}$$

where $p(\theta, \varphi) = 4\pi C_{\text{sca}}^{-1}(dC_{\text{sca}}/d\Omega)$ is the phase function defined in equation (2.9).

Another tack to study and quantify the scattering anisotropy is to limit systematically the integral of equation (8.10) to restricted solid angles. This would be a fraction of the total scattering cross section in the restricted solid angle range. One way to do this is to start the integration at the forward direction, $\theta = 0$, and limit the integration to a finite θ. At any given θ, the entire φ range integration of 0 to 2π is performed. With this in mind, we define the *partial scattering cross section*, $\partial C_{\text{sca}}(\theta_p)$ (Sorensen *et al* 2017), as

$$\partial C_{\text{sca}}(\theta_p) = \int_0^{\theta_p} \int_0^{2\pi} dC_{\text{sca}}/d\Omega \, d\varphi \sin\theta d\theta \tag{8.11}$$

Figure 8.20 shows a diagram of the $\partial C_{\text{sca}}(\theta_p)$ region. The partial scattering cross section shows how the total scattering accumulates when one starts the integration in

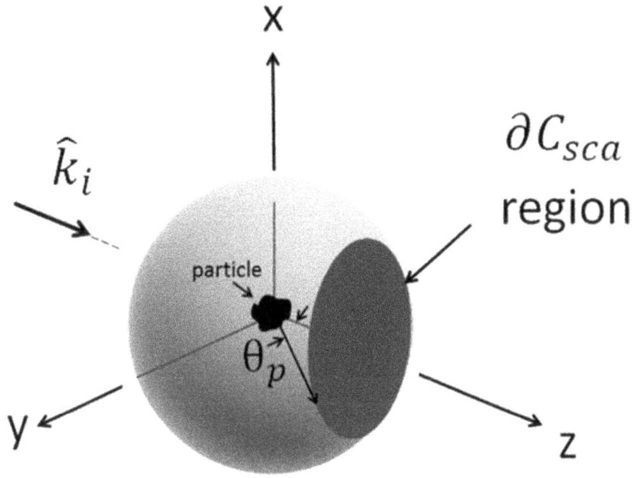

Figure 8.20. Diagram of light incident upon an arbitrary particle. The light scattered at angles θ_p and less symmetrically around the z-axis (darkened area) are included in the partial scattering cross section, $\partial C_{\text{sca}}(\theta_p)$. Reprinted with permission from Sorensen *et al* (2017). © The Optical Society.

the forward direction at $\theta = 0$ and then includes more scattering angles θ into the integral. It follows that $\partial C_{\text{sca}}(180°) = C_{\text{sca}}$, the total scattering cross section.

Figure 8.21 considers scattering of unpolarized light with wavelength $\lambda = 1000\pi/6 = 524$ nm by spherical particles with three different real refractive indices and three different radii such that all three have the same internal coupling parameter $\rho' = 100$. The lower half of figure 8.21 contains the Rayleigh normalized differential scattering cross section versus qR. The calculations were performed with the Mie equations. The quasi-universal behavior for the same ρ' is seen. For this value of ρ', the characteristic patterns of Q-space analysis are present: the forward scattering lobe when $qR < 1$, followed by, with increasing qR, a brief 2d Fraunhofer diffraction regime with a $(qR)^{-3}$ envelop the 2d to 3d diffraction crossover hump centered near ρ' leading to roughly join the RDG, 3d diffraction limit. The upper half of figure 8.21 shows the normalized partial scattering cross section, $\partial C_{\text{sca}}/C_{\text{sca}}$, versus qR, with $q = 2k\sin(\theta_p/2)$. The qR scales are matched between the two halves.

Figure 8.21 shows that the normalized partial scattering cross section initially increases rapidly with qR, i.e. it increases rapidly as more of the forward scattering is included. Comparison to the lower half shows that this rapid increase occurs in the forward scattering lobe. The increase levels off at a value quite close to ½ near $qR = \pi$, which corresponds to an angle of $\theta_p = \lambda/2R = \lambda/D$. After that, $\partial C_{\text{sca}}/C_{\text{sca}}$ remains roughly constant until qR reaches a value at the end of the -3 power law of the 2d circular aperture diffraction. As indicated in the figure's lower half, this is where the hump begins. Then the partial differential cross section increase again throughout the hump. When the hump ends, the increase ends with $\partial C_{\text{sca}}/C_{\text{sca}}$ approaching unity as qR enters the post hump regime close to 3d spherical particle diffraction.

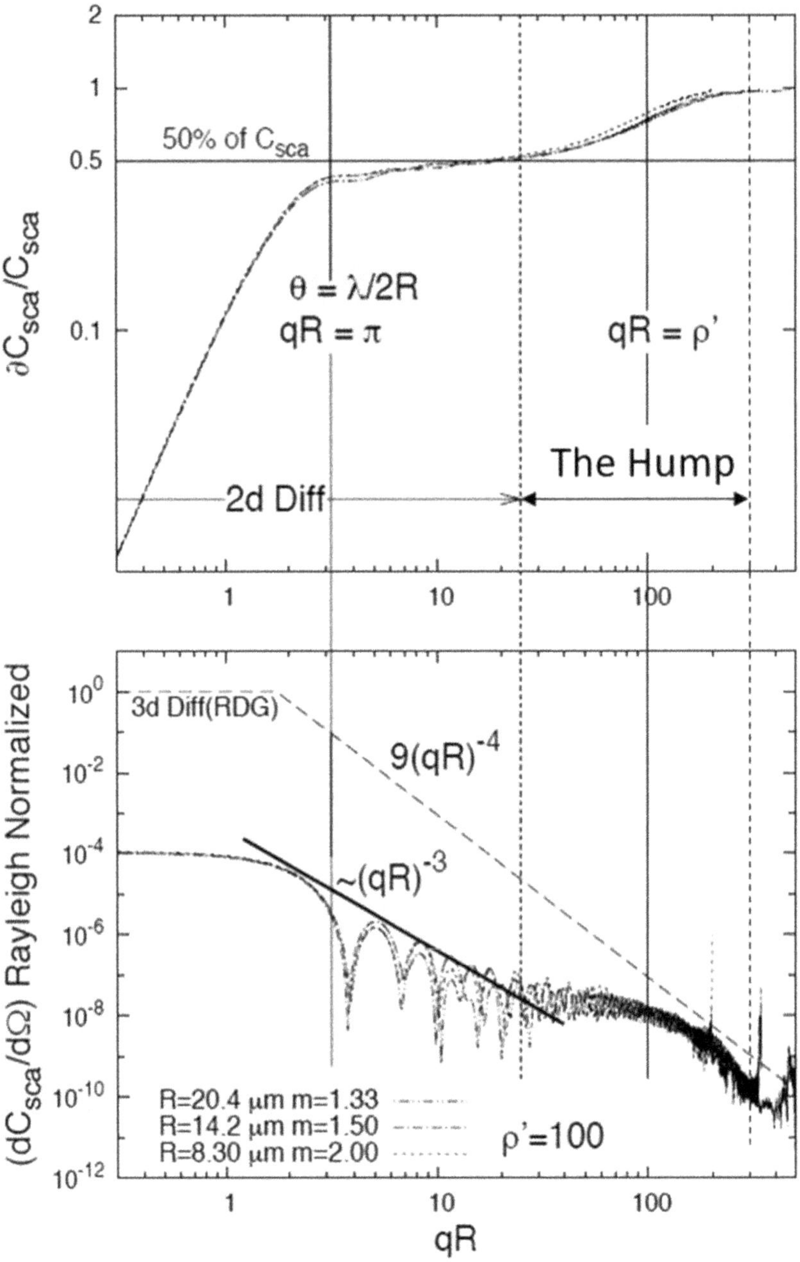

Figure 8.21. (Bottom) A q-space plot of the Rayleigh normalized differential scattering cross section versus the dimensionless qR with $q = 2k\sin(\theta_p/2)$ for three spheres all with a real refractive index and the same internal coupling parameter $\rho' = 100$. The incident light is unpolarized. (Top) The normalized partial scattering cross section for the same three spheres. Regimes of 2d and 3d diffraction are indicated as well as where the scattering angle $\theta = \lambda/2R$. Reprinted with permission from Sorensen *et al* (2017). © The Optical Society.

In summary:

1. The partial scattering cross section, ∂C_{sca} is approximately 50% the total scattering cross section by end of forward scattering lobe near $qR = \pi$, corresponding to $\theta = \lambda/2R = \lambda/D$.
2. The second ~50% is accumulated in the crossover hump, which is centered near $qR \simeq \rho'$.
3. The partial scattering cross section behavior with q (hence θ) is essentially universal with ρ'.

Item 1 above has been well recognized before. In his studies of the extinction paradox, Brillouin (1949) found that to within a few percent half the light energy was scattered at angles of $\theta \leqslant 3\lambda/R$. The physical optics approximation for particles of arbitrary shape (Borovoi 2013) and the geometric optics approximation demonstrate that half the light energy scatters within the Fraunhofer diffraction forward cone (Bohren and Huffman 1983, Mishchenko *et al* 2002). Recognition of the importance of the hump and universality with ρ' are new with these results.

Figure 8.22 shows the partial scattering cross section for a smaller sphere with a radius of $R = 3$ microns, hence a diameter of 6 microns, and a real refractive index of $m = 1.5$. The wavelength is $1000\pi/6 = 524$ nm. For such a particle, the 2d and 3d diffraction regions are not as well defined as for the larger particles examined in figure 8.21. Nevertheless, the conclusions are the same: approximately half the light

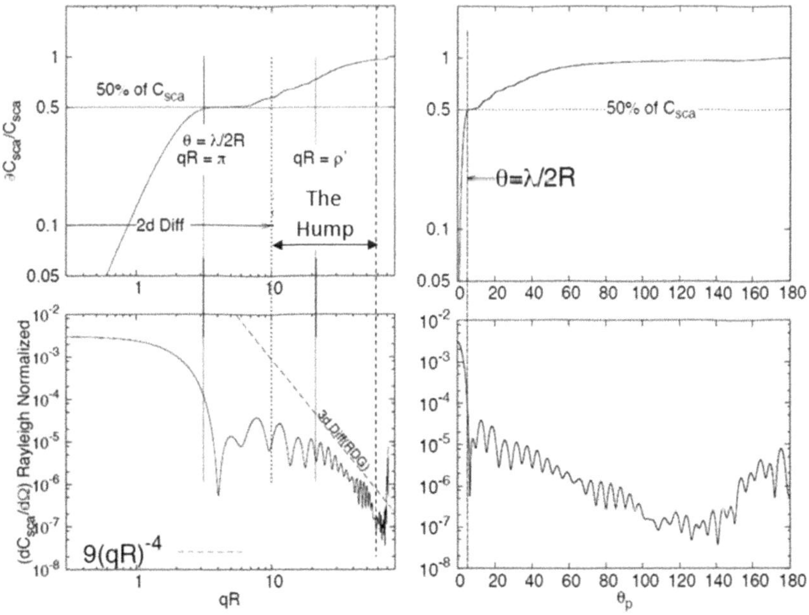

Figure 8.22. Example of scattering of unpolarized light with a wavelength of $1000\pi/6 = 524$ nm by a sphere of radius $R = 3$ microns and real refractive index $m = 1.5$ hence $\rho' = 21.2$. (Bottom) Plots of the Rayleigh normalized differential scattering cross section versus the dimensionless qR (left) or the scattering angle θ (right). (Top) The normalized partial scattering cross section for the same independent variables. Reprinted with permission from Sorensen *et al* (2017). © The Optical Society.

is scattered in the forward lobe and the other half accumulates in the hump. Figure 8.22 compares the q-space analysis to the conventional plot versus the scattering angle θ and thereby explicitly demonstrates the crucial message that about half of all the scattered light is in the narrow forward scattering lobe for which $\theta \leqslant \lambda/D$. In this example of $D = 6$ microns, a typical coarse mode value, and $\lambda = 524$ nm, approximately half of all the scattered light is in angles $\theta \leqslant 5°$.

Close examination of the differential scattering cross section in the bottom of figures 8.21 and 8.22 shows that when ρ' is large, the scattering is broken up into three major regimes listed here in order of increasing qR: 2d Fraunhofer diffraction (which includes the forward scattering lobe), the hump, enhanced backscattering. The partial scattering cross section indicates that approximately half of the total scattering occurs in each of the first two regimes.

We remark that the distinctive nature, uniform scattering across a disk with a sharp boundary to darkness, of the lunar and solar coronas seen when viewing these objects through thin clouds is explained by the forward scattering lobe of constant intensity when $qR \lesssim 1$. In light of the results, here we now recognize that for large, refractive particles, such as water mist droplets, about half the scattered light is in the corona. This is illustrated in figure 8.23.

Figure 8.23. The lunar corona. Although the clouds are not uniform, the scattering across the disk of the corona surrounding the Moon would have been save for the lack of uniformity of the clouds. This corona is a manisfestation of the forward scttering lobe; uniform scattering when $qR \lesssim 1$. The edge of the corona, that is the edge of the forward scattering lobe, is at $qR \simeq 1$. Given this, can the reader explain why the edge is reddish? Given that the apparent angular diameter of the Moon is 1/2°, can the reader give a good estimate of the cloud droplet diameter? Finally, can the reader explain why the sky is so dark beyond the edge? Does the scattering fall off rapidly there? This Lunar Corona.jpg image has been obtained by the authors from the Wikimedia website where it was made available by Wing-Chi Poon under a Creative Commons—Attribution-ShareAlike 2.5 Generic—CC BY-SA 2.5 licence. It is included within this article on that basis. It is attributed to User:Wingchi.

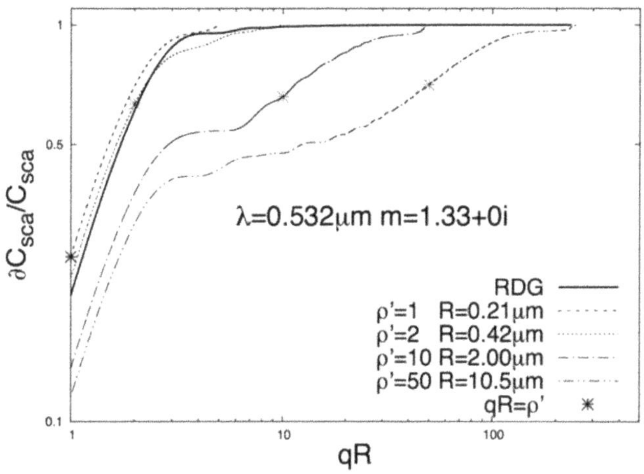

Figure 8.24. Normalized partial cross section versus qR for spheres with a real refractive index $m = 1.33$ and variety of sizes hence different internal coupling parameters ρ' to demonstrate the behavior as $\rho' \to 0$. The asterisks mark where $qR = \rho'$. Reprinted with permission from Sorensen *et al* (2017). © The Optical Society.

Figure 8.24 explores the functionality of the partial scattering cross section on ρ'. This figure shows that because $qR \simeq \rho'$ at the center of the hump and because the hump carries about half the scattered light energy, as ρ' decreases toward unity, the hump half and the forward scattering half of the total energy merge as $\rho' \to 1$. Then in this limit, which is the RDG, 3d diffraction limit, essentially all the light is scattered in the forward lobe.

8.3.1 The partial scattering efficiency

A *partial scattering efficiency* can be defined as

$$\partial Q_{\text{sca}} = \partial C_{\text{sca}}/G \tag{8.12}$$

where G is the projected area of the particle onto a plane perpendicular to the incident direction. With this, one can consider the effects of absorption due to a non-zero imaginary part of the refractive index.

Figure 8.25 (top) shows the partial scattering efficiency plotted versus qR for a large, refractive particle such that $\rho' = 141$. Note that the total extinction efficiency of a particle with this internal coupling parameter is approximately equal to two, and if absorption is minimal, the scattering efficiency is equal to the extinction efficiency. The bottom plots the normalized differential scattering cross section versus the same qR. Three different values of the skin depth parameter κkR are plotted; small, crossover and large (and saturated), $\kappa kR = 0.03$, 0.3 and 3, respectively. The bottom shows how the refraction hump disappears with increasing κkR. Recall that approximately half the scattered light comes from the hump region. Thus, with increasing κkR, the loss of the crossover hump will cause the total scattering to decrease by factor of two. This is confirmed in the far right side of the top of this figure. On the far right side of figure 8.25, θ_p of equation (8.11) is equal to

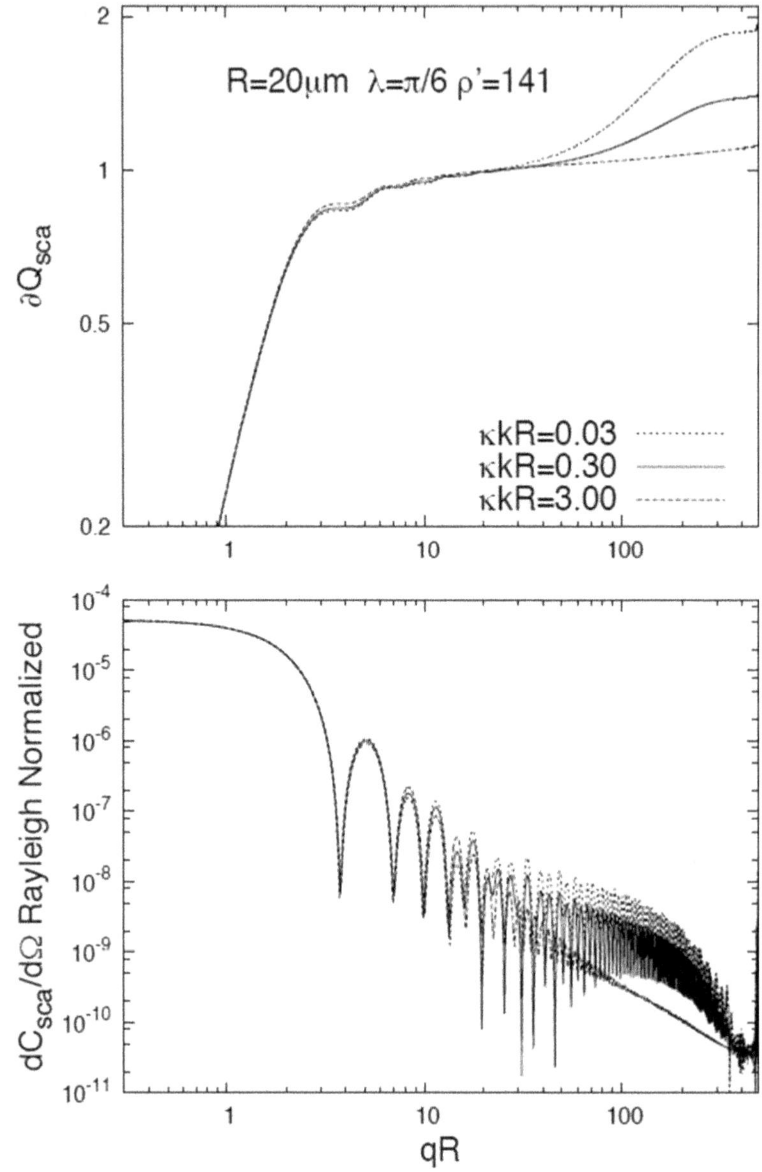

Figure 8.25. (Top) The partial scattering efficiency, $\partial Q_{\mathrm{sca}}$, versus qR for three spheres with $\rho' = 141$ but different values of κkR. (Bottom) Differential scattering cross section versus qR (qR scale matched with the top) for the same three spheres. As κkR increases, the hump disappears in the bottom to cause the total scattering efficiency to decrease from ~2 to ~1.

180°, hence the partial efficiency equals the *total* scattering efficiency. There we see that this total scattering efficiency evolves from a value of ~2 to a value of ~1 as κkR evolves from small to large.

A moment's thought sees the logical consequence to this. A large, refractive particle has an *extinction* efficiency of 2. The extinction efficiency is the sum of the scattering and absorption efficiencies. When absorption is small, as indicated by $\kappa kR \ll 1$, the scattering and extinction efficiencies are equal for such a particle. However, when the absorption is large, as indicated by $\kappa kR > 1$, the absorption efficiency is 1. Thus, the scattering efficiency must fall from 2 to 1 so that the sum of the efficiencies remains equal to 2.

In general, the efficiency can have a wide range of values from 0 to more than 4.5, see figures 7.33–7.35, and an example is given in the lower half of figure 8.26.

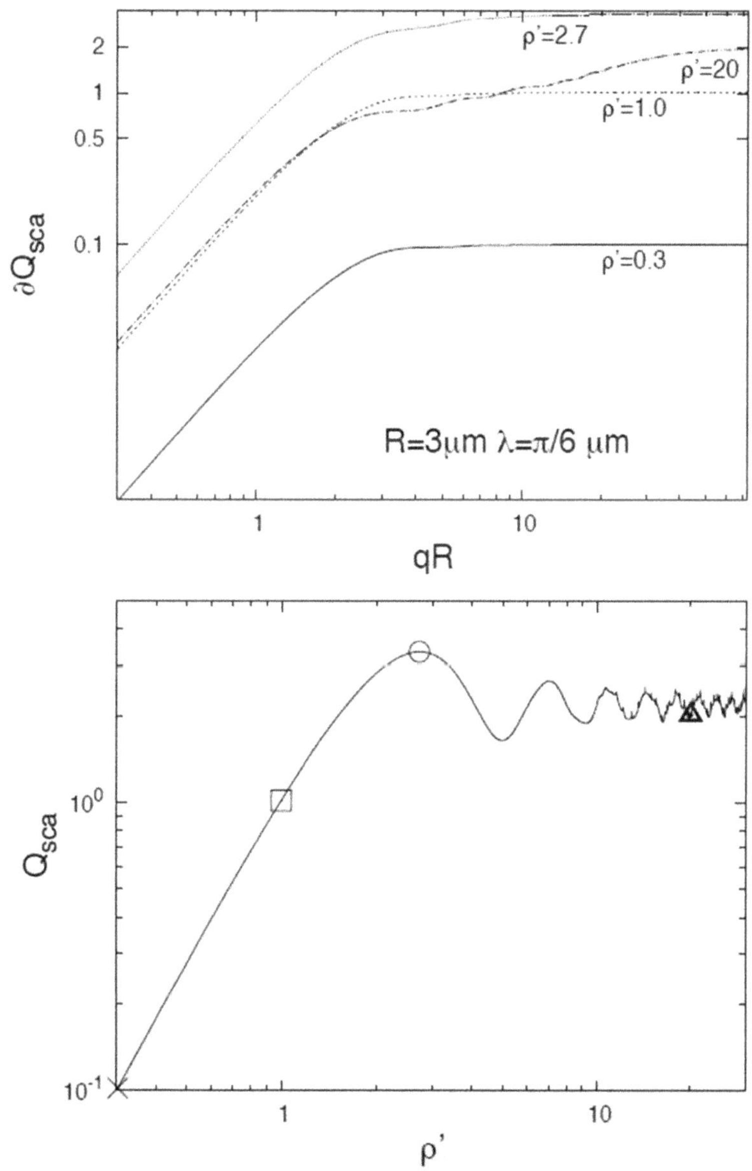

Figure 8.26. (Top) The partial scattering efficiency, $\partial Q_{\rm sca}$, versus qR for spheres with size parameter $kR = 36$ but different internal coupling parameters, ρ'. (Bottom) The total scattering efficiency for the same $kR = 36$ sphere as a function of ρ'. The particular values of ρ' for the spheres in the top are marked.

Remember that the total scattering efficiency is a quasi-universal function of ρ'. The total scattering efficiency is, of course, equal to the partial efficiency at the maximum value of qR, which is $2kR$ and equal to 72 for the particle chosen. For $\rho' \leqslant 3$ one again finds that bulk of the extinction, which for the particle chosen is entirely scattering, is due to the scattering before $qR \simeq \pi$ which corresponds to $\theta \simeq \lambda/2R$. Then the additional contribution beyond $qR \gtrsim \pi$ is miniscule, and indeed there is no hump for this range of ρ' values. From another point of view, when $\rho' \leqslant 3$, nearly all the scattering is in the forward lobe, beyond which is a rapid fall of the differential scattering cross section governed by $(qR)^{-4}$.

8.4 The extinction paradox

Consider the situation in which a beam of light is incident on a particle. How much light energy is removed from the light beam? Formally, the answer is given by the extinction cross section, which accounts for both the light scattered and absorbed by the particle. The extinction cross section is the sum of the scattering and absorption cross sections (equation (2.9))

$$C_{\text{ext}} = C_{\text{sca}} + C_{\text{abs}} \tag{8.13}$$

The extinction cross section can be both measured and calculated, the latter, for example for a sphere, with the Mie equations. Remarkably, one finds for any refractive particle (if it's big, we often call it an object) much larger than the wavelength of the light, the extinction cross section is twice the projected geometric cross section. Figure 7.34 in the previous chapter indicates this limit as Q_{ext} approaches 2 as both the phase shift parameter and the internal coupling parameter grow very large. For a sphere of radius R, the projected area is a circle of the same radius hence $G = \pi R^2$. This value of 2 for the extinction efficiency is truly remarkable because simple physical observation of a large obstacle, a baseball for example, illuminated by a parallel beam of light, shows a shadow behind the obstacle equal to the projected geometric cross section; indeed, we can define 'shadow' in the same way. The shadow tells us that the projected area removes energy from the beam so that the extinction cross section is apparently equal to the projected area, which is the geometric cross section, not twice that value. This observation also makes intuitive sense. Nevertheless, the Mie equations for a sphere, which are based on the fundamental Maxwell equations, predict the extinction cross section for a large, refractive particle is twice the geometric cross section, i.e. $C_{\text{ext}} = 2G$. How can a very large particle remove more energy from a beam than it geometrically intercepts? This counter intuitive result is called the *extinction paradox* (Born and Wolf 1975, Bohren and Huffman 1983). The extinction paradox is not limited to spheres; it applies to any shape so long as it is sufficiently large and refractive.

The paradox becomes more curious when one considers a particle small compared to the wavelength of light, but highly refractive such that ρ' (or ρ) $\gg 1$. Such an object would not have a shadow. Regardless, in this limit, the extinction cross section is twice the geometric cross section.

The canonical explanation of the extinction paradox considers only the large object (van de Hulst 1957). This explanation proposes that with a large object, there is both a geometric optics shadow and a wave optics diffraction around the edge of the object. Each contributes precisely one geometric cross section G to the extinction cross section for a total of twice the geometric cross section, $2G$. Our studies of the partial scattering cross section above, in which we found that approximately half the light is in the forward diffraction lobe and the other half is in the hump, appear to support this explanation. It is also an appealing physical explanation, but it does not account for the small, highly refractive case.

We have offered an alternative explanation (Berg *et al* 2011, Berg and Sorensen 2018) based on the Ewald–Oseen (EO) extinction theorem. This explanation with some elaboration is described below, but in preparation, first we give a brief discussion of the EO extinction theorem (Born and Wolf 1975).

8.4.1 The Ewald–Oseen extinction theorem

As Democritus said (and I paraphrase), 'nothing exists but atoms and the void' (Durant 1954). With this perspective, we can envision the situation in figure 8.27 where an incident electromagnetic wave from the vacuum enters a semi-infinite medium (e.g. a flat slab of glass). If the medium does not react to the presence of the wave, it continues to propagate in the same direction, unaltered, at the speed of light. It is, in some manner, still present in the medium because the medium is mostly 'void', a vacuum. Thus, from the EO point-of-view, this wave passes unperturbed

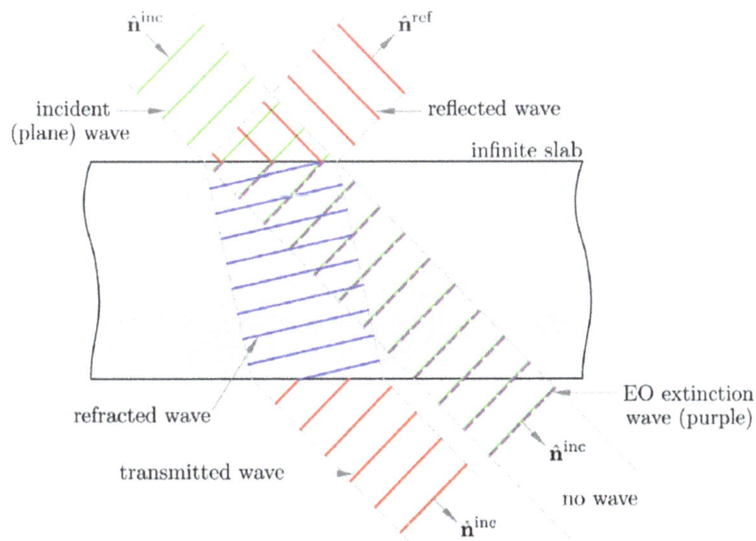

Figure 8.27. An incident wave (green) enters a semi-infinite slab of glass. To aid discussion, the incident wave is imagined to have a finite width much larger than the wavelength such that it can be thought of qualitatively as a plane wave. This wave causes polarization in the glass (grey), which radiates an EO extinction wave (dashed purple) 180° out of phase with the incident wave both inside the slab and after where both propagate out the exit side. The interior polarization wave (blue) also radiates the exterior reflected and transmitted waves, which are scattered waves (red). Subsequent reflections and refractions past the first two are not shown. Reprinted from Berg and Sorensen (2018), Copyright (2018), with permission from Elsevier.

into the glass, through it, and back out the exit side. However, the medium also contains responsive 'atoms', which for us are the electrons in the medium. These electrons feel a force due to the incident electric field, and are thereby perturbed from the equilibrium position to create a polarization wave with concomitant electron acceleration to cause re-emission of electromagnetic waves. One of these re-radiated waves propagates along with the incident wave having the same magnitude but 180° out of phase with the incident wave. This causes destructive interference exactly annihilating, extinguishing, the incident wave in the medium. This new re-radiated wave is the *EO extinction wave*. It is important to note that both the incident wave and the EO extinction wave continue to propagate together as they pass out of the slab forever extinguishing the incident wave. One might say that on the exit side of the slab, the EO wave destructively interferes with the incident wave to create a shadow. In addition, the interaction creates new waves leading to the reflected (from the interface) and refracted waves obeying the laws of reflection and refraction of geometric ray optics.

For a particle the reflected wave from the medium's interface is part of the particle's scattered wave, the medium's refracted wave is the analogue to the particle's interior wave. Subsequent transmissions of this interior wave add to the scattering in the manner of the Debye series. Our new understanding of the extinction paradox makes use of EO and these concepts.

8.4.2 Babinet's principle revisited

A useful example of the EO extinction theorem that will lead us directly to particulate scattering is Babinet's' principle. I invite you to review Babinet's principle described in section 1.5.4. Figure 8.28 shows an obstacle, a cylindrical fiber of width $2R$, whose complimentary screen is a single slit of the same width in an opaque barrier. A light beam (green) with wavelength λ and finite width $2w_0 \gg 2R$ is incident upon it. In the far field we expect, via Babinet's principle, a single slit, Fraunhofer diffraction pattern with a first minimum at an angle $\theta = \lambda/2R$. How do the incident waves diffract around the edges of the obstacle to form the single slit pattern? What is really happening in this experiment?

According to Ewald–Oseen, the incident beam is everywhere within the boundaries of the beam including inside the fiber, unaltered, as if the fiber was not there; see figure 8.28. However, the incident beam also accelerates the electrons in the fiber. Thus, the incident beam excites an oscillating electric polarization into the material of the fiber and this oscillating electric polarization radiates other, secondary electromagnetic waves. One wave, the EO extinction wave, travels along with the incident beam and has the same field magnitude but is exactly 180° out of phase with it. This causes complete destructive interference within and immediately behind the fiber, the latter creates the shadow.

Now consider the incident and extinction waves in the far field where we observe the single slit, Fraunhofer diffraction pattern. The wider incident beam diffracts less than the narrower extinction wave. Thus, if we view the diffraction pattern at any angle other than within the narrow diffraction of the incident beam (i.e. we view where

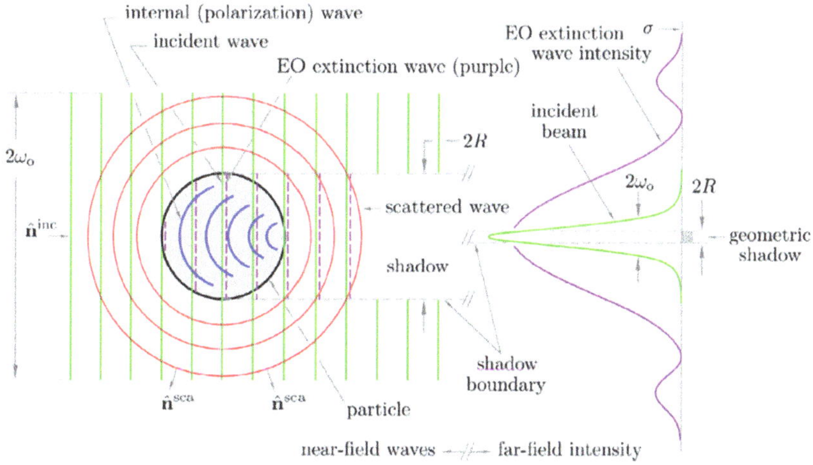

Figure 8.28. Diagram of a beam of light with width $2w_0$ propagating to the right, incident upon a long, cylindrical fiber of radius R and whose axis is perpendicular to the plane of the figure. The diagram is split with the left portion showing the particle interior and near-field zone with component *waves*, and the right portion shows the far-field zone *intensities* at a much larger scale. The component waves inside and outside of the particle are spaced by an arbitrary integer number of wavelengths and indicated by the color coding as follows: incident wave green, internal (polarization) wave blue, EO extinction wave dashed purple, and scattered wave (outside) red. Note that the shadow region shown may, or may not, constitute a complete field-free region. If the fiber is large (compared to λ) and strongly absorbing, i.e. if $\kappa kR > 1$, it will, whereas if the particle is smaller and/or partly absorbing, it will not. The intent here is to emphasize the extinction, via destructive interference, of the incident wave by the EO extinction wave inside and immediately behind the particle and to contrast that to the lack of extinction—due to the significantly different amounts of diffraction—between these same two waves in the far field. The far-field intensities on the screen σ show the classic diffraction pattern from the fiber, while the remainder of the incident beam is seen as the brighter central peak, i.e. the Drude spot (Drude 1959). Reprinted from Berg and Sorensen (2018), Copyright (2018), with permission from Elsevier.

the incident beam amplitude is essentially zero, recall the condition $E_0 = 0$ described in section 1.5.4), we will see only the extinction wave. However, the extinction wave has emanated from a slit-shaped source, so it should have a diffraction pattern consistent with the single slit, Fraunhofer diffraction pattern, which it does. Furthermore, the cylinder is narrower than the incident beam so it will diffract to wider angles where the incident beam amplitude is essentially zero. Thus, the light we observe diffracted by a fiber on a distant screen is, in fact, the extinction wave of Ewald and Oseen.

For the fiber, the situation differs from that of the slab because the fiber diameter is smaller than the width of the incident beam. Thus, as the two waves propagate into the far field, the narrower EO extinction wave will diffract (spread) more than the incident wave to the point where wave overlap, and hence destructive interference, will be minimal. On the other hand, the slab is wider than the incident beam so that the EO extinction wave and the incident wave are the same width and remain so as they diffract to the far field, forever precisely destructively interfering with each other. The end result for the fiber is the diffraction pattern of a single slit composed of the EO extinction wave on the distant screen with a bright central spot due to the incident wave.

With this explanation, *the incident light beam is not 'blocked'* by the fiber. Electromagnetic waves, unlike, e.g. water waves or sound waves, do not propagate in a material medium. Thus, there is no medium in which one can place an object to block the waves. Instead, electromagnetic waves propagate into material media unperturbed. If the material medium has electric charges, these charges will respond to create secondary waves that make it *appear* that the incident wave was diffracted or scattered. Rather than blocking the electromagnetic wave, which it cannot do, to create a shadow, the fiber radiates a wave, the extinction wave, to cancel the incident beam and create the shadow immediately behind the fiber. Furthermore, note that with this explanation, *the incident beam does not diffract around the edges* of the fiber as a water wave would do; rather the slit-shaped fiber radiates a wave, the extinction wave, into the far field where it takes on the intensity profile of single slit diffraction. This is a radically different description of diffraction than that encountered in sophomore physics. It is, however, the correct description because it is based on Maxwell's equations.

In our study of diffraction (Berg and Sorensen 2018), we also considered a screen with a hole the Babinet compliment to an obstacle, which, via Babinet's principle, should have approximately the same diffraction pattern in the far field. Then the incident wave passes unperturbed through the screen. At the screen, the EO extinction wave is created. Immediately behind the screen, the destructive interference between the incident wave, which passed unperturbed through the screen, and EO waves creates a shadow. Immediately behind the aperture, there is no EO wave, and hence, no destructive interference for the incident wave which then propagates into the far field to create the diffraction pattern. This is the same diffraction pattern as for the complimentary obstacle. Note, however, that for the screen there is no Drude spot.

To summarize, we have found that the wave that is responsible for the diffraction pattern of the aperture is the remainder of the incident wave, whereas the obstacle's diffraction pattern is due to the EO extinction wave, which is 180° out of phase with the incident wave. This phase difference appears in the simple derivation of Babinet's principle in section 1.5.4, the negative sign in equation (1.61).

8.4.3 The extinction paradox for particles

Two limiting situations are useful to begin an understanding of the extinction paradox for large, refractive particles: the perfect reflector and the perfect absorber (Berg *et al* 2011).

8.4.3.1 The perfect reflector

Recall the EO concept that the incident field is everywhere until it gets cancelled or altered by the field radiated by the polarized scatterer. When the light, or electromagnetic wave, encounters a particle, the particle becomes polarized at a frequency equal to that of the incident light and emits light itself. In the case of a large, perfect reflector the emitted light is comprised of both the wave reflected off its front, illuminated surface, and, according to Ewald–Oseen, a field that is exactly 180° out of phase with the incident light and hence cancels it, via destructive interference, to create the shadow. In each case, it is straightforward to see that the scattering cross section due to each is precisely equal to the geometric cross section of the object. Thus, the total scattering cross section is twice the geometric cross section. Since there is no absorption, this total scattering is the extinction.

8.4.3.2 The perfect absorber

Again recall the EO concept. Now rather than reflected, the light is absorbed and turned into another form of energy at the surface with a cross section equal to the geometric cross section. In addition, as for the reflected case, the light emitted from the oscillating polarization of the particle is exactly 180° out of phase with the incident light and hence cancels it to create the shadow, to yield a scattering cross section precisely equal to the geometric cross section. Thus, the extinction cross section, which is the sum of the two cross sections, equals twice the geometric cross section to create the extinction cross section.

From a macroscopic point of view wherein material objects are solids that by their physical presence can block the light, the two explanations above appear to be a sleight-of-hand. Indeed, if the perfect reflector is a non-atomic, solid object, the light reflects off its surface with a cross section of G and this reflected light is removed from the beam to create the shadow too. Then, one could argue, reflection and shadow making are the same event, so don't count it twice and the total cross section is G. However, in reality the incident beam cannot see the perfect reflector, it passes right through it, but it can excite electron oscillations, or, equivalently, an oscillating electric polarization, which radiates waves that are, in fact, the reflected wave and another wave that extinguishes the incident wave within and behind the particle. The cross section for each of these waves is G.

In each of these 'perfect' situations above, the fact that a well-defined shadow exists immediately behind the particle implies that the scattering cross section must have a component with a value precisely equal to the geometric cross section. Thus, the formation of the well-defined shadow behind the particle is a consequence of EO cancellation within the interior. This is the crucial recognition missed in the traditional explanations since the extinction is not formulated in terms of its source in the particle's internal field. Moreover, this shows that even a large, glass particle will have an extinction cross section twice the geometric cross section due to the fundamental requirement that its incident wave be canceled within. With this, there is no need to rely on the concept of diffraction around the particle's edge.

8.4.3.3 An arbitrary sphere

The partial total scattering cross section and the associated partial extinction efficiency presented in sections 8.4.1 and 8.4.3 can yield further insight to the extinction paradox for an arbitrary particle.

Figure 8.29 plots the Rayleigh normalized differential cross section and the partial scattering efficiency for a sphere with size parameter $kR = 1000$ and internal coupling parameter $\rho' = 410$. Such a particle would have a diameter of ~170 microns for green light and a refractive index $m = 1.33$. The plots are stacked on top of each other with a common abscissa with the variable qR. For such a particle Mie theory predicts a total scattering cross section of $C_{sca} = 2.02\pi R^2$, hence the scattering efficiency is 2.02.

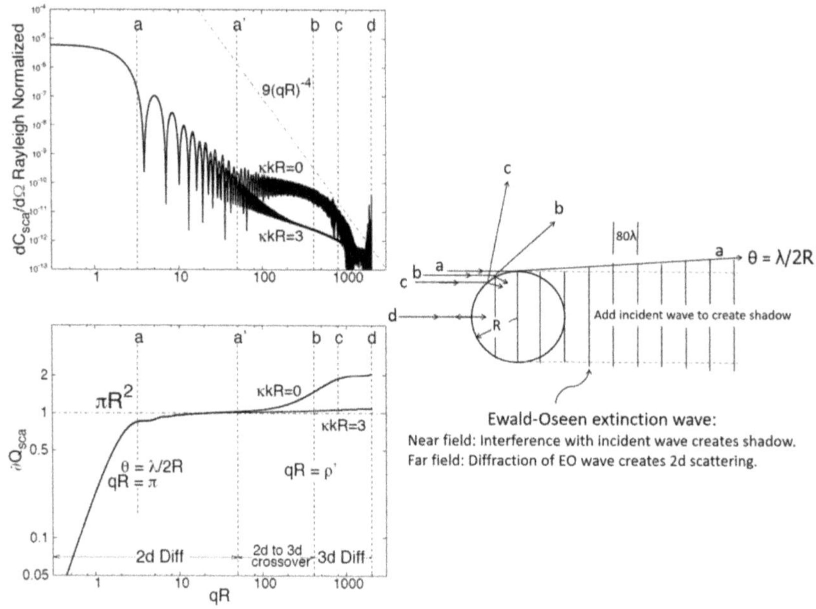

Figure 8.29. (Top left) The differential scattering cross section for a spherical particle with a size parameter of kR = 1000 and internal coupling parameter ρ' = 410 normalized by the Rayleigh differential cross section plotted versus qR for both κkR = 0 and 3. The positions of the rays in the diagram to the right are indicated. (Bottom left) The partial scattering efficiencies plotted versus qR matched with the upper plot. (Right) Diagram of the scattering with light incident (not shown) from the left. A few selected light geometric rays (a, b, c and d) are drawn. Also drawn is the EO extinction wave, which accounts for the 2d diffraction pattern, traveling to the right.

Behavior very similar to figures 8.21, 8.22 and 8.23 is seen. The third part of the figure (at the right) is a ray diagram for light incident upon the particle. Four rays (a, b, c and d) are marked on the diagram and the positions of these rays in qR space are marked on the plots to the left. Ray a is a reflected, geometric ray at the diffraction angle $\lambda/2R$, where R is the radius of the particle, and not drawn is Ray a′, which is at the end of the 2d diffraction regime, but still at very small angles. These rays are reflected in an angular range within the 2d diffraction pattern; they are not diffracted rays. They are relatively faint because the circular area from which they reflect is very thin due to the glancing angle. Other geometric rays encountering the surface are drawn, and when reflected and refracted, these rays lie beyond the 2d diffraction regime. Also included in figure 8.29 are plots for the same particle but with $\kappa kR = 3$. When κkR is this big, the hump disappears to imply that the reflected and refracted rays a, a', b, c and d are eliminated. Then, 2d diffraction dominates the differential cross section. This in turn causes the total scattering cross section to fall from ca. $2\pi R^2$ to πR^2 while the absorption cross section increases from zero to πR^2. Throughout the transition from non-absorptive to very absorptive, the extinction efficiency remains ca. $2\pi R^2$, which is the paradoxical result.

The virtue of figure 8.29 is that it explicitly demonstrates that the contributions to the total extinction cross section (or efficiency) divide into two regimes that each contributes $G = \pi R^2$ to the total scattering cross section: the forward, 2d diffraction regime and the geometric, refractive hump regime. The forward, 2d diffraction regime (marked as '2d Diff' in the lower left figure) is a scattered wave due to the oscillating polarization induced into the sphere by the incident field. In physical space, as drawn to the right, in the near field it is the EO extinction wave, which creates the shadow immediately behind sphere. Furthermore, this same extinction wave continues into the far field to create the 2d diffraction part of the particle's forward scattering. Notice that we have drawn no rays that diffract around the edge of the sphere. The hump dominates the scattering at angles greater than the 2d diffraction regime. It includes incident rays that reflect and refract to angles in the hump region, or rays that get absorbed, or some combination of reflection, refraction and absorption depending on the refractive index and the parameter κkR. These hump regime rays, regardless of the degree of absorbance, *do not create the shadow*, but they do contribute one factor of G to the total extinction. Therefore, it is incorrect to say that the object blocks the rays that are reflected, refracted and/or absorbed and thereby creates the shadow.

To continue, the concepts of getting blocked and diffracting around the edges of the blocking object go together. But Maxwell equations know nothing about fantasies such as blocked rays, shadows and diffraction around objects! Maxwell's equations show that the incident wave polarizes the particle. This polarization is oscillating so that a new wave, the EO extinction wave, forms. In the near field close to behind the particle, this EO wave cancels the incident wave to create the shadow. In the far field, this same wave forms a diffraction pattern consistent with the projected geometric cross section of the particle. This extinction wave contributes one G to the extinction. The other factor of G comes mostly from the hump, and this hump is also due to the oscillating polarization induced into the sphere. It manifests itself in the large particle limit as rays that endure some combination of reflection, refraction and absorption.

References

Abramowitz M and Stegun I A 1972 *Handbook of Mathematical Functions, with Formulas, Graphs, and Mathematical Tables* (New York: Dover)

Berg M J and Sorensen C M 2018 A review and reassessment of diffraction, scattering, and shadows in electrodynamics *J. Quant. Spectrosc. Radiat. Transf.* **210** 225–39

Berg M J, Sorensen C M and Chakrabarti A 2005 Patterns in Mie scattering: evolution when normalized by the Rayleigh cross section *Appl. Opt.* **44** 7487–93

Berg M J, Sorensen C M and Chakrabarti A 2011 A new explanation of the extinction paradox *J. Quant. Spectrosc. Radiat. Transf.* **112** 1170–81

Bohren C F and Huffman D R 1983 *Absorption and Scattering of Light by Small Particles* (New York: Wiley)

Born M and Wolf E 1975 *Principles of Optics: Electromagnetic Theory of Propagation, Interference, and Diffraction of Light* (Oxford: Pergamon)

Borovoi A G 2013 Light scattering by large particles: physical optics and the shadow-forming field *Light Scattering Reviews 8: Radiative Transfer and Light Scattering* ed A A Kokhanovsky vol 8 (Berlin: Springer-Praxis) pp 115–38

Brillouin L 1949 The scattering cross section of spheres for electromagnetic waves *J. Appl. Phys.* **20** 1110–25
Debye P 1908 Das elektromagnetische Feld um einen Zylinder und die Theorie des Regenbogens *Phys. Z.* **9** 775–78
Drude P 1959 *The Theory of Optics* (New York: Dover)
Durant W 1954 *The Story of Civilization: Part 1. Our Oriental Heritage* (New York: Simon and Schuster)
Gautam P and Sorensen C M 2021 A wide range (0.32 degrees-177.6 degrees), multi-angle light scattering setup and concomitant analysis method *Rev. Sci. Instrum.* **92** 113105
Hecht E 2002 *Optics* (Reading, MA: Addison-Wesley)
Heinson W R, Chakrabarti A and Sorensen C M 2014 Crossover from spherical particle Mie scattering to circular aperture diffraction *J. Opt. Soc. Am.* A **31** 2362–4
Maughan J B 2020 Q-space analysis of the orientationally averaged light scattering by particles of various shapes *PhD Thesis* Kansas State University, Manhattan, KS
Mishchenko M I, Travis L D and Lacis A A 2002 *Scattering, Absorption, and Emission of Light by Small Particles* (Cambridge: Cambridge University Press)
Sorensen C M and Fischbach D J 2000 Patterns in Mie scattering *Opt. Commun.* **173** 145–53
Sorensen C M, Maughan J B and Chakrabarti A 2017 The partial light scattering cross section of spherical particles *J. Opt. Soc. Am.* A **34** 681–84
van de Hulst H C 1957 *Light Scattering by Small Particles* (New York: Wiley)
Wang G, Chakrabarti A and Sorensen C M 2015 Effect of the imaginary part of the refractive index on light scattering by spheres *J. Opt. Soc. Am.* A **32** 1231–5

Chapter 9

Light scattering and absorption by fractal aggregates

Small, solid particles often stick together, for one reason or another, to form aggregates (or clusters, agglomerates[1]) and hence are common in Nature. In many cases, the aggregate structure appears rather random and thus difficult to describe and quantify[2]. However, Witten, Forest and Sanders (Forrest and Witten 1979, Witten and Sander 1981) discovered a certain class of aggregates that have a quantifiable fractal morphology. These *fractal aggregates* (more precisely fractal-like aggregates, see below) are formed when smaller particulate components of the aggregate come together via a random process and stick together. This is not an uncommon process, so fractal aggregates represent a major subset of aggregates and are found in many natural and practical situations. Because of this, we desire to know how they scatter and absorb light. Furthermore, since scattering and absorption depends on structure, we need to understand fractal aggregate structure as well as possible.

9.1 Fractals

Mandelbrot coined the term *'fractal'* to describe a *scale invariant* object (Mandelbrot 1977, Mandelbrot and Mandelbrot 1982). Scale invariance means the object appears the same regardless of the scale at which it is viewed. Said differently, the object looks the same regardless of whether you zoom in or zoom out on it. An equally useful term to describe fractals is that they are *self-similar*, which means that any part has a structure similar to any other part at any scale including the whole. Figure 9.1 shows a common mathematical example. Notice that each square has five smaller squares in it for a few iterations. If this continued *ad infinitum*, the example

[1] Usage takes an agglomerate as an assembly of smaller particles loosely joined, e.g. via van der Waals forces, whereas an aggregate is an assembly more strongly joined, e.g. via sintering.
[2] Note that small, liquid particles are usually spheres that coalesce on contact to form larger spherical particles.

doi:10.1088/978-0-7503-5157-7ch9

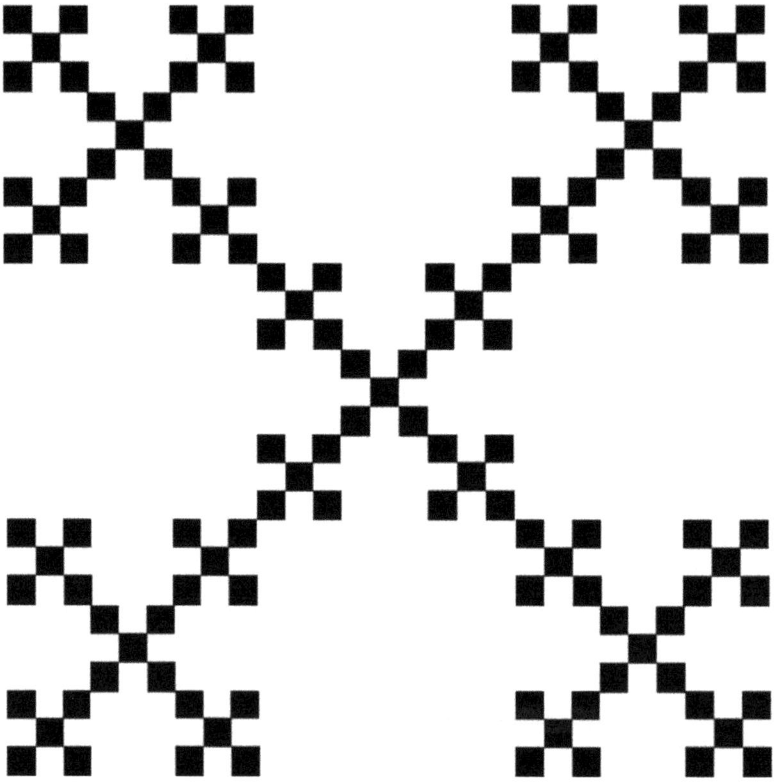

Figure 9.1. An example of a fractal-like object. Starting at the smallest scale of one dark square, whenever the linear dimension is increased by a factor of three, the mass increases by a factor of five. Thus, this object has a mass scaling dimension of $= \ln(5)/\ln(3) = 1.465\ldots$. This dimension is also called the fractal dimension. Note, however, that this object resides in two-dimensional space.

would be scale invariant, hence a fractal. Notice the self-similarity as any part has the same structure as any other. The squares in figure 9.1 iterate only three times, so it is 'fractal-like' over a finite range, as will be all the aggregates we encounter in Nature because they will be composed of finite sized monomers and have an overall limit to the aggregate size. However, I will use the term 'fractal aggregate' for these realistic objects for simplicity of expression.

The fundamental numerical parameter to describe as a scale-invariant, fractal object is its scaling dimension. A scaling dimension is the exponent of a power law that relates two variables. For example, a sphere's volume is proportional to its radius cubed. This proportionality is a power law with a scaling dimension of three. We would say that a sphere's volume scales with its linear size to a dimensionality of three. Now recognize that a sphere's surface area scales with a dimension of two with its size. Most uniform objects have volume and surface scaling dimensions of three and two, respectively. A moment's thought leads us to realize that our body mass also scales with our linear size, our height for instance, to the third power if we remember to keep our proportions the same; as we rescale our height, we must rescale our width and depth. This scaling dimension of three is what we mean when we say 'three-dimensional' object[3]. The scaling relationship is

$$\text{mass} \sim (\text{linear size})^{D} \tag{9.1}$$

where D is the *scaling dimension*. In the scaling relationship, we are careful to use the term 'linear size' because there are many ways to measure size including by mass. We also use mass because apparent volume might be ill-defined for 'fluffy' aggregates. For many geometric or human-made objects D is an integer. However, for scale invariant fractals D is a non-integer; a remarkable fact!

As an example, return to figure 9.1. The mass of the example is proportional to the number of dark squares, N. At the smallest scale is one square, a monomer, with $N = 1$. Let the side length of the monomer be unity. At the next iteration up in size, the side length is 3 and $N = 5$. The next iteration yields side length 9 and $N = 25$. The proportionality of equation (9.1) yields the same value for the scaling exponent with any of these length-N combinations, $D = \ln 5/\ln 3 = 1.465\ldots$. Thus, the scaling dimension for the example in figure 9.1 is $D = 1.465\ldots$. This is certainly a non-integer, and it is also not equal to the spatial dimension ($d = 2$) in which it is embedded. It is common to call these scaling dimensions for fractal objects '*fractal dimensions*'.

Mandelbrot pointed out that many objects in Nature have a fractal morphology. The perimeters of clouds and coastlines, the tributary systems of rivers, the shapes of mountain ranges, trees, and so on. Pull a twig off a tree, and it looks like a miniature tree, see figure 9.2, scale invariance; trees are fractals.

9.2 Fractal aggregate structure

Fractal aggregates form when solid particles randomly meet and stick together without coalescing. The initial particles are called the monomers or primary particles. If monomers move, hit and stick to a stationary aggregate one at a time, a roughly spherically symmetric aggregate forms, and the process is called the particle-cluster model. If the motion of the monomers is diffusive, the particle-cluster model is often called *diffusion limit aggregation* (DLA) (Witten and Sander 1981). If an ensemble of monomers moves and sticks when they hit, forming larger aggregates that can also move, hit and stick, a cluster–cluster model is at hand. In this case, if the motion is diffusive, the process is called *diffusion limited cluster aggregation* (DLCA) (Kolb *et al* 1983, Meakin 1983). Variants on these two main models include reaction limited situations in which the sticking probability is less than one, so-called *reaction limited cluster aggregation* (RLCA), and ballistic aggregation wherein the motion between collisions is a straight-line, so-called *ballistic limited cluster aggregation* (BLCA) (Jullien and Botet 1987). If, on the other hand, upon collision, the colliding particles can continue to move until they find the most energetically favorable position, then equilibrium is achieved with a dense, non-fractal aggregate with a scaling dimension equal to the spatial dimension. An example of a dense,

[3] This scaling dimension is different that the dimensionality of space, which allows us degrees of freedom.

Figure 9.2. A tree, an example of a scale invariant fractal object in Nature.

non-fractal, equilibrium aggregate is a crystal. During crystal growth, the atoms or molecules collide but can continue to move until they find the lowest energy position in the growing lattice. There are also intermediate situations in which dense equilibrium growth occurs over short length scales and then non-dense fractal growth over larger length scales (Skjeltorp 1987, Chakrabarti *et al* 2004). Fractal aggregates can occur in both colloidal and aerosol systems, and the resulting fractal structures are univeral, independent of the material details of the particulate system (Lin *et al* 1989). Examples are given in figure 9.3.

The scale invariance of fractal aggregates is demostrated for soot in figure 9.4. On the left, there are two micrographs of the soot. The upper picture was taken with a

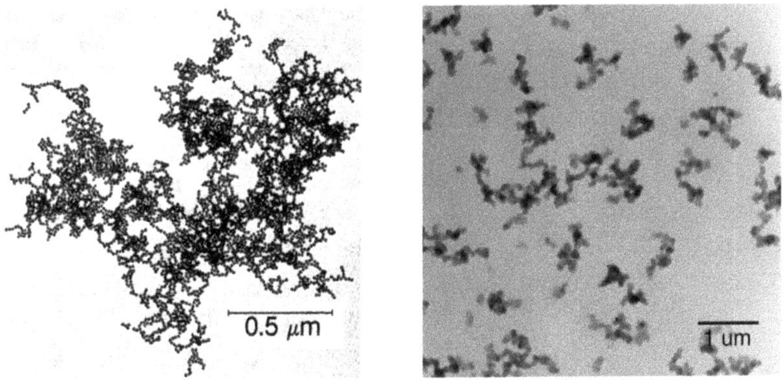

Figure 9.3. (Left) TEM image of a gold colloid aggregate. This cluster contains 4739 gold monomer particles (go ahead, count them) and has a fractal dimension of $D = 1.75$. Reprinted figure with permission from Weitz and Oliveria (1984), Copyright (1984) by the American Physical Society. (Right) TEM image of soot aggregates from an acetylene in air diffusion flame. These aggregates have a fractal dimension of $D = 1.8$.

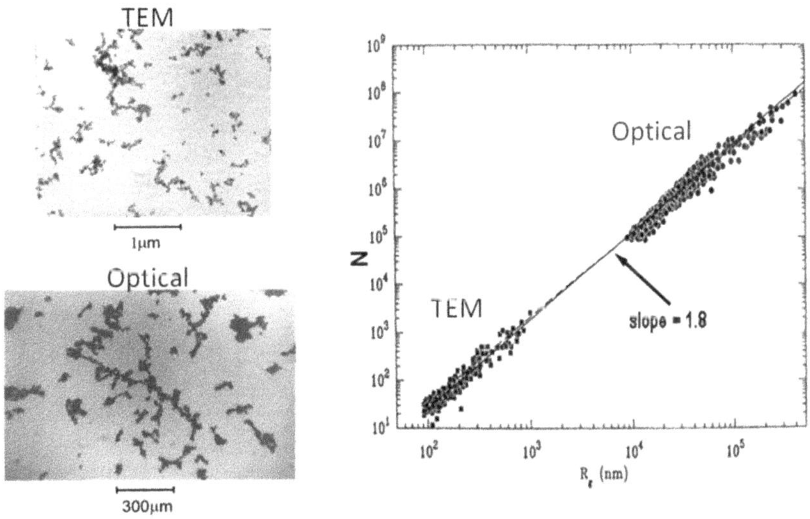

Figure 9.4. (Top left) TEM image of soot aggregates sampled low from an acetylene in air diffusion flame; (bottom left) optical microscope image of soot captured high in the same flame where larger aggregates reside the result of continuing aggregation with height in the flame. (Right) Plot of the number of monomers per aggregate N versus aggregate radius of gyration R_g for soot sampled both low and high in the flame. Note the scaling remains over three and a half orders of magnitude in linear size. Picture and data from Sorensen and Feke (1996), reprinted by permission of the publisher (Taylor & Francis Ltd, http://www.tandfonline.com).

transmission electron microscope (TEM) and the scale bar represents 1 micron. The lower picture was taken with an optical microscope and the scale bar is 300 microns, two-and-a-half orders of magnitude larger than the upper picture. Regardless of the drastically different scales, the general features of the morphology are the same in each picture; this is scale invariance. On the right side of figure 9.4, the number of monomers is plotted versus the cluster radii of gyration for both the TEM and optical images. There the straight line on this log–log plots that fits to both sets of data indicating the same morphology for the two sets of aggregates. The power law of equation (9.1) applies. The slope of this line is the fractal dimension of the ensemble of aggregates, which is $D = 1.8 \pm 0.1$.

When dealing with aggregates, the radius of gyration R_g (a root-mean-square radius, see section 4.7) is a very useful measure of linear size since the aggregate has an indefinite boundary, although other measures have been used such as longest apparent length. The radius of gyration is given by equations (4.41) and (4.42)

$$R_g^2 = \langle r^2 \rangle = \frac{\int n(r) r^2 d\mathrm{r}}{\int n(r) d\mathrm{r}} \tag{9.2}$$

Remember, the vector $\boldsymbol{r}$ is measured relative to the center of mass of the aggregate. The mass is very difficult to measure because it is so small, but to a good approximation it is proportional to the number of monomers in the aggregate, N which can be seen in microscope images. Then following relation (9.1) the *general scaling relation* for a fractal aggregate is

$$N = k_0 (R_g / a)^D \tag{9.3}$$

where k_0 is the *scaling prefactor* and a is the monomer radius. A sketch of a fractal aggregate created by a computer simulation is shown in figure 9.5.

Numerous experiments in both colloids and aerosols and computer simulations wherein the aggregates have formed by moving with Brownian, random walk motion, meet and then stick with high probability have yielded essentially the same fractal dimension in the range of $D = 1.8 \pm 0.10$. These are perhaps the most

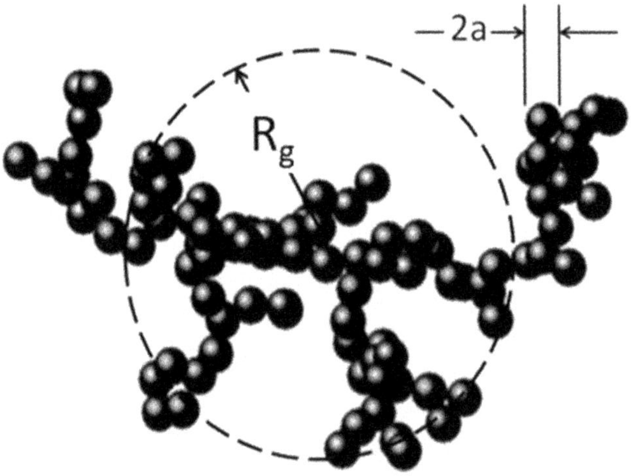

Figure 9.5. A computer generated diffusion limited cluster aggregation (DLCA) fractal aggregate of radius of gyration R_g, monomer diameter $2a$ and fractal dimension 1.8.

common conditions under which aggregates (or clusters) form. This situation is the diffusion limited cluster–cluster aggregation mentioned above; DLCA for short. Empirically, we can conclude that the fractal dimension value of 1.8 can be considered universal for this type of aggregation process.

There are very few measurements of the prefactor in the literature and those that exist are for soot. In our laboratory, two different flame soot studies using TEM yielded $(D, k_0) = (1.74, 1.23)$ (Cai *et al* 1995) and $(1.8, 1.7 \pm 0.4)$ (Sorensen and Feke 1996) with about 5% uncertainty except as indicated. In contrast, others (Koylu and Faeth 1994a, Koylu and Faeth 1994b, Koylu *et al* 1995a, 1995b) also using analysis of TEM images, measured similar fractal dimensions but k_0 values ranging from 2.0 to 2.7. Of course, soot formation and growth likely has more than spherical monomer DLCA involved and this might be some source of the discrepancy.

DLCA can be simulated on a computer to yield a facile 'explanation' for the fractal dimension observed in experiments and values for both D and k_0. With simulations, one can readily vary the embedding spatial dimension d and the conditions for the aggregation. The particle diffusion coefficient is a function of the properties of both the particle and the medium, and the functionality depends on the Knudsen number, which is the ratio of the mean free path of the medium molecules to the effective diameter of the particle. When the Knudsen number is small, the Stokes–Einstein diffusion coefficient applies; when large, the Epstein diffusion coefficient applies; and when intermediate, the Cunningham corrected Stokes–Einstein coefficient applies (see Oh and Sorensen 1997, Seinfeld and Pandis 2006). Simulations that determined both D and k_0 (Oh and Sorensen 1997, Sorensen and Roberts 1997, Lattuada *et al* 2003, Heinson *et al* 2010, 2012) yield within error the same fractal dimension of $D = 1.78 \pm 0.04$ with a prefactor of $k_0 = 1.30 \pm 0.10$ for a spatial dimension of $d = 3$. Moreover, simulations that delve into the Epstein regime yield very similar results regardless of the Knudsen number (Pierce *et al* 2006). Finally, a simple analytical theory, the restricted hierarchy model, for the morphology of DLCA aggregates yields $D = 1.815$ and $k_0 = 1.3$ (Sorensen and Oh 1998, Heinson *et al* 2015). See appendix C.

Our lab has shown that if DLCA is allowed to proceed until the aggregates occupy an appreciable fraction of the space, a situation which we have called the cluster-dense limit, superaggregates occur (Sorensen *et al* 2003, Kim *et al* 2004, 2006, Dhaubhadel *et al* 2006). These superaggregates have a hybrid structure with a fractal dimension of $D = 2.5$ at large length scales but are composed of smaller DLCA aggregates with $D = 1.8$. When the smaller DLCA aggregates are treated as monomers, the prefactor is $k_0 = 0.6$. As described in Heinson *et al* (2017), these values are the same as those obtained for a 'static' percolation cluster (Stauffer and Aharony 1994). Super-aggregation can lead to aerosol gels (Dhaubhadel *et al* 2007, 2012, Sorensen and Chakrabarti 2011) in which the entire aerosol forms a percolated network, 'frozen smoke'. The light scattering properties of superaggregates are described in detail below in section 9.5.

It is important to quote both the fractal dimension and the prefactor when an ensemble of aggregates is analyzed via equation (9.3) because they are anticorrelated in the fit of N versus R_g. This is illustrated in figure 9.6 where the open circles

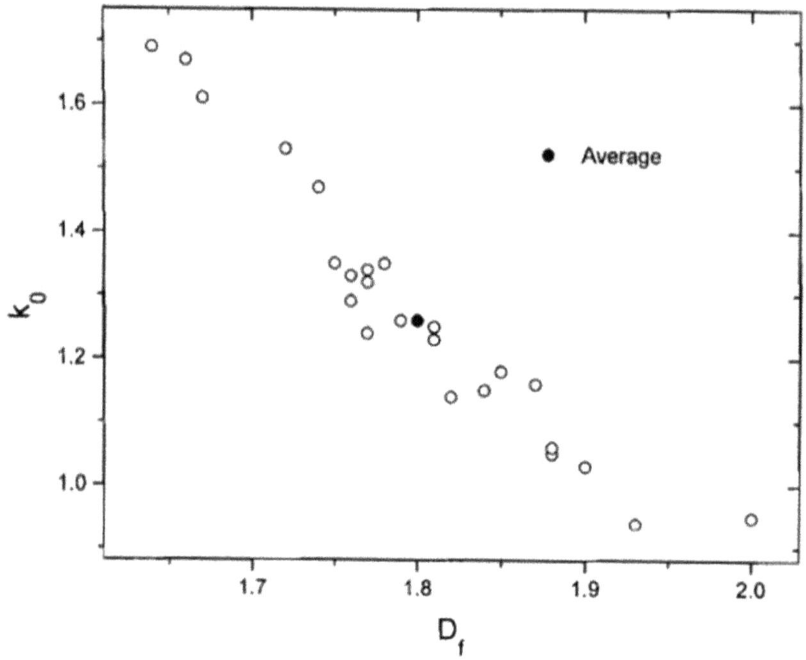

Figure 9.6. An example of the anticorrelation of the fractal dimension D and the prefactor k_0 that occurs when fitting simulation data to equation (9.3). The open circles are fractal dimension/prefactor, (D, k_0), ordered pairs determined from individual, repetitive DLCA simulation runs fitting to equation (9.3) (including all shapes). Reprinted from Sorensen and Roberts (1997), Copyright (1997), with permission from Elsevier.

represent (D, k_0) ordered pairs from a series of DLCA simulation runs all operating under the same conditions. The anticorrelation is readily apparent.

A detailed description of DLCA fractal aggregate morphology is necessary to determine their scattering and absorption properties. Such a description can be obtained from a bottom up approach to yield the monomer pair correlation function (Heinson *et al* 2012). To start, consider the number of spherical monomers of radius a that can be packed into a volume of radius r in a D dimensional space

$$N(r) = \varphi(r/a)^D \tag{9.4}$$

Equation (9.4) defines the monomer packing fraction φ. With this foundation, one can determine the monomer pair correlation function to be

$$G(r) = \varphi\frac{D}{4\pi a^D}r^{D-3}\exp\left[-(r/\xi)^\gamma\right] \tag{9.5}$$

The correlation function of equation (9.5) has three parts related to the three aspects of the structure:

(1) The coefficient, which involves φ, a and D with φ describing the structure at the monomer level;

(2) The spatial power law decay involving D, which is the fractal part of the structure between the monomer and aggregate length scales; and
(3) The stretched exponential cutoff of the power law decay with stretching exponent γ describing the boundary of the aggregate with size specified by ξ.

Beyond the sizes of the monomer and the aggregate, there are three parameters, φ, D and γ involved in $G(r)$. Since the correlation function completely describes the morphology of the aggregate, these three parameters provide a necessary and sufficient description of the aggregate morphology.

The correlation function of equation (9.5) has spherical symmetry due to the assumptions under which it was derived. Since individual aggregates lack spherical symmetry, equation (9.5) represents an orientationally averaged aggregate or ensemble of aggregates with random orientations. The stretching exponent γ is a measure of how sharply the pair correlation cuts off at the edge of the aggregate; an infinite γ represents a sharp boundary, finite γ implies a diffuse boundary, more diffuse with decreasing γ. Real aggregates have both diffuse boundaries and non-spherical shape. Realize that a non-spherical object with a sharp boundary, a rugby ball for example, when rotationally averaged would appear to have a diffuse boundary hence have a cutoff function with a finite γ. Thus, the stretching exponent γ for a rotationally averaged aggregate has both an intrinsic source due to the diffuse nature of the surface and a source due to its non-spherical shape.

The shape of an aggregate can be described in at least two different ways. One way is to find the principal radii of gyration by diagonalizing the rotational inertia tensor (Fry *et al* 2004)

$$T = \sum_{i=1}^{N} \begin{pmatrix} y_i^2 + z_i^2 & -x_i y_i & -x_i z_i \\ -x_i y_i & x_i^2 z_i^2 & -y_i z_i \\ -x_i z_i & -y_i z_i & x_i^2 y_i^2 \end{pmatrix} \xrightarrow{\text{diagonalization}} T = N \begin{pmatrix} R_1^2 & 0 & 0 \\ 0 & R_2^2 & 0 \\ 0 & 0 & R_3^2 \end{pmatrix} \tag{9.6}$$

The principal radii of gyration $R_1 \geqslant R_2 \geqslant R_3$ are related to the radius of gyration by $R_g^2 = (1/2)(R_1^2 + R_2^2 + R_3^2)$. One can define an aggregate anisotropy as the ratio

$$A_{13} = (R_1/R_3)^2 \tag{9.7}$$

If $A_{13} = 1$, the aggregate has spherical symmetry, and if $A_{13} > 1$, it does not, and the symmetry decreases with increasing A_{13}. That is the talent of A_{13}. However, A_{13} does a poor job of indicating the shape. For example, consider a rectangular solid with dimensions 1 by 2 by 4. Note that the middle dimension is the geometric mean of the large and small dimensions; the dimensions are in a geometric proportion. On the other hand, the ratios of the principal radii of this rectangular solid are $R_1/R_2 = 1.085$, $R_2/R_3 = 1.84$, and $R_1/R_3 = 2.00$, which is not a geometric progression. This indicates that the magnitudes of the principal radii are useful but poor descriptors of shape.

The other method to describe anisotropy is to enclose the aggregate with the smallest circumscribing rectangular solid (Heinson *et al* 2015). This method has the advantage of preserving the correct proportions between the three sides of a rectangular solid (obviously). It also proves essential in confirming a simple analytical theory for DLCA fractal aggregates structure that shows that DLCA aggregate shape is described by the Divine proportion, in $d = 2$, and its generalization in $d = 3$ (Heinson *et al* 2015). See appendix appendix C.

Simulation work (Heinson *et al* 2012) indicates a packing fraction of $\varphi = 0.68 \pm 0.02$ for the DLCA. The simulations yielded a distribution of shapes hence a distribution of stretching exponents, the average of which was $\langle\gamma\rangle = 2.02$. This value is consistent with previous simulation studies and empirical light scattering measurements (Sorensen *et al* 1992); the so-called Gaussian cutoff. If $\gamma = 2$, one can show that $\xi^2 = (4/D)R_g^2$. Furthermore, the anisotropy A_{13} was found to be directly related to and interchangeable with γ. Intuitive sense of this can be made by rotationally averaging a long object to yield spherical symmetry with a very diffuse boundary. The fractal dimension was found to be $D = 1.8$ as many times before, but interestingly, independent of shape; a result consistent with Heinson *et al* (2010). Thus, long stringy and short stocky nearly spherical aggregates have the same D provided that they were formed under the same conditions of spatial dimension, aggregate motion and sticking probability. This is perhaps counterintuitive, but understandable when we keep in mind that the fractal dimension only describes how the mass scales with the linear size and this scaling can be the same regardless of shape. Now recall the general scaling relation for a fractal aggregate, equation (9.3). For a constant N if the aggregate gets stringier, R_g will increase, so k_0 must decrease. This is in fact true as illustrated in figure 9.7. Thus, whereas D is not a function of shape, k_0 is. It was shown (Heinson *et al* 2012) that while k_0 appears as a fundamental property in the scaling relation, equation (9.3), it is actually dependent on the more fundamental parameters φ, D and γ via

$$k_0 = \left[\frac{2\Gamma(D/\gamma)}{\Gamma[(D+2)/\gamma]}\right]^{D/2} \frac{\varphi D \Gamma(D/\gamma)}{\gamma} \tag{9.8}$$

where $\Gamma(x)$ is the Gamma function.

9.3 Fractal aggregate structure factor

The structure factor describes diffraction from an object. Diffraction is the refractive index $m \to 1$ limit of light scattering. What follows is a derivation of the structure factor for fractal aggregates. If we assume or can show that light scattering by fractal aggregates is in this limit, then what is to follow below is often called the Rayleigh–Debye–Gans fractal aggregate (RDG-FA) theory for light scattering by fractal aggregates.

The fractal aggregate structure factor can be obtained by Fourier transforming the pair correlation function $G(r)$ of equation (9.5) following equation (4.29) without the additive factor of 1 and normalized by N. $G(r)$ is spherically symmetric so the angular integrations of the Fourier transform simplify analogous to the steps that led equation (4.34) to equation (4.35)

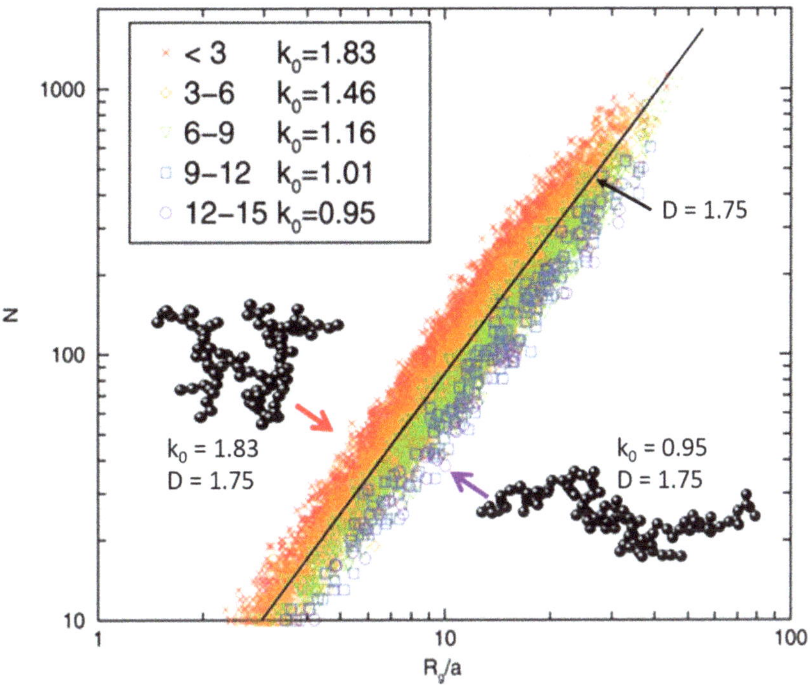

Figure 9.7. Shape dependent prefactors for DLCA fractal aggregates. Number of monomers, N, versus reduced cluster size, R_g/a, for aggregates of different shape as indicated by the aggregate anisotropy A_{13}. The legend gives A_{13} ranges and the prefactor k_0 value for that range. Note that results for each A_{13} range run parallel to the black guideline with slope $D = 1.75$, the fractal dimension. Hence, the fractal dimension is independent of shape. Also shown are typical aggregates for two different anisotropies. Adapted from Heinson *et al* (2010), reprinted by permission of the publisher (Taylor & Francis Ltd, http://www.tandfonline.com).

$$S(q) = 4\pi \int G(r)\left[\frac{\sin qr}{qr}\right]r^2 dr \tag{9.9}$$

Note that equation (9.9) contains the scattering wave vector

$$q = (4\pi/\lambda)\sin\left(\frac{\theta}{2}\right) \tag{9.10}$$

The integrand of equation (9.9) can be viewed from two perspectives: small q and large q (the details are described in Heinson *et al* 2012). Recall from section 4.7 that the expansion of the $\sin qr$ term for small qr led to the Guinier formula and an expression for the radius of gyration, R_g, equations (4.39)–(4.42). Application of appendix equation (D.1) to the correlation function equation (9.5) yields an expression relating R_g and ξ

$$R_g^2 = \xi^2\left[\frac{\Gamma((D+2)/\gamma)}{2\Gamma(D/\gamma)}\right] \tag{9.11}$$

The large q part of the transformation deals directly with the power law decay, the 'fractal part', of the correlation function. This power law in space Fourier transforms into a power law in reciprocal space, q-space. Given these two parts, we write

$$S(0) = 1 \tag{9.12a}$$

$$S(q) = 1 - \frac{1}{3}\left(qR_g\right)^2 \quad \text{for} \quad q < R_g^{-1} \tag{9.12b}$$

$$= C\left(qR_g\right)^{-D} \quad \text{for} \quad R_g^{-1} < q < a^{-1} \tag{9.12c}$$

where C is a coefficient to be determined below.

Equation (9.12c) cannot apply for *all* $qR_g > 1$, because a real aggregate has a small length scale limit to its fractal nature and that is the monomer size a. Remember that the correlation function in equation (9.5) is the 'monomer-monomer pair correlation function', and as such does not contend with length scales less than the monomer size a. For length scales less than the monomer size a, the cluster structure factor is multiplied by the monomer structure factor that is often called the form factor and designated as either $f(q)$ or $p(q)$. For example, if the monomers are spheres, the form factor is the Rayleigh–Debye–Gans (RDG) formula for a sphere, equation (4.47).

The full fractal part of the aggregate structure factor is obtained from the spherically symmetric Fourier transform in equation (9.9). The result is (9.12c) with the coefficient

$$C = \left[\frac{\Gamma(D-1)}{\Gamma(D/\gamma)}\right]\left[\frac{\Gamma((D+2)/\gamma)}{2\Gamma(D/\gamma)}\right]^{D/2} \gamma \sin\left[\frac{\pi}{2}(D-1)\right] \tag{9.13}$$

In my 2001 review on fractal aggregate light scattering (Sorensen 2001), this structure factor and this coefficient were also derived for the correlation function (9.5) and displayed with a few other attempts to describe scattering from that era. When $\gamma = 1$, the cutoff of the correlation function is exponential, and the result is the exponential cut off structure factor. Substitution of $\gamma = 1$ into equations (9.11) and (9.13) yields results identical to those given in table 2 of that review. When $\gamma = 2$, the cutoff of the correlation function is Gaussian, and the result is the Gaussian cut off structure factor. Substitution of $\gamma = 2$ into equation (9.11) again yields results identical to that given in table 2 of that review. However, the coefficient C *appears* different. Fortunately, substitution of $\gamma = 2$ into the alternate equation (9.13) and application of Euler's inversion formula and the duplication formula for gamma functions (Abramowitz and Stegun 1972) shows that the two expressions are, in fact, equivalent. Note, however, equation (9.13) is more general because it is a function of γ as well as D.

In the 2001 review, I described reasons to believe that the Gaussian cut off structure factor worked best for single sized (monodisperse) DLCA aggregates. This conclusion has been substantiated through the intervening years; most recently in the three parameter work that has been the main source for our discussion so far that concluded $\langle\gamma\rangle = 2.02$ for a DLCA derived ensemble of aggregates (Heinson *et al* 2012). Other groups have performed aggregation simulations to find a Gaussian cutoff for the spatial correlation function (Nicolai *et al* 1994, Haw *et al* 1995, Lattuada *et al* 2003). Experimental evidence for the Gaussian cutoff is solid but not abundant. Our lab noticed a 'hump' in the structure factor near the Guinier regime measured optically for soot in a premixed flame (Sorensen *et al* 1992). This hump strongly favors the Gaussian over the exponential cutoff, as a close examination of figures 9.8 and 9.9, both below, will show. We also used TEM analysis of soot from a flame to measure the spatial correlation function directly and found a Gaussian cutoff (Cai *et al* 1995). Fractal aggregate mobility data also imply a Gaussian cutoff (Sorensen 2011a).

Proceeding now with some assurance that the monomer pair correlation function for a DLCA aggregates has a Gaussian cutoff, the Fourier transform of equation (9.9) applied to the correlation function of equation (9.5) yields (table 2 of Sorensen 2001)

$$S(qR_g) = \exp\left[-\left((qR_g)^2/D\right)\right]F_{1,1}\left[\frac{3-D}{2}, \frac{3}{2}; -\frac{(qR_g)^2}{D}\right] \tag{9.14}$$

where $F_{1,1}$ is the Kummer or hypergeometric function. This function is plotted in figure 9.8. Note the dashed line in the figure indicates a power law described by equation (9.12c) with slope $-D = -1.8$ and coefficient $C = 1.0$. More will be said of this below.

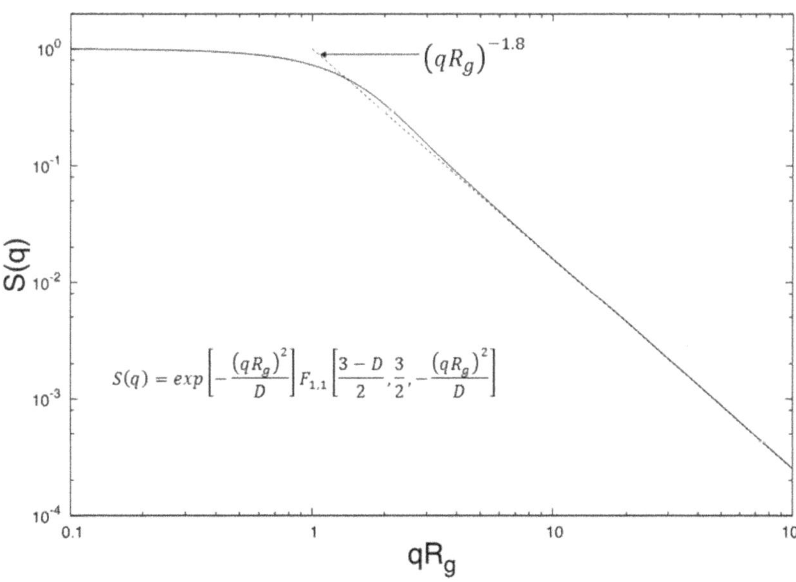

Figure 9.8. Plot of equation (9.14), the structure factor for DLCA aggregates that have a Gaussian cutoff monomer pair correlation function. The dashed line indicates the power law of equation (9.12c) with $D = 1.8$ and $C = 1.0$. Note the small hump in the region near $qR_g \approx 2$.

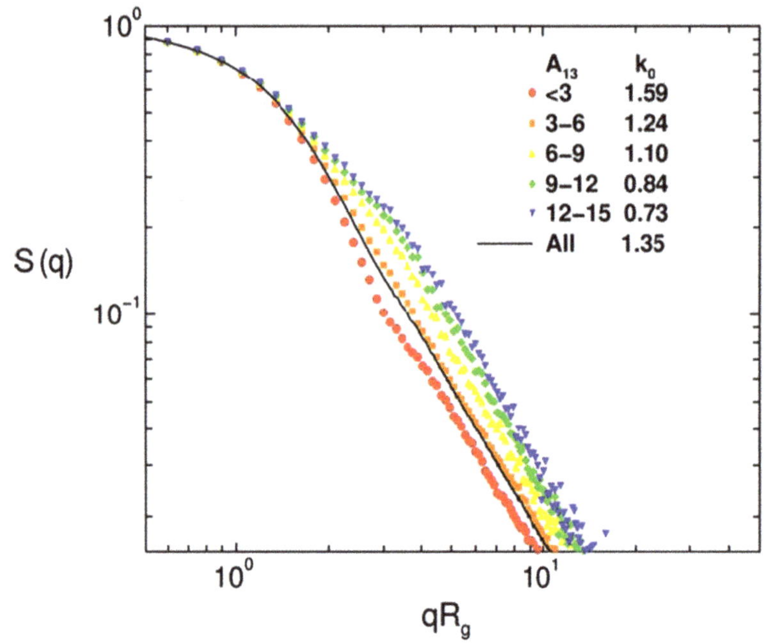

Figure 9.9. Structure factors for DLCA fractal aggregates (also studied in figure 9.7). Aggregates were binned according to anisotropy A_{13} as indicated by points. The structure factor for all the aggregates, hence the true structure factor for the DLCA simulation is indicated by the black line. Reprinted from Heinson *et al* (2012), Copyright (2012), with permission from Elsevier.

Figure 9.9 shows the structure factor obtained from a DLCA simulation in which the aggregates were sorted into bins and then their collective structure factor was obtained directly by Fourier transforming each aggregate, squaring that, and then averaging across the bin. The bins were determined by the aggregate anisotropy A_{13}. There we see that the Guinier regime and the slope of the large qR_g power law, which is equal to $-D$, remain unaffected by the changes in shape. On the other hand, as the anisotropy A_{13} increases, hence the stretching exponent γ deceases, the level, as determined by the coefficient C, of the large qR_g power law decreases. These simulation results are consistent with the predictions, equations (9.12) and (9.13), above. Notice that as the aggregates become less anisotropic, hence more globular, the structure factor developes a small hump in the Guinier regime. Finally, notice in the limit of small anisotropy A_{13}, which corresponds to large γ, a kink occurs in $S(q)$ near $qR_g \simeq 4$. This is the beginning of a ripple structure due to a sharply defined surface. Recall that the first minimum of the structure factor for a uniform sphere, equation (4.49) is at $qR = 4.49$ which is equivalent to $qR_g = 3.48$.

The reciprocal space, figure 9.9, complements the real space, figure 9.7. Each one retains the power law with the fractal dimension D independent of the shape, but each one shows that the coefficient of the power law, C or k_0, respectively, changes with shape. This is predicted by the formalism above by combining equations (9.8) and (9.13) to find

$$C = D\Gamma(D - 1)\sin[\pi(D - 1)/2]\varphi k_0^{-1} \tag{9.15}$$

For DLCA, $D = 1.8$ and $\varphi = 0.68$, then equation (9.12c) becomes

$$S(q) = \frac{1.35}{k_0}\left(qR_g\right)^{-D} \text{ for } R_g^{-1} < q < a^{-1} \tag{9.16}$$

This equation ties together figures 9.7 and 9.9. Reciprocal spaces: one with $+D$ power law, the other with $-D$, and their coefficients k_0 and C inversely related. How else could it be? We conclude that for the ensemble of DLCA aggregates with all shapes produced by the simulation, the coefficients of the real space and inverse space power laws are related via $Ck_0 = 1.35$, and the averages for the entire ensemble are $C = 1.0$ and $k_0 = 1.35$. And because the simulation is an attempt to imitate nature, we expect the same coefficient in nature as well.

The exponential cutoff was actively used in the early days of fractal aggregate structure factors. There are a number of reasons for this including the simplicity of the cutoff, the fact that it fit the data well, and that it was similar to the, at that time well-known, Fisher–Burford structure factor (Dietler *et al* 1986) used to describe scattering from fluids near their critical points. The Fisher–Burford structure factor is

$$S(q) = \left(1 + 2q^2R_g^2/3D\right)^{-D/2} \tag{9.17}$$

After dealing with the equations above, this structure factor has appealing simplicity. Moreover at $D = 2$, it yields $C = 3$ for the coefficient of the power law, equation (9.12*c*), identical to the C coefficients predicted by equations (9.13) and (9.14) for $D = 2$ and $\gamma = 1$. The trouble is, is that it is wrong. This was well demonstrated in the 2001 review. Fisher–Burford fits the data well only if one ignores the polydispersity of the aggregates (perhaps why it worked so well for critical fluids). The better course of action is to use the proper monodisperse structure factor and account for polydispersity directly (Sorensen *et al* 1992) as will be described below in section 9.4.3.

One enigma to be overcome is how the power law exponent $-D$ for the fractal structure factor crosses over as $D \to d$, the spatial dimension, so that when $D = d$ the power law exponent becomes $-(d + 1) = -4$ and not $-d = -3$. The answer, simply said, is that 'it's in the math' (you're welcome). Instead of following the math explicitly, we demonstrate the crossover as $D \to d$ with two graphs; one for $\gamma = 1$ (exponential, figure 9.10, left side) and one for $\gamma = 2$ (Gaussian, figure 9.10, right side) cutoffs. One can see that it is not until $D > 2.9$ that the power laws begin to leave the $-D$ values and precede to the limit of -4 for larger D. Notice for the Gaussian cutoff the exponential functionality $\exp[-(qR_g)^2/3]$ in equation (9.14) with $D = 3$ becomes more prominent as $D \to d = 3$. We can conclude that for these two cutoffs and for $D < 2.9$ the asymptotic $S(q) \sim q^{-D}$ behavior begins immediately after the Guinier regime. Finally, we recall that the scaling analysis, equation (5.10), predicts the exponent to be $-(2D_m - D_s)$ so that as $D_m \to d$, $D_s \to d - 1$.

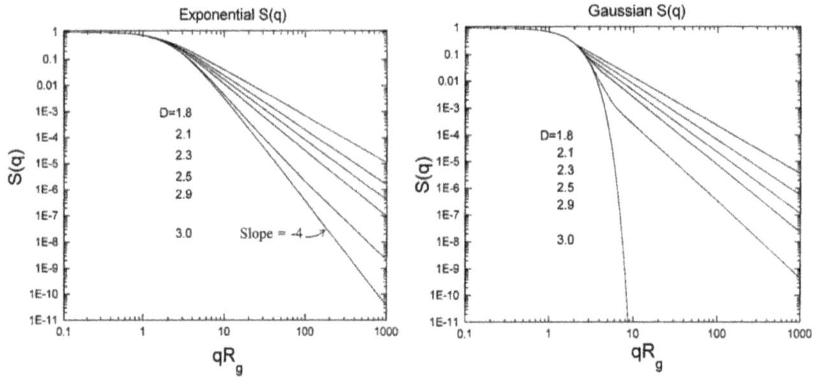

Figure 9.10. (Left) Structure factors for an exponential cut-off ($\gamma = 1$), left, and a Gaussian cut-off ($\gamma = 2$), right, for the pair correlation function, equation (9.5), and a variety of fractal dimensions D. Slopes at large qR_g are $-D$, except when $D = 3.0$, and then the slope is -4 for the exponential cut-off, whereas for Gaussian cut-off the functionality becomes $\exp[-(qR_g)^2/3]$. From Sorensen (2001), reprinted by permission of the publisher (Taylor & Francis Ltd, http://www.tandfonline.com).

9.4 Light scattering and absorption by fractal aggregates

9.4.1 The Rayleigh–Debye–Gans approximation for fractal aggregates

The structure factor of an object is revealed by scattering waves that interact only weakly with it. For example, x-rays that, as Superman knows, pass readily through matter, are used to determine the structure factor of condensed matter systems. Ordinary matter's refractive index for x-rays is very close to unity. In such a situation, the incident waves penetrate nearly uninhibited into the object. With small but finite probability, the waves interact with and scatter, at wave vector q, from each small subvolume of the object and then the scattered waves continue uninhibited out of the object to the detector. At the detector, the scattered waves sum together, keeping track of their relative phases, to form the q-component of the Fourier transform of the object. The detector squares this sum to yield the structure factor of the object, $S(q)$.

This simple, heuristic description of scattering that leads to a structure factor has a key characteristic that we have emphasized throughout this book: each subvolume of the object sees only the incident wave. Corollaries follow: a given subvolume is not affected by scattering from another subvolume; the internal coupling (measured by the internal coupling parameter ρ') is small; there is no internal multiple scattering. This is, as we have described in many of the chapters above, the Rayleigh–Debye–Gans (RDG) limit. And that is diffraction.

For fractal aggregates under the RDG approximation, because each monomer sees only the incident wave, the scattering at a given q (or angle θ) is proportional to the monomer differential scattering cross section. It is usually assumed that the monomers are small so that this cross section is scattering angle independent when the polarization is perpendicular to the scattering plane. Aggregate scattering in the forward direction, defined by $q < R_g^{-1}$, will see the scattered waves from all N

monomers add together in phase to yield a scattered intensity proportional to N^2. At non-forward directions, the relative phases of the monomer scattered waves must be accounted for and that leads to the aggregate structure factor, $S(q)$. Thus, the RDG differential scattering cross section of light with polarization perpendicular to the scattering plane for an aggregate is

$$\frac{dC_{\text{sca}}^{\text{agg}}}{d\Omega} = N^2 \frac{dC_{\text{sca}}^{m}}{d\Omega} S(q) \tag{9.18}$$

where $dC_{\text{sca}}^{m}/d\Omega$ is the monomer differential scattering cross section. Recall that $S(0) = 1$. Equation (9.18) is consistent with equation (7.1). If the monomers are in the Rayleigh regime,

$$\frac{dC_{\text{sca}}^{m}}{d\Omega} = k^4 R^6 F(m) \tag{9.19}$$

Recall $F(m)$ is the square of the Lorentz–Lorenz function equation (6.6).

The total scattering cross section is obtained by integrating the differential cross section over the entire solid angle. This is dependent upon the incident polarization so proper attention must be paid to that. We assume incident light linearly polarized perpendicular to the scattering plane. In the RDG limit the result is analogous to Rayleigh scattering described in section 6.3. For fractal aggregates the integration is messy so that in the past various approximations have been made. One good approximation (Dobbins and Megaridis 1991) leads to

$$C_{\text{sca}}^{\text{agg}} = N^2 C_{\text{sca}}^{m} G(kR_g) \tag{9.20}$$

where, for monomers in the Rayleigh regime,

$$C_{\text{sca}}^{m} = (8\pi/3) k^4 R^6 F(m) \tag{9.21}$$

and

$$G(kR_g) = \left(1 + \frac{4}{3D} k^2 R_g^2\right)^{D/2} \tag{9.22}$$

For absorption under the RDG approximation, each of the N monomers absorbs the incident light independent of all the other monomers in the aggregate (no internal coupling). Thus, the total absorption cross section is simply N times the monomer absorption cross section

$$C_{\text{abs}}^{\text{agg}} = N C_{\text{abs}}^{m} \tag{9.23}$$

If the monomers are in the Rayleigh regime,

$$C_{\text{abs}}^{m} = 4\pi k R^3 E(m) \tag{9.24}$$

Recall $E(m)$ is the Imaginary part of the Lorentz–Lorenz function, equation (6.20).

Table 9.1 gives a summary of all the cross sections for a fractal aggregate.

Table 9.1. Summary of DLCA fractal aggregates light scattering and absorption under the RDG approximation. The monomers are spheres scattering in the Rayleigh regime with radius a and refractive index m. The incident polarization is perpendicular to the scattering plane.

Differential scattering:	monomer	$\frac{dC^{m}_{sca}}{d\Omega} = k^4 a^6 F(m)$		(9.19)
	aggregate	$\frac{dC^{agg}_{sca}}{d\Omega} = N^2 \frac{dC^{m}_{sca}}{d\Omega} S(q)$		(9.18)
where		$F(m) = \left\lvert\frac{m^2-1}{m^2+2}\right\rvert^2$		(9.25)
The structure factor		$S(0) = 1$	$qR_g \ll 1$	(9.12a)
		$= 1 - \frac{q^2 R_g^2}{3}$	$qR_g \lesssim 1$	(9.12b)
		$= C(qR_g)^{-D}$	$qR_g > 1$	(9.12c)
where C = 1.0				
Total scattering:		$C^{m}_{sca} = \frac{8}{3}\pi k^4 a^6 F(m)$		(9.21)
		$C^{agg}_{sca} = N^2 C^{m}_{sca} G(kR_g),$		(9.20)
where, for a fractal aggregate		$G(kR_g) = \left(1 + \frac{4}{3D} k^2 R_g^2\right)^{-D/2}$		(9.22)
Absorption:	monomer	$C^{m}_{abs} = 4\pi k a^3 E(m)$		(9.24)
	aggregate	$C^{agg}_{abs} = N\, C^{m}_{abs}$		(9.23)
where		$E(m) = Im\left[\frac{m^2-1}{m^2+2}\right]$		(9.26)

9.4.2 Light scattering by an ensemble of monodisperse aggregates

Table 9.1 presents equations describing light scattering by a single fractal aggregate. However, experiments very often deal with a system, an ensemble, of aggregates, and so we address that situation now. To start, we next consider an ensemble of same sized aggregates, a monodisperse ensemble.

The many equations in table 9.1 are, perhaps, daunting, but they can be tamed somewhat by the realization that we rarely make absolute measurements of the scattered intensity. Thus, typically one need only be concerned with the scattered intensity proportionality. If the number of aggregates (clusters) in the scattering volume is N_c, this proportionality goes as

$$I(q) \sim N_c N^2 (a^6/\lambda^4) S(q) \tag{9.27}$$

Don't fail to notice that equation (9.27) is written under the assumption that all of the aggregates have the same size, as designated by N, and that all of the aggregates are composed of the same sized monomers, as designated by a. Now recognize that because $S(q = 0) = 1$, equation (9.12a), one can simply write

$$I(q) = I(0)S(q) \tag{9.28}$$

If the system conserves the number of monomers per unit volume, i.e. if the total number of monomers, N_m, in the system is constant, then $N_m = N_c N$ and from the coefficient of equation (9.27)

$$I(0) \sim N_m N \tag{9.29}$$

Equation (9.29) is the Tyndall effect (see section 6.4.3) for aggregates, not necessarily fractal, showing that the forward scattering increases as the aggregates grow in size as measured by the number of monomers per aggregate, N. This applies to many aggregating systems. For any q, however, $I(q) = NN_m\, S(q)$. When $q > R_g^{-1}$, $S(q)$ takes the form of equation (9.12c). Recalling the fundamental scaling relation for fractal aggregates, equation (8.3), we obtain with the addition of equations (9.28) and (9.29)

$$I(q) \sim N_m C k_0 (qa)^{-D};\, qR_g > 1 \tag{9.30}$$

This relation shows that the scattering in the power law region does not change during aggregation. Thus, there is a nuance for the Tyndall effect not found for Rayleigh scatterers because Rayleigh scatterers have no power law regime. Also, note that via equation (9.16) $Ck_0 = 1.35$ for DLCA aggregates. Figure 9.11 gives a

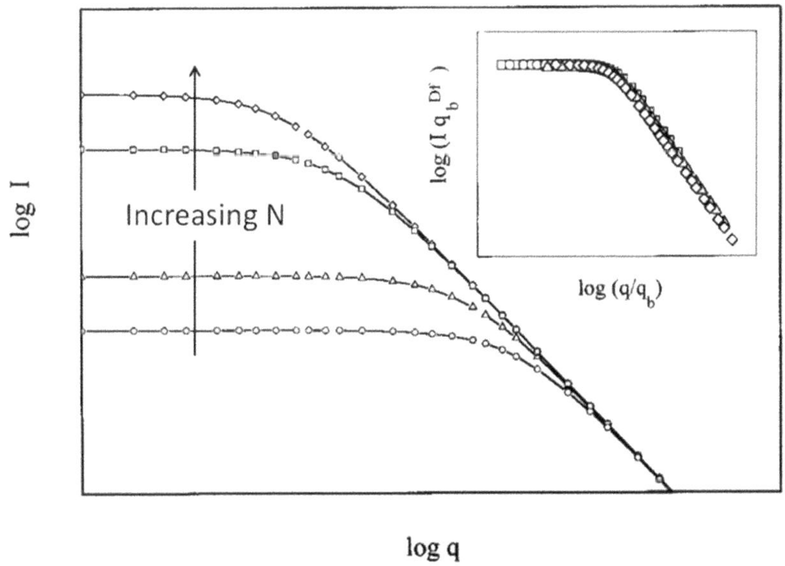

Figure 9.11. The Tyndall effect for an ensemble of fractal aggregates. The scattered intensity versus the wave vector q for an aggregating system at various times after the onset of aggregation as the number of monomer per aggregate N increases. The inset shows single curve scaling of these intensities, where q_b is the value of q at the bend in $S(q)$, which is the Guinier regime, and D is the cluster fractal dimension. Adapted from Sorensen (2001), reprinted by permission of the publisher (Taylor & Francis Ltd, http://www.tandfonline.com).

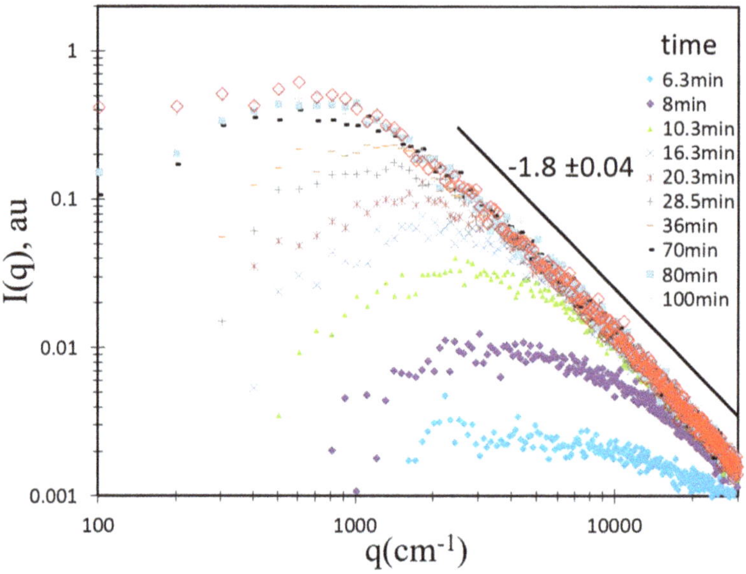

Figure 9.12. Scattered light intensity from an aggregating colloid of 40 nm diameter polystyrene nanospheres in water versus the scattering wave vector q; an example of the Tyndall effect for an aggregating ensemble. Shown are plots at times after the colloid was destabilized and aggregation began. The aggregation was DLCA to yield aggregates with a fractal dimension of $D = 1.8$. The decrease in scattering a smallest q is due to cluster–cluster anticorrelation as explained in section 5.5.2.

graphical description of the Tyndall effect for fractal aggregates. One can show that the Tyndall effect holds for a polydisperse system of aggregates as well. This is demonstrated in figure 9.12, which shows the scattered light intensity from an aggregating colloid of 40 nm diameter polystyrene nanospheres in water.

9.4.3 Light scattering by an ensemble of polydisperse aggregates

Aggregates are a result of aggregation, which gives a finite width to the aggregate size distribution. Hence, any real experiment detecting scattered light from an ensemble of fractal aggregates will involve a polydisperse ensemble. Here, we discuss the effects of that polydispersity on the functional characteristics of the scattered light. Much of this section follows my 2001 review (Sorensen 2001).

Polydispersity causes the observed shape of the functional behavior of the scattered light intensity, $I(q)$ versus q, to be different than that of the structure factor of any single cluster in the distribution (Nicolai *et al* 1994, Sorensen and Wang 1999). The single cluster structure factor, dependent on the cutoff function and D, was described above, and table 9.1, with equations (9.18) and (9.19), gives the light scattered intensity for a single sized aggregate in the RDG limit. Note that there are two ways to specify size, the number of monomers in the aggregated, N, and the radius of gyration, R_g. These are connected via equation (9.3). For aggregates it is common to use N as the size variable and $n(N)$ as the size distribution function. The distribution $n(N)$ is a number density with dimensions of inverse volume, and $n(N)dN$ is the number of aggregates per unit volume in the size range N to $N + dN$. With all this, we write the forward normalized scattered intensity from an ensemble of aggregates as

$$I(q)/I(0) = \int_1^\infty S[qR_g(N)]N^2n(N)dN / \int_1^\infty N^2n(N)dN \tag{9.31}$$

Equation (9.31) contains moments of the size distribution. The ith moment of the size distribution is defined as

$$M_i = \int_1^\infty N^i n(N)dN \tag{9.32}$$

The size distribution is normalized such that the zeroth moment, $i = 0$, is the total number of aggregates (clusters) N_c in the ensemble

$$N_c = M_0 = \int_1^\infty n(N)dN \tag{9.33}$$

Then the first moment, $i = 1$, is the total number of monomers N_m in the ensemble

$$N_m = M_1 = \int_1^\infty Nn(N)dN \tag{9.34}$$

With equation (9.3) and equations (9.12), equation (9.31) becomes

$$I(q)/I(0) = 1, \ qR_{g,z} \ll 1 \tag{9.35a}$$

$$I(q)/I(0) = 1 - (q^2/3)\, R_{g,z}^2 \ qR_{g,z} \leqslant 1 \tag{9.35b}$$

$$I(q)/I(0) = k_0(M_1/M_2)C(qa)^{-D}, \ qR_{g,z} \gg 1 \tag{9.35c}$$

Equation (9.35b) contains the so-called *z-averaged radius of gyration*, $R_{g,z}$, a term which applies whenever the size distribution used to calculate the average is weighted by the square of the mass, which for aggregates is N^2. The explicit form for $R_{g,z}$ is

$$R_{g,z}^2 = \int_1^\infty R_g^2(N)N^2n(N)dN / \int_1^\infty N^2n(N)dN \tag{9.36}$$

Substitution of $R^2_g = a^2k_0^{-2/D}N^{2/D}$, from equation (9.3), and application of equation (9.32) yields

$$R_{g,z}^2 = a^2k_0^{-2/D}M_{2+2/D}/M_2 \tag{9.37}$$

Note that equation (9.37) shows that $R_{g,z}$ is strongly weighted by the large size end of the distribution because it depends on high order moments, e.g. ~ M_3/M_2 when D ~ 2.

The question now arises, what is the size distribution? There are certainly lots of options to respond to this question, but the process of aggregation leads to self-preserving, i.e. scaling, size distributions (Friedlander and Wang 1966, Wang and Friedlander 1967, van Dongen and Ernst 1985), and that fact mostly answers the question. However, the answer comes with the caveat that the aggregates have only grown via aggregation and have not seen any other process that could modify the size distribution.

A self-preserving distribution does not change shape as the system aggregates. Thus, its parameters, such as most probable size and distribution width, can be re-scaled to dimensionless variables; hence the term 'scaling distribution'. The scaling distribution for an aggregating system is (van Dongen and Ernst 1985)

$$n(N) = Ax^{-\tau}e^{-\alpha x} \tag{9.38}$$

where x is the relative size

$$x = N/s \tag{9.39}$$

and s is a mean size. The exponent τ is a measure of the width of the distribution with large τ implying a broad distribution. For example, $\tau = 0$ for the DLCA process and $\tau = 1.5$ for the RLCA process (van Dongen and Ernst 1985, Ball *et al* 1987). Equation (9.38) is valid for the large size end of the distribution when $x > 1$, the small x form being different. Since, as we have seen, light scattering strongly weights large sizes, the small x part of the distribution can be ignored.

Other forms for the size distribution exist. One popular form is the zeroth order logarithmic distribution (ZOLD) see e.g. (Kerker 1969) a form of log normal distributions that uses a parameter σ to describe the geometric width of the distribution. The relation $(2 - \tau)/(1 - \tau) = \exp[\ln^2 \sigma_N]$ holds, to some degree of accuracy for $\tau < 1$ and large x, where σ_N is the geometric width in N-space. However, caution must be exercised in the use of distributions other that equation (9.38) for systems that have experienced only aggregation. For example, we have shown (Sorensen *et al* 1992) that the log-normal distribution yields erroneous values for distribution moments higher than the second when compared to the exact scaling distribution. Since scattering involves higher moments, such as $M_{2+2/D} \simeq M_3$ for $D \simeq 2$, see equation (9.37), this would lead to errors.

Equation (9.35b) suggests that a plot of its inverse[4],

$$I(0)/I(q) = 1 + (q^2/3)\, R_{g,z}^2 \quad qR_{g,z} \leqslant 1 \tag{9.35d}$$

$I(0)/I(q)$ versus q^2 would be linear so long as $qR_{g,z} \leqslant 1$, hence $I(0)/I(q) \leqslant 4/3$, and the slope of the plot would be $R_{g,z}^2/3$. This is the classic Guinier analysis to determine particle size. An example is shown in figure 9.13.

For the most part the data used in figure 9.13 fit the criterion of $qR_g \lesssim 1$ hence $I(0)/I(q) \lesssim 4/3$. Through the years, however, we have seen that the Guinier analysis of real data often yields linear slopes well beyond these limits. Figure 9.14 gives an example. We have shown (Sorensen *et al* 1995) that the curvature, or lack of it, of these Guinier plots is related to the fractal dimension D, the density correlation function cutoff exponent γ, equation (9.5), and the size distribution exponent τ. This

[4] We use the binomial theorem result that $1 - x \approx 1 + x$ when $x < 1$. We use $x \approx qR_g \approx qR_{g,z}$ and given all the approximations, use these two forms interchangeably here.

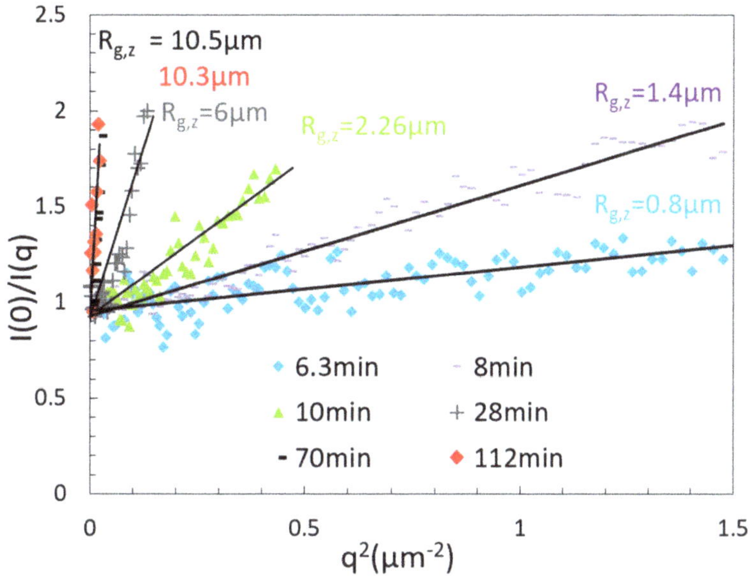

Figure 9.13. Inverse normalized scattered light intensity from an aggregating colloid of 40 nm diameter polystyrene nanospheres in water versus the square of the scattering wave vector; an example of Guinier analysis. These data are from figure 9.12. Shown are plots at six different times after the colloid was destabilized and aggregation began. These plots are labeled with the z-averaged radius of gyration, $R_{g,z}$, inferred from equation (9.35d). The aggregation was DLCA to yield aggregates with a fractal dimension of $D = 1.8$. Growth ceased near 70 min when the system gelled.

is demonstrated with theoretical calculations in figure 9.15 for a system for which $D = 1.75$ and $\gamma = 2.0$, the Gaussian cutoff. There we see that once beyond the true Guinier regime, especially when $I(0)/I(q) \gtrsim 3$, all the lines for different τ and a monodisperse system separate. This affords the opportunity to measure τ given D and γ from other measurements or assumption. The data included in figure 9.15 for methane soot indicate $\tau \simeq 0.25 \pm 0.3$ for the soot ensemble. This figure also implies that given the realistic fact that most systems are not monodisperse using the Guinier formula beyond its mathematically determined bounds will very likely cause no significant error in particulate size measurement.

A size distribution also affects the power law regime and this takes us to consider equation (9.35c) in detail. The functionality on monomer radius a in equation (9.35c) can be useful when a is known, as we will see. However, the monomer size is typically small and not detectable within the q-range of light scattering. On the other hand, the aggregate z-averaged radius of gyration is readily measured, as we have just seen. The experimentalist in the laboratory will typically measure an uncalibrated $I(q)$ and normalize it with $I(0)$ and then use Guinier analysis to determine $R_{g,z}$. Having $R_{g,z}$ and not a and k_0, leads us to solve equation (9.37) for $k_o a^{-D}$ and then substitute on $k_o a^{-D}$ into equation (9.35c) to find

$$I(q)/I(0) = (M_1/M_2)(M_{2+2/D}/M_2)^{D/2} C\left(qR_{g,z}\right)^{-D}, \quad qR_q \gg 1 \tag{9.40}$$

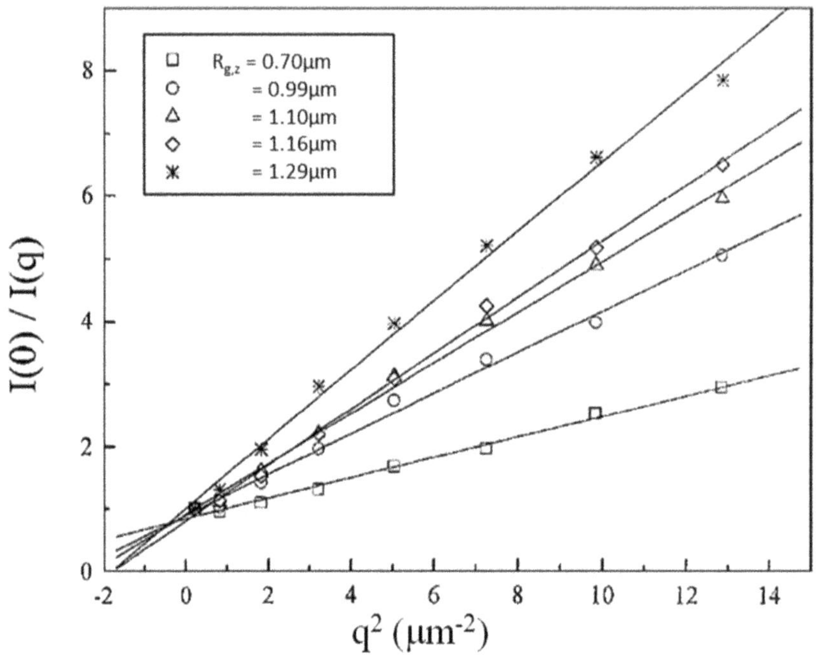

Figure 9.14. Inverse normalized scattered light intensity from an aggregating aerosol of 40 nm mean diameter TiO_2 quasi-spherical particles (the monomers) in air at room conditions versus the square of the scattering wave vector; an example of Guinier analysis. Shown are plots at five different times during the aggregation of the aerosol. The z-averaged radius of gyration, $R_{g,z}$, inferred from equation (9.35d), is given in the inset. The aggregation was DLCA to yield aggregates with a fractal dimension of D = 1.7. Adapted from Sorensen (2001), reprinted by permission of the publisher (Taylor & Francis Ltd, http://www.tandfonline.com).

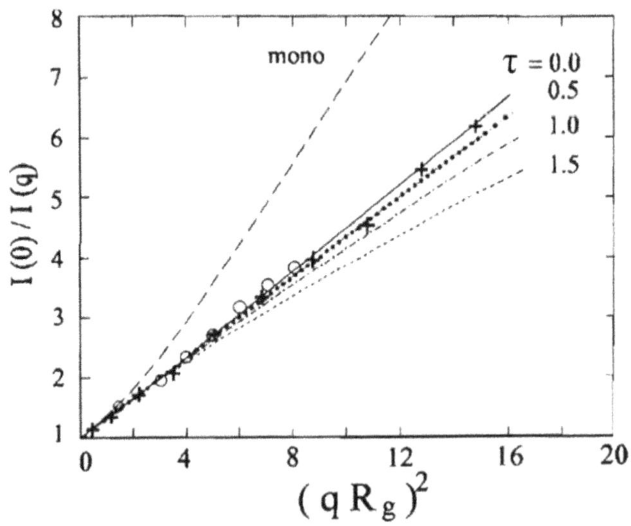

Figure 9.15. Inverse normalized scattered intensity versus $(qR_g)^2$. Lines are calculated using equation (9.31) for a Gaussian structure factor (γ = 2, equation (9.5)) with D = 1.75 convoluted with various size distributions: dashed line monodisperse, and the scaling distribution, equation (9.38), with τ = 0.0, 0.5, 1.0, and 1.5 as labeled. Data are from a soot aerosol in a premixed methane/oxygen flame. Adapted from Sorensen *et al* (1995), Copyright (1995), with permission from Elsevier.

Equation (9.40) shows that the coefficient of the power law is modified by the polydispersity of the ensemble. This alerts us to the fact that if this modifying factor is significantly different from unity, the use of single cluster structure factors on scattering data could yield erroneous results. The result also opens an opportunity to measure the polydispersity of the ensemble.

We have defined the polydispersity factor C_p in equation (9.40) as (Sorensen and Wang 1999)

$$C_p = (M_1/M_2)(M_{2+2/D}/M_2)^{D/2} \tag{9.41}$$

Then equation (9.40) becomes

$$I(q)/I(0) = C_p C\left(qR_{g,z}\right)^{-D}, \quad qR_q \gg 1 \tag{9.42}$$

The polydispersity factor can be calculated with any known size distribution. Here we are dealing with aggregates formed via aggregation so we will again use the scaling distribution, equations (9.38) and (9.39). With these, we find the polydispersity factor to be

$$C_p = \frac{1}{2 - \tau}\left[\frac{\Gamma(3 - \tau + 2/D)}{\Gamma(3 - \tau)}\right]^{D/2} \tag{9.43}$$

where $\Gamma(x)$ is the gamma function. C_p is plotted as a function of the width parameter τ for a variety of fractal dimensions D in figure 9.16. There we see that, in general, C_p

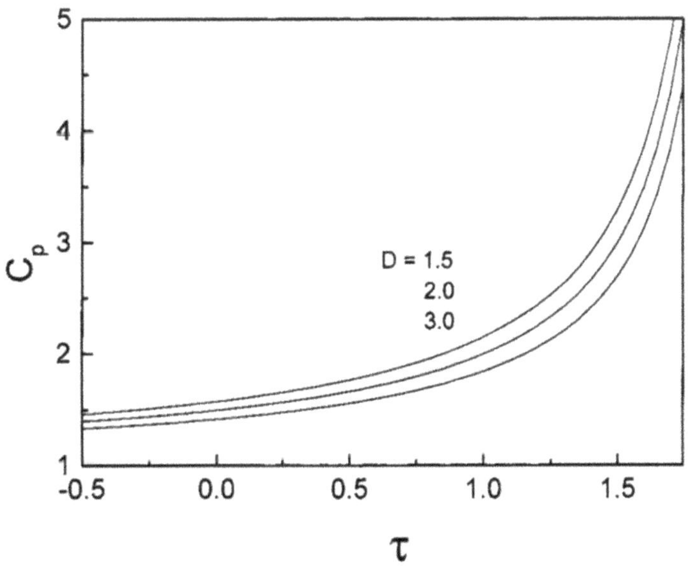

Figure 9.16. The polydispersity factor C_p, equation (9.43), which is a coefficient of the power law regime for light scattering from an ensemble of fractal aggregates, equation (9.42), versus the polydispersity exponent τ for various fractal dimensions D. Adapted figure with permission from Sorensen and Wang (1999), Copyright (1999) by the American Physical Society.

is significantly greater than unity. It increases with τ, thus with the width of the distribution. It is not strongly dependent on the fractal dimension, D. In particular, for DLCA in the continuum regime where $\tau = 0$ and $D = 1.78$, $C_p = 1.53$. For RLCA in the continuum regime $\tau = 1.5$ and $D = 2.15$, $C_p = 2.9$. Furthermore, for aerosols, which aggregate via DLCA, τ varies with the Knudsen number, the ratio of medium molecule mean free path to particle radius, and hence C_p would vary as well.

We applied this method to determine the τ exponent for two aggregating systems using light scattering: a titania aerosol aggregating via DLCA kinetics and a polystyrene colloid aggregating via RLCA kinetics (Sorensen and Wang 1999). Fractal dimensions of 1.7 and 2.15 were found for these two systems, respectively, as expected (Jullien and Botet 1987). Figure 9.17 plots in $(qR_{g,z})^D I(q)/I(0)$ versus $qR_{g,z}$. By equation (9.42) such a log–log plot will level off when $qR_{g,z} > 1$ to the value CC_p. From our considerations of the single cluster structure factor we expect $C = 1.0$; thus the level in figure 9.17 is C_p. Our analysis of several sols found $\tau = 0.24 \pm 0.23$ for the DLCA aerosol and $\tau = 1.35 \pm 0.2$ for the RLCA colloid, values consistent with previous work, see Sorensen and Wang (1999).

We conclude with yet another reproduction from the 2001 review. Figure 9.18 gives an overview of scattering by an ensemble of fractal aggregates.

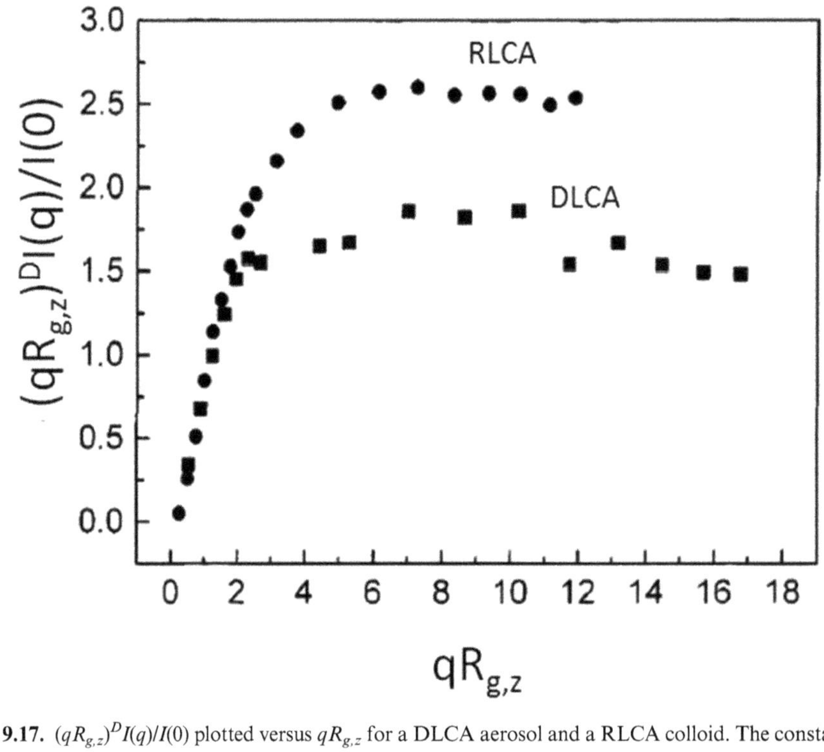

Figure 9.17. $(qR_{g,z})^D I(q)/I(0)$ plotted versus $qR_{g,z}$ for a DLCA aerosol and a RLCA colloid. The constant level at large $qR_{g,z}$ is, by equation (9.42), equal to C_pC. Adapted figure with permission from Sorensen and Wang (1999), Copyright (1999) by the American Physical Society.

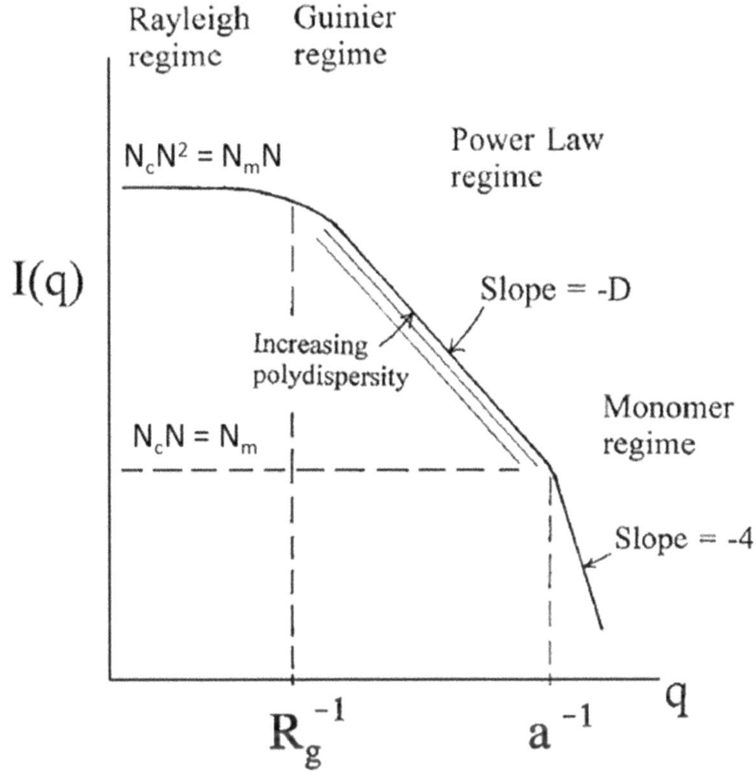

Figure 9.18. Schematic plot of the scattered light intensity *I*(*q*) from an ensemble of fractal aggregates versus *q* on a log–log plot. There are N_m total monomers of radius *a* in the ensemble. There are *N* monomers per aggregate and there is a total of N_c aggregates (clusters). Hence, $N_m = N_cN$. The fractal dimension of the aggregates is *D*. Adapted from Sorensen (2001), reprinted by permission of the publisher (Taylor & Francis Ltd, http://www.tandfonline.com).

9.4.4 The internal coupling parameter for fractal aggregates

From our studies of spheres (and other shapes, see chapter 10) we expect scattering to evolve away from the RDG limit with increasing internal coupling as parameterized by the internal coupling parameter ρ'. In section 10.2.2, the general expression for ρ' for any shape will be derived as (equation (10.8))

$$\rho' = 2\pi k \frac{V}{A_{\text{proj}}} |\alpha(m)| \tag{9.44}$$

In (9.44) *V* is the volume of the particle, $\alpha(m)$ is the average volume polarizability, which is dependent on the both the refractive index *m* and the shape of the particle, and A_{proj} is the orientationally averaged projected area.

We propose for an aggregate with *N* spherical monomers of radius *a*

$$V = N\left(\frac{4\pi}{3}\right)a^3 \tag{9.45}$$

and

$$A_{\text{proj}} = N^{0.92}\pi a^2 \text{ for large } N \tag{9.46}$$

Equation (9.46) is an empirical result found in many situations from TEM observations to mobility considerations for DLCA aggregates with $D \approx 1.8$ (Sorensen 2011b). To have a physical sense for this, realize that if the exponent of N was not 0.92, but instead 1.0, the projected area of the aggregate would be the simple sum of the projected areas of all the monomers, assumed to be spherical, in the aggregate. The fact that the exponent is less than 1.0 is due to monomer-monomer screening (occultation) during the projection. Substitution of (9.45) and (9.46) into (9.44) leads to an expression for the internal coupling parameter for scattering by a DLCA aggregate (the exponent 0.92 is specific to DLCA aggregates)

$$\rho'_{\text{agg}} = \rho'_{\text{mon}} N^{0.08} = \rho'_{\text{mon}} k_0^{0.08} (R_g/a)^{0.14} \tag{9.47}$$

Equation (9.47) indicates that the internal coupling for aggregates is directly related to the internal coupling parameter for the monomers. This latter parameter is typically small due to the monomer size. Moreover, the dependence on aggregate size is increasing but very weak. It also suggests, surprisingly, that the value of the prefactor is not very important.

The argument made in my 2001 review based on Maxwell–Garnett averaging of the monomer refractive index led to a phase shift parameter, ρ, with a $R_g^{D-2} = R_g^{-0.2}$ dependence. Here, we now use the same argument applied to the internal coupling parameter, ρ'. The Maxwell–Garnett theory predicts that the effective index of refraction is related to the refractive index of the dispersed phase and its volume fraction f by

$$\left(\frac{m^2 - 1}{m^2 + 2}\right)_{\text{eff}} = f\left(\frac{m^2 - 1}{m^2 + 2}\right) \tag{9.48}$$

The volume fraction of monomers can be approximated as

$$f = Na^3/R_g^3 \tag{9.49}$$

The internal coupling parameter for an aggregate can be written as

$$\rho'_{\text{agg}} = 2kR_g\left(\frac{m^2 - 1}{m^2 + 2}\right)_{\text{eff}} \tag{9.50}$$

Substitution of equations (9.48) and (9.49) into (9.50) yields (with $D = 1.8$)

$$\begin{aligned}\rho'_{\text{agg}} &= \rho'_{\text{mon}} k_0 (R_g/a)^{D-2} = \rho'_{\text{mon}} k_0 (R_g/a)^{-0.2} \\ &= \rho'_{\text{mon}} k_0^{2/D} N^{(D-2)/2} = \rho'_{\text{mon}} k_0^{1.1} N^{-0.10}\end{aligned} \tag{9.51}$$

This Maxwell–Garnett formulation shows a decreasing internal coupling with increasing aggregate size, although the dependence is weak. It also suggests that the prefactor is important. Both these conclusions are contrary to the formulation that led to equation (9.47).

To summarize, the similarity between these two formulations is that both depend linearly on the monomer internal coupling, and the functionality on aggregate size, while contrary, is weak. Equation (9.47) implies the internal coupling will increase with increasing aggregate size, while equation (9.51) implies the internal coupling will decrease with increasing size. The functionality on the prefactor, k_0, however, is quite different and could be tested. I know of no such tests.

A qualitative explanation for the weak (and upon average of both derivations, lack of) size dependence and linear dependence on the monomer internal coupling of the internal coupling for DLCA fractal aggregates might reside with the fact that $D < 2$. Then projection of the three-dimensional aggregate onto a two-dimensional plane can be achieved with very little monomer-monomer occultation. In fact, for $D = 1.8$ only about 8% of the monomers are occulted upon projection, as indicated by equation (9.46) and the exponent 0.92. When $D > 2$, however, plenty of projectional occultation occurs. From this perspective when $D < 2$, the cluster is transparent, whereas when $D > 2$, it is opaque. When $D < 2$, the aggregates are open and porous. The monomers are seen and do not couple well with each other.

9.4.5 Experimental examples of scattering by fractal aggregates

Figure 9.12, above, gives an example of scattering from an experiment in our lab watching a colloid aggregate with figure 9.13 a Guinier analysis of the data. Figure 9.19 contains

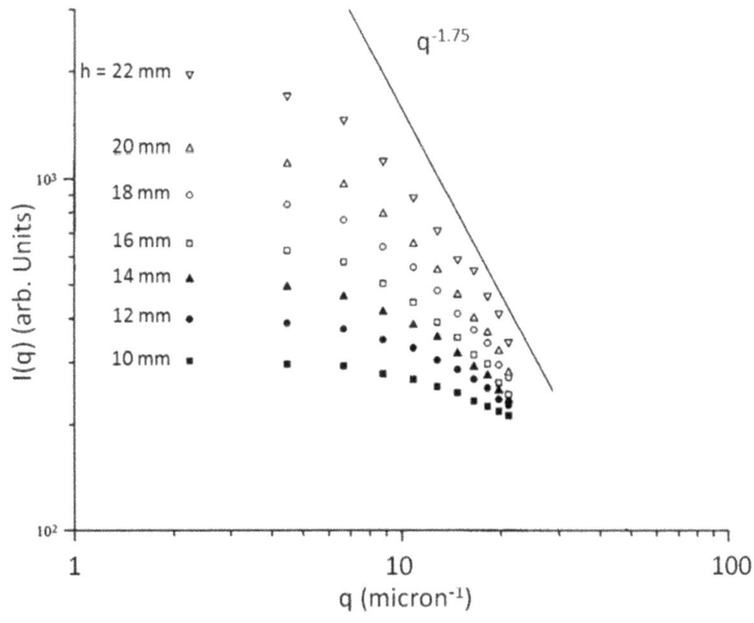

Figure 9.19. Scattered light intensity versus q for in-flame soot as a function of height above the burner surface, h. The flame was a methane/oxygen premixed flame with a carbon/oxygen ratio of 0.75. Reprinted figure with permission from Sorensen *et al* (1998), Copyright (1998) by the American Physical Society.

some data from our lab showing light scattering from soot in a methane/oxygen premixed flame (Sorensen *et al* 1998). Soot is typically composed of roughly spherical monomers with diameters of approximately 30 nm. Values for the refractive index of soot have a significant history with a range of values including the popular $m = 1.57 + i0.56$ (Dalzell and Sarofim 1969) to a more current and equally popular value of $m = 1.95 + i0.79$ (Bond and Bergstrom 2006). In our work, a flat flame burner with a circular porous frit and an overhead stagnation plate was used. Such a flame is quasi-one-dimensional so that the soot is uniform in the horizontal direction across the flame but grows in size due to aggregation with distance above the flame, the height above burner, h. Early in the flame, e.g. at $h = 10$ mm, the scattering is small and nearly isotropic indicating small particles. With increasing height above the burner, the scattering both increases and becomes more anisotropic, and the characteristic shape of the structure factor for a fractal aggregate (see figures 9.8, 9.9, etc) appears. In all, one sees a constant forward scattering lobe at small q, followed by a Guinier regime and finally a developing power law regime at largest q. The power law is approaching $q^{-1.75}$ to imply, if the structure factor is a correct interpretation for the light scattering (to a good approximation, it is, see below), a fractal dimension of $D = 1.75$ for the soot aggregates in the flame. Both the Guinier regime progressing to smaller q and the increasing intensity of the forward lobe imply that the particles are growing with increasing h. For example, when $h = 10$ mm, the Guinier regime is near $q \simeq 10$ micron^{-1} to imply $R_g \simeq 0.1$ micron.

Figure 9.20 uses some of the data in figure 9.19 plotted for Guinier analysis via equation (9.35d). With this, figure 9.20 plots the normalized inverse intensity versus q^2 to find approximately linear graphs with slopes of $R_{g,z}^2/3$. Note that the linearity versus q^2 goes well beyond $I(0)/I(q) = 4/3$ as discussed above in section 9.4.3.

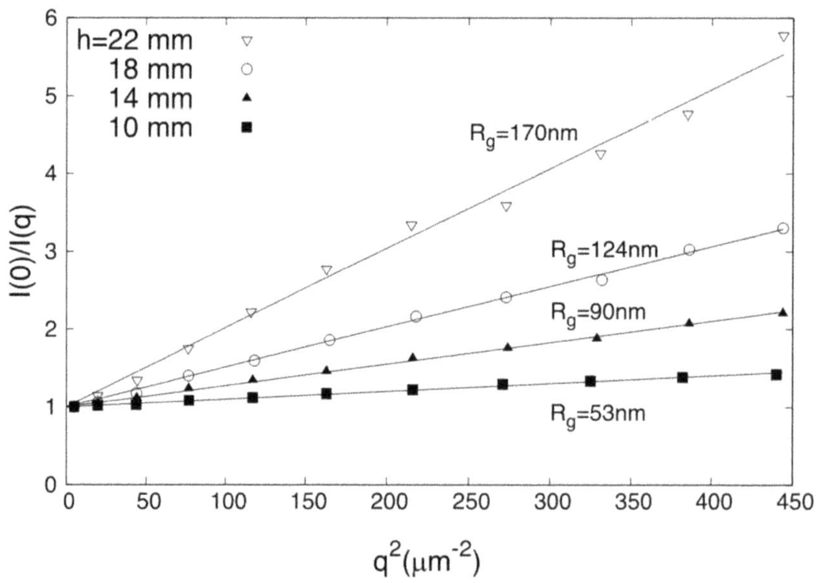

Figure 9.20. Guinier analysis of the scattering data in figure 8.12 for soot in a premixed methane/oxygen flame; normalized inverse intensity versus q^2. By equation (9.35d) the slopes of these plots are equal to $R_{g,z}^2/3$.

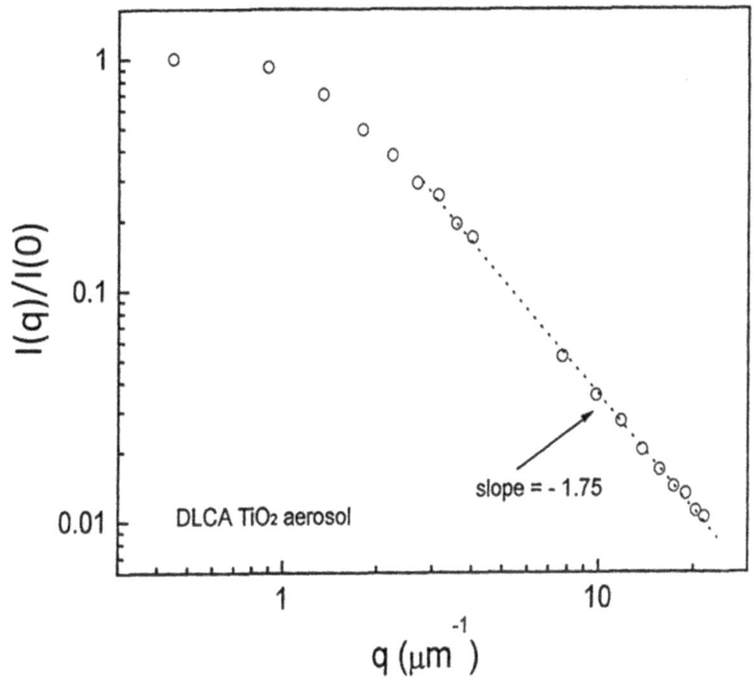

Figure 9.21. Scattered light intensity (normalized structure factor $S(q)$) versus q for an aerosol of TiO_2 aggregates. Can the reader estimate the mean aggregate size?

Figure 9.21 shows yet another example; scattering from TiO_2 aggregates that have formed via DLCA. This example has an extensive power law regime with an exponent equal to the negative fractal dimension of the aggregates.

The fact that the experimental data in figures 9.12, 9.19 and 9.21 show scattering profiles consistent with the RDG formulation is an indication that this formulation has some validity.

9.4.6 Fractal aggregate albedo

The albedo ω is the ratio of the scattering to the total extinction, as defined in section 2.4. Thus, from equations (2.9) and (2.10)

$$\omega = C_{sca}/C_{ext} = C_{sca}/(C_{sca} + C_{abs}) \tag{9.52}$$

It is also called the single scattering albedo, SSA, when there is the chance that multiple scattering might be present. Substitution of equations (9.20)–(9.24) into (9.52) yields

$$\omega_{agg} = \left\{1 + \frac{3E(m)}{2F(m)}(ka)^{-3}\left[NG(kR_g)\right]^{-1}\right\}^{-1} \tag{9.53}$$

This result was given in the 2001 review (Sorensen 2001). Following that review, this result is plotted in figure 9.22 versus the aggregate radius of gyration for

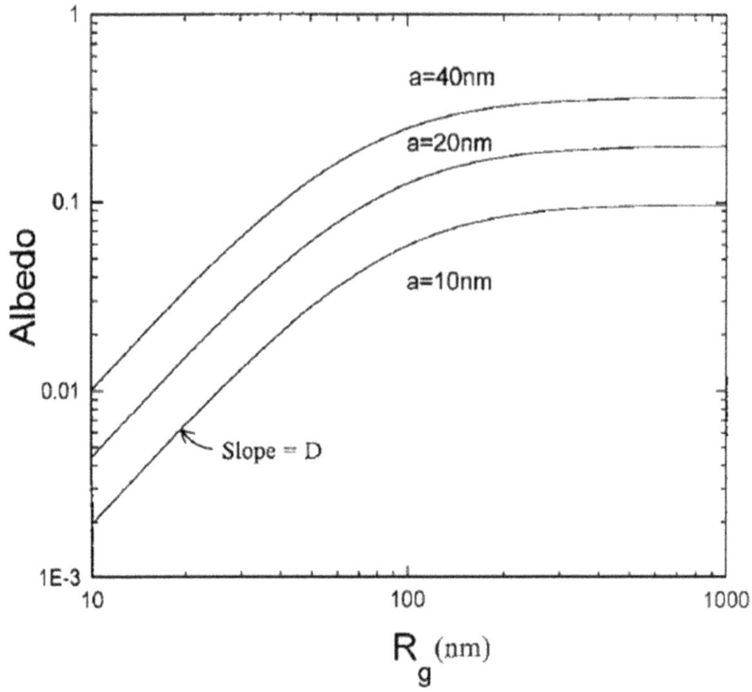

Figure 9.22. The albedo plotted versus the radius of gyration for a soot-like fractal aggregate with a fractal dimension of $D = 1.75$ and light with wavelength 500 nm. Results for three different monomer radii are shown. From Sorensen (2001), reprinted by permission of the publisher (Taylor & Francis Ltd, http://www.tandfonline.com).

$(3E(m)/2k_0F(m)) = 1$, a value consistent with a rough mean of the widely ranging values for the refractive index of soot available in the literature[5].

The crossover near $R_g \simeq 100$ nm for the albedo in figure 9.22 occurs when the size parameter for the aggregate is $kR_g \simeq 1$. With this, figure 9.22 displays two asymptotic regimes for the small and large limits of kR_g. Equations (9.53) and (9.22) yield these limits as

$$\omega = \frac{2F(m)}{3E(m)}(ka)^3 N = \frac{2F(m)}{3E(m)} k^3 k_0 a^{3-D} R_g^D \text{ for } kR_g \ll 1 \tag{9.54}$$

and

$$\omega = \left\{ \left[\frac{3E(m)}{2F(m)} \right] \left[\frac{4}{3D} \right]^{D/2} \right\}^{-1} k_0 (ka)^{3-D} \text{ for } kR_g \gg 1 \tag{9.55}$$

The asymptotic functionalities on the monomer size a and the overall aggregate size R_g are apparent in figure 9.22.

[5] The factor of k_0 comes from the factor of N in equation (9.53) and that figure 9.22 uses the variable R_g for plotting.

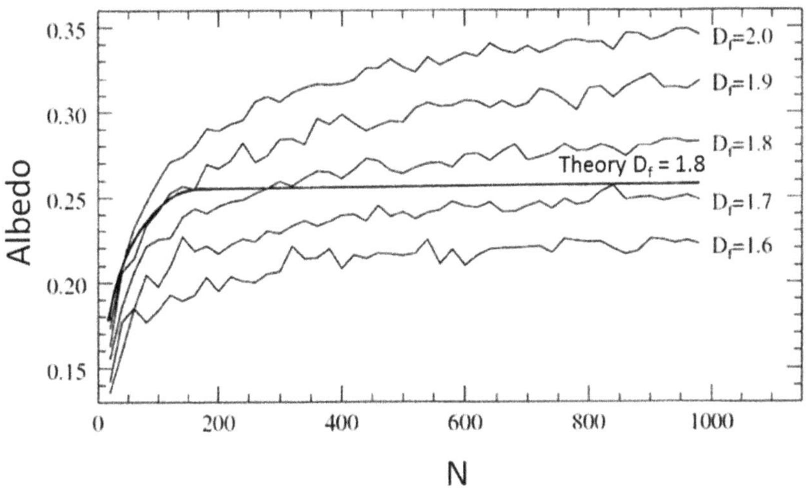

Figure 9.23. The single scattering albedo for soot fractal aggregates as a function of the number of monomers per aggregate, N, parameterized with the fractal dimension, D_f (Smith and Grainger 2014). The solid line marked 'Theory $D_f = 1.8$' is the prediction of equation (9.53).

The 2001 review demonstrated that equation (9.53) gave an excellent fit to calculation data for simulated DLCA soot aggregates with three different monomer size parameters. Figure 9.23 presents a comparison of equation (9.53) to more recent calculations using the T-Matrix method (Smith and Grainger 2014). The aggregates had monomer radii of 25 nm, a refractive index of $m = 1.95 + i0.79$, a prefactor of $k_0 = 1.18$, and selectable fractal dimension ranging from 1.6 to 2.0. Figure 9.23 shows the results. Although not as successful as the comparison in the review, equation (9.53) yields a semi-quantitative fit to the calculations.

9.4.7 Validity of the Rayleigh–Debye–Gans approximation for fractal aggregate light scattering and absorption

In my 2001 review of fractal aggregate scattering and absorption (Sorensen 2001), the literature up to that time regarding RDG validity for fractal soot aggregates was reviewed. The overall conclusion from both comparisons to other theories and experimental measurements was that the RDG approximation compared quite well in predicting magnitudes of the scattering and absorption, usually better than 10% differences but sometimes as large as 20%, in comparison with not necessarily perfect alternatives. Furthermore, the functionalities of N^2 and q^{-D} for scattering and N for absorption were essentially 'spot on'. The magnitude discrepancies were that RDG underestimated scattering and absorption by about 10% for small to intermediate size aggregates.

Comparison work since then has not changed this conclusion. We made perhaps the only calibrated, absolute value measurement of the scattering cross section using two fractal aggregate aerosols (Wang and Sorensen 2002). These aerosols were composed of vitreous silica, SiO_2, with a refractive index of 1.46, and anatase, TiO_2, with a refractive index of 2.61. Calibration was achieved by scattering from gases with known Rayleigh ratios. Aerosol mass densities were measured by collecting measured volumes of the aerosol. The aggregates of both aerosols had radii of gyration of approximately 280 nm, fractal dimensions of 1.75, and monomer sizes of approximately 20 nm with approximately 150 monomers per cluster. We found good agreement, ~ 20%–30%, with the Rayleigh–Debye–Gans prediction for the scattering cross section of fractal aggregates.

Soewono and Rogak (2013) used the discrete dipole approximation (DDA) to calculate scattering and absorption by soot fractal aggregates and compared to RDG. Aggregates ranged in size as $15 < N < 600$ monomers with either $D = 1.78$ or 2.1 with $k_0 = 2.24$ and 2.26, respectively. The primary particle size parameter was $ka = 0.177$ with a refractive index of $m = 1.6 + 0.6i$. As in many previous studies, both scattering and absorption cross sections were 10% to 15% higher than that predicted by RDG, but came into agreement with increasing size.

We (Berg and Sorensen 2013) compared RDG to calculations using the DDA for simulated soot fractal aggregates in fixed and random orientations. For a DLCA aggregate ($D = 1.8$) with aggregate size parameter of $kR_g = 5.4$, monomer size parameter of $ka = 0.42$, monomer number $N = 99$, and refractive index $m = 1.57 + 0.56i$ with a fixed orientation, the internal field magnitude deviated from the incident field over a range of ±25%. Recall that RDG assumes the internal field equals the incident field. Orientational and ensemble averaging lessened these deviations somewhat. Despite internal field deviations, the far-field scattered intensity from aggregates agreed well with RDG, better than 10%. Thus, while RDG does a poor job of representing the internal field, it still does a good job of representing the scattered field. We proposed that the explanation for the agreement was that the positive and negative deviations of the internal field in some manner negated each other to yield agreement for the external fields. Furthermore, the agreement between the DDA and RDG improved with increasing aggregate size provided the fractal dimension was less than two. This supports the formulation that led to equation (9.51).

Given all this, it is reasonable to conclude that RDG does a very good job of predicting fractal aggregate scattering and absorption for soot aggregates even though the internal field might no longer be described by the incident field. Soot aggregates typically have primary particles with complex index of refraction and small size parameters and the aggregate fractal dimension is less than 2. It is reasonable to conclude the same degree of validity for $D < 2$ aggregates of any material so long as the primary particles, the monomers, have a small size parameter.

As the reader has very likely become aware, there is a strong preoccupation of scattering and absorption by soot fractal aggregates despite the fact that the title of this chapter does not contain the modifier 'soot'. Certainly, given the importance of soot as a by-product of combustion including fossil fuels and biomass burning, it is not surprising that the bulk of previous work concerning scattering and absorption of light by fractal aggregates has been limited to soot. Early work at the beginning of the 'fractal revolution' scattered light from gold and silica colloids, but the authors

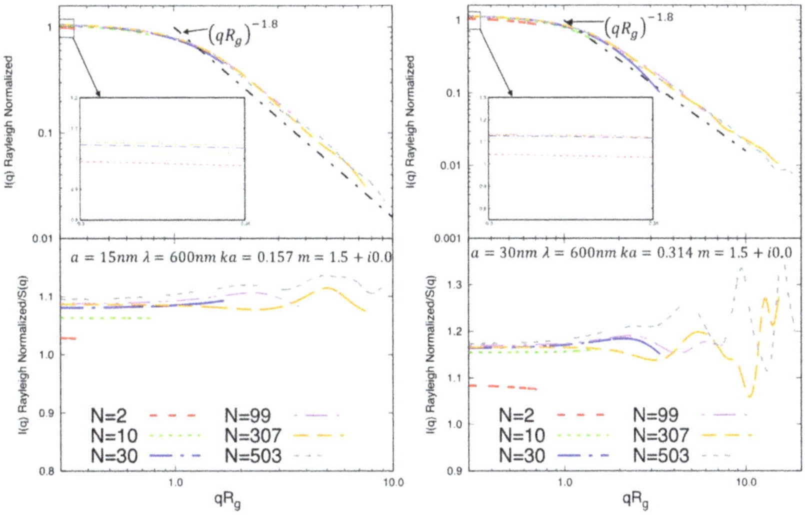

Figure 9.24. (Top) Rayleigh normalized scattered intensity for DLCA fractal aggregates of various number of monomers, N, versus qR_g. The monomer index of refraction is $m = 1.5 + i0.0$ and the size parameters are $ka = 0.157$ (left plot) and 0.314 (right). The marked dashed line in each shows the functionality $C(qR_g)^{-D}$ of equation (9.12c) with $C = 1.0$ and fractal dimension $D = 1.8$. Insets show the behavior for very small qR_g, hence essentially $I(0)$, Rayleigh normalized. (Bottom) Rayleigh normalized scattered intensity divided by the structure factor, equations (9.12) and (9.30). If this ratio is unity, there is no deviation from RDG. Insets show the behavior of the forward scattering. Adapted from Sorensen *et al* (2018), Copyright (2018), with permission from Elsevier.

of those studies were, it seems, happy just to see the general features of the scattering and did not test the RDG theories. Unfortunately, by concentrating on soot, the opportunity to detect systematic trends in the scattering and absorption and thereby develop physical intuition for these processes for fractals in general is lost.

This deficiency was addressed to some degree in Sorensen *et al* (2018), who studied scattering and absorption by DLCA aggregates with fractal dimension $D = 1.78 \pm 0.04$ and prefactor $k_0 = 1.35 \pm 0.10$. The variables were number of monomers per aggregate N, ranging from $N = 1$ to 1000, and monomer refractive index, $m = n + i\kappa$. These were systematically varied for aggregates of two different monomer sizes with size parameters of $ka = 0.157$ and 0.314. The orientationally averaged light scattering calculations for the aggregates were performed using the multiple sphere T-matrix code (MSTM) developed by Mackowski and Mishchenko (2011).

In general, the results for both the Rayleigh normalized forward scattering and absorption showed that the deviations from RDG ranged at most from −20% to + 30% and less so for the smaller monomers. The functionality with N was very weak, with a mild increase when $1 < N < 100$, consistent with equation (9.47), then slowly decreasing when $N > 100$, consistent with equation (9.51). The deviations depended on the refractive index with the real part, n, causing positive deviations whereas the imaginary part, κ, caused negative deviations, and these two counteracted each other. The insets of figures 9.24 and 9.25 show some of this behavior.

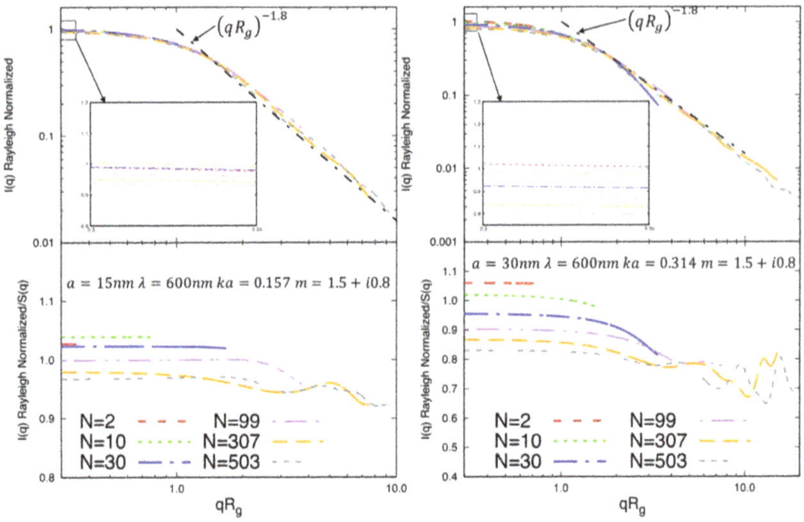

Figure 9.25. (Top) Rayleigh normalized scattered intensity for DLCA fractal aggregates of various number of monomers, N, versus qR_g. The monomer index of refraction is $m = 1.5 + i0.8$ and the size parameters are $ka = 0.157$ (left plot) and 0.314 (right). The marked dashed line in each shows the functionality $C(qR_g)^{-D}$ of equation (9.12c) with $C = 1.0$ and fractal dimension $D = 1.8$. Insets show the behavior for very small qR_g, hence essentially $I(0)$, Rayleigh normalized. (Bottom) Rayleigh normalized scattered intensity divided by the structure factor, equations (9.12) and (9.30). If this ratio is unity, there is no deviation from RDG. Insets show the behavior of the forward scattering. Adapted from Sorensen *et al* (2018), Copyright (2018), with permission from Elsevier.

The Rayleigh normalized forward scattered intensity was also studied as a function of ρ'_{agg}, equation (9.47). There was an undeniable and notable similarity of the functionality for the aggregates compared to that for spheres as seen in figure 7.6. This included large fluctuation near $\rho'_{\text{agg}} = 1$, and when $\rho'_{\text{agg}} \gtrsim 1$, the a precipitous decrease limiting to a ρ'^{-2}_{agg} functionality.

Beyond forward scattering, the q-dependence was also explored. Figures 9.24 and 9.25 display the results.

These figures show that light scattering does not significantly affect the shape of the plots, i.e. the $I(q)$ versus q functionality, compared to the RDG limit, which is the structure factor, $S(q)$. Most notably, this uniformity implies that the slopes of the light scattering plots reflect the fractal dimension well. Also of note is the fact that the coefficient C of the $C(qR_g)^{-D}$ power law can need adjustments from the RDG value of $C = 1.0$ by as much as ±15%.

9.5 Superaggregates

Among the joys of research are surprises in the lab. One surprise for us was the discovery of aerosol gelation and superaggregates (Sorensen *et al* 1998, 2003, Sorensen and Hageman 2001, Fry *et al* 2004, Kim *et al* 2004, 2006, Dhaubhadel *et al* 2006, 2012). A superaggregate is so named not so much because it is super, but because it has a hybrid, hierarchical, super-structure. It is an aggregate composed of aggregates; it is an aggregate with a global fractal dimension composed of smaller aggregates with a different fractal dimension. The superaggregates we discovered have a fractal dimension of $D \simeq 2.5$ to 2.6 at a large length scale and are composed of smaller aggregates with fractal dimensions of $D = 1.8$ at small scales.

Figure 9.26 shows electron microscope pictures of graphene soot collected from a detonation of acetylene and oxygen in a 17 liter chamber. On the left is a large aggregate roughly 100 microns in diameter. Structure factor analysis, achieved by Fourier transforming the image, yielded a fractal dimension for this aggregate of $D = 2.50$. A perimeter analysis yielded 2.42 to corroborate. On the right is a zoom into the large aggregate to a scale roughly 100 times smaller. On this scale, structure factor analysis yielded a fractal dimension of $D = 1.80$, while the perimeter analysis gave 1.78. This aggregate with different fractal dimensions at different scales is a superaggregate (Dhaubhadel *et al* 2006). Figure 9.27 shows a sketch of a superaggregate.

Superaggregates were discovered in flames of fuels with large sooting indices, i.e. heavily sooting flames. Subsequently, they have been observed in other situations involving dense aerosols, such as fuel pool fires (Kearney and Pierce 2012), forest fires (Chakrabarty *et al* 2014a) and inverted flames (Chakrabarty *et al* 2014b). They occur because DLCA aggregation can proceed from cluster-dilute, where normal $D = 1.8$ DLCA aggregates form, to cluster-dense, so that eventually the aggregate volume fraction becomes so close to unity that the $D = 1.8$ DLCA aggregates jam together to form the $D = 2.5$ superaggregates (Fry *et al* 2004). Ultimately, these kinetics yield macroscopic gels as reviewed in Sorensen and Chakrabarti (2011) and recently substantiated in Liu *et al* (2019) and in Ebini and Sorensen (2019). Superaggregates formed via DLCA kinetics are isomorphic with static percolation clusters (Heinson *et al* 2017), a result consistent with gel formation. Figure 9.28 gives an example of an in-flame gel.

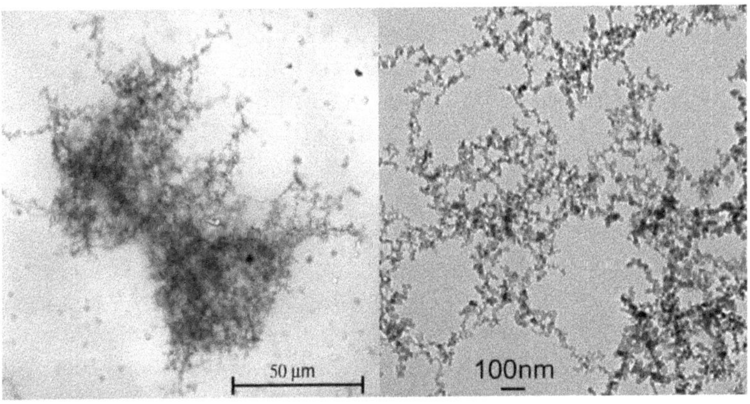

Figure 9.26. (Left) Transmission electron microscope picture of a graphene soot cluster formed after the explosion of acetylene gas mixed with oxygen. The fractal dimension of this cluster is $D \simeq 2.46$. (Right) A zoom into the cluster on the left by a factor of ~100. The fractal dimension at this scale is $D \simeq 1.78$. Adapted figure with permission from Dhaubhadel *et al* (2006), Copyright (2006) by the American Physical Society.

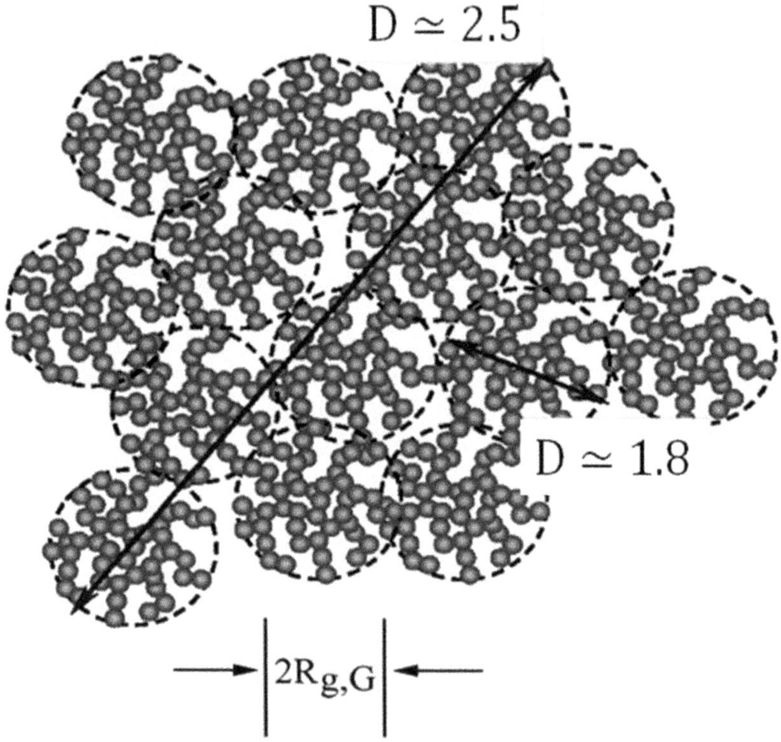

Figure 9.27. Sketch of a superaggregate with a fractal dimension of $D \simeq 2.5$ over large scales and $D \simeq 1.8$ over small scales. The smaller aggregates were formed via DLCA and grew until they reached a size of $R_{g,G}$, the radius of gyration at the gel point. Adapted from Mokhtari *et al* (2008), Copyright (2008), with permission from Elsevier.

Light scattering from superaggregates appears to be well described by diffraction, i.e. the RDG limit. This is surprising given the fact that one of the fractal dimensions is greater than 2. Perhaps it's the small scale dimension of 1.8 that is relevant. Figure 9.29 shows results from some of the first light scattering experiments (Kim *et al* 2006). One sees slopes of −2.5 and −1.8 at small and large q, respectively, to imply $D = 2.5$ and 1.8 at large and small length scales, respectively. All TEM analyses that I am aware of, such as those in figures 9.26 and 9.29, yield the same dimensions and indicate sizes consistent the scattering; thus bolstering the diffraction claim. This issue could use further work, see below.

Simulations of DLCA indicate that the classic $D = 1.8$ DLCA aggregates form when the system is cluster-dilute. 'Cluster-dilute' is defined by when the aggregate (i.e. cluster) size is much less than the mean aggregate nearest neighbor separation. However, because the aggregate scaling dimension D (the fractal dimension) is less that the spatial dimension (the scaling dimension of space), the separation-to-size ratio decreases as the aggregation continues and the aggregates grow. Another way to say this is that because $D < d$, the aggregate size grows faster than the aggregate separation. Thus, eventually, the size overtakes the separation and the system enters

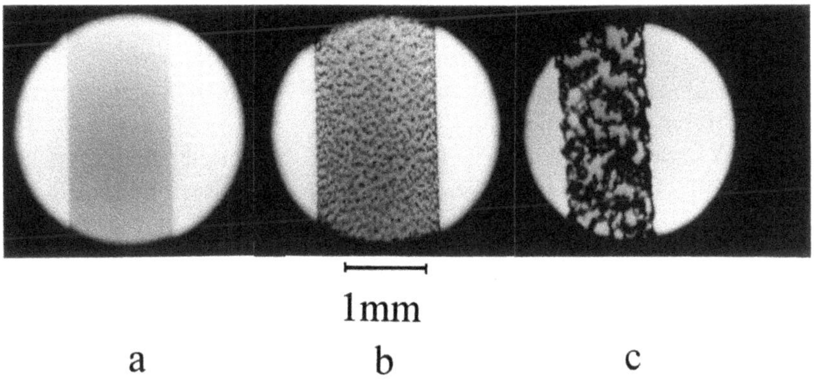

Figure 9.28. Microsecond photographs of soot aerosols at a height above burner of 10 cm in an acetylene/air diffusion flame. (a) Acetylene flow rate = 0.83 $cm^3\ s^{-1}$ yields submicron soot. (b) Acetylene flow rate = 1.7 $cm^3\ s^{-1}$ yields supramicron soot. (c) Acetylene flow rate = 3.4 $cm^3\ s^{-1}$ yields an annular gel network of soot. Reprinted figure with permission from Sorensen *et al* (1998), Copyright (1998) by the American Physical Society.

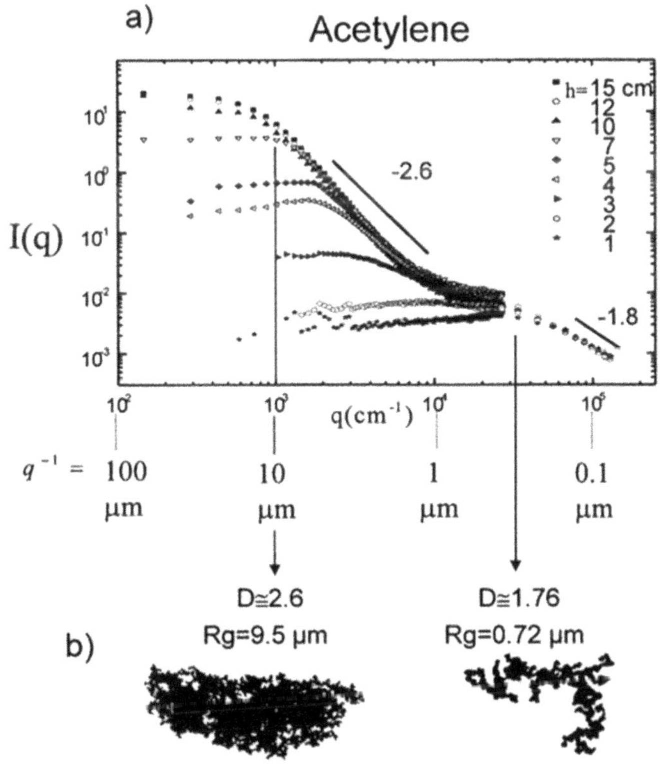

Figure 9.29. (a) Static light scattered intensity as a function of scattering wave vector q from an acetylene/air flame at various heights above burner h. Inverse q values represent length scales as indicated. (b) On the left is a soot superaggregate with fractal dimension $D \simeq 2.6$ and radius of gyration $R_g = 9.5$ micron collected from the flame at h = 12 cm. On the right is a soot aggregate with $D \simeq 1.76$ and $R_g = 0.72$ micron collected from the flame at h = 2 cm. Arrows from (a) to (b) indicate correspondence between the aggregates and the scattered light. Reprinted from Kim *et al* (2006), Copyright (2006), with permission from Elsevier.

the cluster-dense regime where the separation to size ratio is small. Then superaggregates form and these quickly jam together to form a gel (Sorensen and Chakrabarti 2011).

Figure 9.30 shows the structure factor of a simulated superaggregate. The simulation involved over 10 000 monomers in a cubical box with size adjusted so that the volume fraction of the monomers was $f_v = 0.01$. Note the $q^{-2.5}$ and $q^{-1.8}$ power law regimes indicating the hybrid superaggregate structure of this simulated aggregate. This is, of course (if you've been paying attention), the RDG, $\rho' = 0$ limit for light scattering. There is a Guinier regime near $qR_g \simeq 1$ indicating the overall size of the superaggregate and a crossover between the two power laws indicating $R_{g,G}$, the radius of gyration of the $D = 1.8$ aggregates when they percolate at the ideal gel point, see figure 9.27.

Figure 9.30 also contains Amsterdam discrete-dipole-approximation (ADDA) (Yurkin and Hoekstra 2011) calculations of the light scattering for the superaggregate. The monomer radius is a = 25 nm. Thus, for a wavelength of $\lambda = 532$ nm, the monomer size parameter is 0.30. There are about 33 dipoles per monomer, and the refractive index is $m = 1.6 + i0.6$. The overall superaggregate size is $R_g = 2.34$ μm. The light scattering agrees very well with the structure factor to imply that the light scattering is in the RDG regime. This supports the experimental results and leads us to conclude that superaggregates of this size scatter light in the RDG regime.

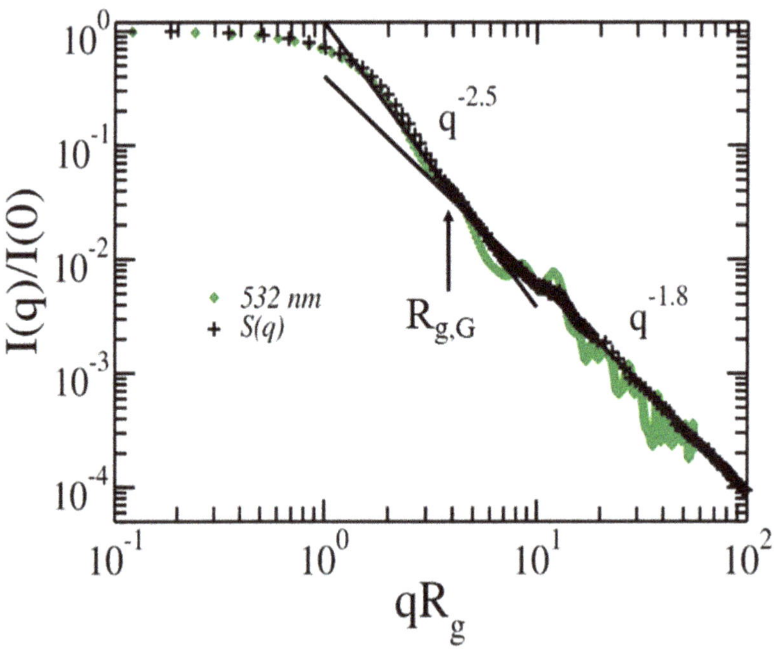

Figure 9.30. Structure factor, black + signs, and light scattering calculations, green diamonds, for a simulated superaggregate with over 10 000 soot-like monomers (credit to W R Heinson).

This result can be understood theoretically because of the non-dense, open structure of the superaggregate. In section 9.4.4, above, the Maxwell–Garnett equation was used to explain how the non-dense nature of fractal aggregates leads to small internal coupling ρ' and thus supports the validity of the RDG approximation for fractal aggregate scattering. However, recall that equation (9.51) predicted that the coupling should increase with size if the fractal dimension is $D > 2$. This is partially true for superaggregates, but not entirely true given their hybrid structure. Here we reconsider the Maxwell–Garnett approach embodied in equation (9.50) and calculate the effective Lorentz–Lorenz term for the superaggregate directly from the volume fraction of the monomers in the superaggregate, f. Then the superaggregate internal coupling parameter is

$$\rho'_{\mathrm{sa}} = 2kR\left|\frac{m^2-1}{m^2+2}\right|_{\mathrm{sa}} = 2kRf\left|\frac{m^2-1}{m^2+2}\right|_{\mathrm{mono}} \tag{9.59}$$

When $m = 1.6 + i0.6$, the Lorentz–Lorenz term is 0.49. For R we use the radius of gyration $R_g = 2.34$ μm of the aggregate in figure 9.30. The volume fraction of monomers in the superaggregate is $f \simeq 7 \times 10^{-3}$. With all this, the internal coupling parameter for the superaggregate is $\rho'_{\mathrm{sa}} = 0.19$. This is well within the RDG regime and so consistent with the results in figure 9.30.

Figure 9.31 shows $I(q)$ measured for soot gel particles created in an inverted flame (Heinson *et al* 2018). Large q slopes of −3 and −1.8 are seen followed by significant

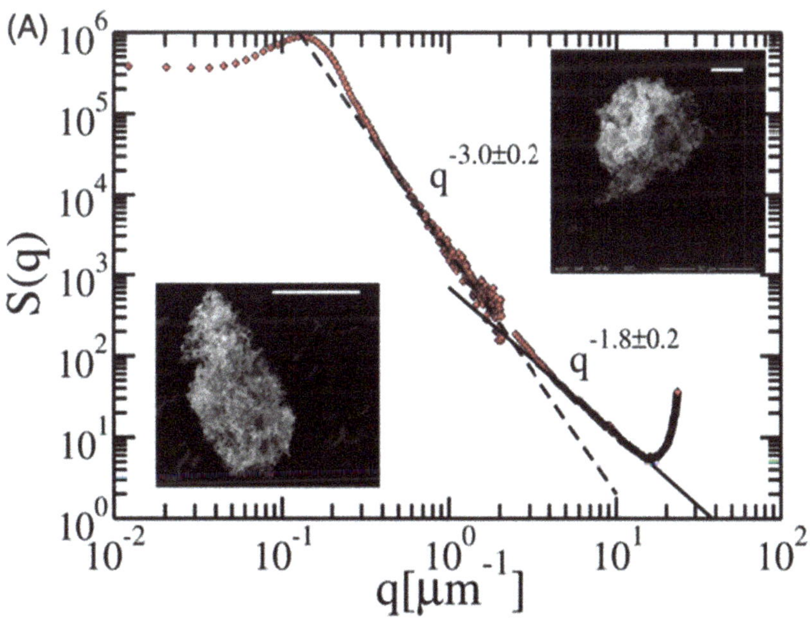

Figure 9.31. Scattered intensity $S(q)$ versus q for soot in an inverted flame that produces superaggregates. Adapted from Heinson *et al* (2018), reprinted by permission of the publisher (Taylor & Francis Ltd, http://www.tandfonline.com).

backscattering. The −3 slope was interpreted as due to superaggregates with mass and surface scaling dimension of $D_m = 2.5$ and $D_s = 2$, respectively. See section 5.3 and equation (5.10). Note that for these fractals $D_m \neq D_s$, which is rather unusual but supported by microscope analysis and simulations. The slope of −1.8 at larger q, hence smaller size regimes, implies the superaggregate structure. Also of note is the significant backscattering, which was not explained and needs further study.

References

Abramowitz M and Stegun I A 1972 *Handbook of Mathematical Functions, with Formulas, Graphs, and Mathematical Tables* (New York: Dover)

Ball R C, Weitz D A, Witten T A and Leyvraz F 1987 Universal kinetics in reaction-limited aggregation *Phys. Rev. Lett.* **58** 274

Berg M J and Sorensen C M 2013 Internal fields of soot fractal aggregates *J. Opt. Soc. Am.* A **30** 1947–55

Bond T C and Bergstrom R W 2006 Light absorption by carbonaceous particles: an investigative review *Aerosol Sci. Technol.* **40** 27–67

Cai J, Lu N and Sorensen C M 1995 Analysis of fractal cluster morphology parameters: structural coefficient and density autocorrelation function cutoff *J. Colloid Interface Sci.* **171** 470–3

Chakrabarti A, Fry D and Sorensen C M 2004 Molecular dynamics simulation of the transition from dispersed to solid phase *Phys. Rev.* E **69** 031048

Chakrabarty R K, Beres N D, Moosmuller H, China S, Mazzoleni C, Dubey M K, Liu L and Mishchenko M I 2014a Soot superaggregates from flaming wildfires and their direct radiative forcing *Sci. Rep.* **4** 5508

Chakrabarty R K, Novosselov I V, Beres N D, Moosmuller H, Sorensen C M and Stipe C B 2014b Trapping and aerogelation of nanoparticles in negative gravity hydrocarbon flames *Appl. Phys. Lett.* **104** 243103

Dalzell W H and Sarofim A F 1969 Optical constants of soot and their application to heat-flux calculations *J. Heat Transfer* **91** 100–4

Dhaubhadel R, Gerving C S, Chakrabarti A and Sorensen C M 2007 Aerosol gelation: synthesis of a novel, lightweight, high specific surface area material *Aerosol Sci. Technol.* **41** 804–10

Dhaubhadel R, Pierce F, Chakrabarti A and Sorensen C M 2006 Hybrid superaggregate morphology as a result of aggregation in a cluster-dense aerosol *Phys. Rev.* E **73** 011404

Dhaubhadel R, Rieker T P, Chakrabarti A and Sorensen C M 2012 Synthesis of silica aerosol gels via controlled detonation *Aerosol Sci. Technol.* **46** 596–600

Dietler G, Aubert C, Cannell D S and Wiltzius P 1986 Gelation of colloidal silica *Phys. Rev. Lett.* **57** 3117–20

Dobbins R A and Megaridis C M 1991 Absorption and scattering of light by polydisperse aggregates *Appl. Opt.* **30** 4747–54

Ebini R H and Sorensen C M 2019 Light scattering studies of the sol-to-gel transition in particulate systems *J. Colloid Interface Sci.* **556** 577–83

Forrest S R and Witten T A 1979 Long-range correlations in smoke-particle aggregates *J. Phys. A: Math. Gen.* **12** L109–17

Friedlander S K and Wang C S 1966 Self-preserving particle size distribution for coagulation by Brownian motion *J. Colloid Interface Sci.* **22** 126–32

Fry D, Chakrabarti A, Kim W and Sorensen C M 2004 Structural crossover in dense irreversibly aggregating particulate systems *Phys. Rev.* E **69** 061401

Fry D, Mohammad A, Chakrabarti A and Sorensen C M 2004 Cluster shape anisotropy in irreversibly aggregating particulate systems *Langmuir* **20** 7871–9

Haw M D, Sievwright M, Poon W C K and Pusey P N 1995 Structure and characteristic length scales in cluster–cluster aggregation simulation *Physica* A **217** 231–60

Heinson W R, Chakrabarti A and Sorensen C M 2015 Divine proportion shape invariance of diffusion limited cluster–cluster aggregates *Aerosol Sci. Technol.* **49** 786–92

Heinson W R, Chakrabarti A and Sorensen C M 2017 Kinetic percolation *Phys. Rev.* E **95** 052109

Heinson W R, Heinson Y W, Liu P and Chakrabarty R K 2018 Breakdown of fractal dimension invariance in high monomer-volume-fraction aerosol gels *Aerosol Sci. Technol.* **52** 953–6

Heinson W R, Sorensen C M and Chakrabarti A 2010 Does shape anisotropy control the fractal dimension in diffusion-limited cluster–cluster aggregation? *Aerosol Sci. Technol.* **44** i–iv

Heinson W R, Sorensen C M and Chakrabarti A 2012 A three parameter description of the structure of diffusion limited cluster fractal aggregates *J. Colloid Interface Sci.* **375** 65–9

Jullien R and Botet R 1987 *Aggregation and Fractal Aggregates* (Singapore: World Scientific)

Kearney S P and Pierce F 2012 Evidence of soot superaggregates in a turbulent pool fire *Combust. Flame* **159** 3191–8

Kerker M 1969 *The Scattering of Light, and Other Electromagnetic Radiation* (New York: Academic)

Kim W, Sorensen C M and Chakrabarti A 2004 Universal occurrence of soot superaggregates with a fractal dimension of 2.6 in heavily sooting laminar diffusion flames *Langmuir* **20** 3969–73

Kim W, Sorensen C M, Fry D and Chakrabarti A 2006 Soot aggregates, superaggregates and gel-like networks in laminar diffusion flames *J. Aerosol Sci.* **37** 386–401

Kolb M, Botet R and Jullien R 1983 Scaling of kinetically growing clusters *Phys. Rev. Lett.* **51** 1123

Koylu U O and Faeth G M 1994a Optical-properties of overfire soot in buoyant turbulent-diffusion flames at long residence times *J. Heat Transfer ASME* **116** 152–9

Koylu U O and Faeth G M 1994b Optical-properties of soot in buoyant laminar diffusion flames *J. Heat Transfer ASME* **116** 971–9

Koylu U O, Faeth G M, Farias T L and Carvalho M G 1995a Fractal and projected structure properties of soot aggregates *Combust. Flame* **100** 621–33

Koylu U O, Xing Y C and Rosner D E 1995b Fractal morphology analysis of combustion-generated aggregates using angular light scattering and electron microscope images *Langmuir* **11** 4848–54

Lattuada M, Wu H and Morbidelli M 2003 A simple model for the structure of fractal aggregates *J. Colloid Interface Sci.* **268** 106–20

Lin M Y, Lindsay H M, Weitz D A, Ball R C, Klein R and Meakin P 1989 Universality in colloid aggregation *Nature* **339** 360–2

Liu P, Heinson W R, Sorensen C M and Chakrabarty R K 2019 Kinetics of sol-to-gel transition in irreversible particulate systems *J. Colloid Interface Sci.* **550** 57–63

Mackowski D W and Mishchenko M I 2011 A multiple sphere T-matrix FORTRAN code for use on parallel computer clusters *J. Quant. Spectrosc. Radiat. Transf.* **112** 2182–92

Mandelbrot B B 1977 *Fractals: Form, Chance and Dimension* (San Francisco, CA: W.H. Freeman)

Mandelbrot B B and Mandelbrot B B 1982 *The Fractal Geometry of Nature* (San Francisco, CA: W.H. Freeman)

Meakin P 1983 Formation of fractal clusters and networks by irreversible diffusion-limited aggregation *Phys. Rev. Lett.* **51** 1119–22
Mokhtari T, Chakrabarti A, Sorensen C M, Cheng C Y and Vigil D 2008 The effect of shear on colloidal aggregation and gelation studied using small-angle light scattering *J. Colloid Interface Sci.* **327** 216–23
Nicolai T, Durand D and Gimel J C 1994 Static structure factor of dilute-solutions of polydisperse fractal aggregates *Phys. Rev.* B **50** 16357–63
Oh C and Sorensen C M 1997 The effect of overlap between monomers on the determination of fractal cluster morphology *J. Colloid Interface Sci.* **193** 17–25
Oh C and Sorensen C M 1997 Light scattering study of fractal cluster aggregation near the free molecular regime *J. Aerosol Sci.* **28** 937–57
Pierce F, Sorensen C M and Chakrabarti A 2006 Computer simulation of diffusion-limited cluster–cluster aggregation with an Epstein drag force *Phys. Rev.* E **74** 021411
Seinfeld J H and Pandis S N 2006 *Atmospheric Chemistry and Physics: From Air Pollution to Climate Change.* (Hoboken, NJ: Wiley)
Skjeltorp A T 1987 Visualization and characterization of colloidal growth from ramified to faceted structures *Phys. Rev. Lett.* **58** 1444–47
Smith A J A and Grainger R G 2014 Simplifying the calculation of light scattering properties for black carbon fractal aggregates *Atmos. Chem. Phys.* **14** 7825–36
Soewono A and Rogak S N 2013 Morphology and optical properties of numerically simulated soot aggregates *Aerosol Sci. Technol.* **47** 267–74
Sorensen C M 2001 Light scattering by fractal aggregates: a review *Aerosol Sci. Technol.* **35** 648–87
Sorensen C M 2011a The mobility of fractal aggregates: a review *Aerosol Sci. Technol.* **45** 755–69
Sorensen C M 2011b The mobility of fractal aggregates: a review *Aerosol Sci. Technol.* **45** 765–79
Sorensen C M, Cai J and Lu N 1992 Light-acattering measurements of monomer size, monomers per aggregate, and fractal dimension for soot aggregates in flames *Appl. Opt.* **31** 6547–57
Sorensen C M, Cai J and Lu N 1992 Test of static structure factors for describing light-scattering from fractal soot aggregates *Langmuir* **8** 2064–69
Sorensen C M and Chakrabarti A 2011 The sol to gel transition in irreversible particulate systems *Soft Matter* **7** 2284–96
Sorensen C M and Feke G D 1996 The morphology of macroscopic soot *Aerosol Sci. Technol.* **25** 328–37
Sorensen C M, Hageman B, Rush T J, Huang H and Oh C 1998 Aerogelation in a flame soot aerosol *Phys. Rev. Lett.* **80** 1782
Sorensen C M and Hageman W B 2001 Two-dimensional soot *Langmuir* **17** 5431–4
Sorensen C M, Kim W, Fry D, Shi D and Chakrabarti A 2003 Observation of soot super-aggregates with a fractal dimension of 2.6 in laminar acetylene/air diffusion flames *Langmuir* **19** 7560–63
Sorensen C M, Lu N and Cai J 1995 Fractal cluster size distribution measurement using static light scattering *J. Colloid Interface Sci.* **174** 456–60
Sorensen C M and Oh C 1998 Divine proportion shape preservation and the fractal nature of cluster–cluster aggregates *Phys. Rev.* E **58** 7545–8
Sorensen C M, Oh C, Schmidt P W and Rieker T P 1998 Scaling description of the structure factor of fractal soot composites *Phys. Rev.* E **58** 4666–72
Sorensen C M and Roberts G C 1997 The prefactor of fractal aggregates *J. Colloid Interface Sci.* **186** 447–52

Sorensen C M and Wang G M 1999 Size distribution effect on the power law regime of the structure factor of fractal aggregates *Phys. Rev.* E **60** 7143–8

Sorensen C M, Yon J, Liu F S, Maughan J, Heinson W R and Berg M J 2018 Light scattering and absorption by fractal aggregates including soot *J. Quant. Spectrosc. Radiat. Transf.* **217** 459–73

Stauffer D and Aharony A 1994 *Introduction to Percolation Theory: Revised* 2nd edn (New York: Taylor & Francis)

van Dongen P G J and Ernst M H 1985 Dynamic scaling in the kinetics of clustering *Phys. Rev. Lett.* **54** 1396

Wang C S and Friedlander S K 1967 Self-preserving particle size distribution for coagulation by Brownian motion: 2. Small particle slip correction and simultaneous shear flow *J. Colloid Interface Sci.* **24** 170–79

Wang G M and Sorensen C M 2002 Experimental test of the Rayleigh–Debye–Gans theory for light scattering by fractal aggregates *Appl. Opt.* **41** 4645–51

Weitz D A and Oliveria M 1984 Fractal structures formed by kinetic aggregation of aqueous gold colloids *Phys. Rev. Lett.* **52** 1433

Witten T A and Sander L M 1981 Diffusion-limited aggregation, a kinetic critical phenomenon *Phys. Rev. Lett.* **47** 1400–3

Yurkin M A and Hoekstra A G 2011 The discrete-dipole-approximation code ADDA: capabilities and known limitations *J. Quant. Spectrosc. Radiat. Transf.* **112** 2234–47

IOP Publishing

Chapter 10

Light scattering and absorption by particles of any shape

The problem of how light scatters from spherical particles was solved more than 100 years ago, and now, with Q-space analysis, a new perspective and understanding has been achieved. Furthermore, the manner in which light scatters from fractal aggregates, a type of non-spherical, irregularly shaped particle is fairly well understood. For both of these situations the Q-space analysis has allowed quantitative descriptions not obtainable in angular θ-space. Given this background, it is now time to apply Q-space analysis to develop a universal description for scattering by particles of any shape.

The term 'any shape' brings me to notation. I propose 'any shape' means just that, any shape. Under 'any shape', I single out spheres. After that, I propose regular and irregular shapes, a safe tautology. Regular shapes include various geometric shapes such as ellipsoids (which include spheres), cylinders, prisms and non-random combinations of these like some ice crystals. Irregular shapes include aggregates, both fractal and not, and objects with some sort of shape randomization such as randomized and perturbed spheres. In some sense, the classification to irregular is the dustpan[1] of this classification scheme to pick up any shape that is not regular. Furthermore, a particle of any shape can have a roughened surface. Note, however, that roughening can be viewed as randomization at length scales small compared to the characteristic dimensions of the particle to distinguish it from shape. The purpose of this chapter is to see how much light[2] can be shown into the final frontier of light scattering problems, the scattering of light from irregularly shaped particles. When this is achieved, the description of scattering and absorption of light by any shape is at hand.

At the start, the very term 'irregularly shaped particles' is problematic. What does this term mean? Can 'irregular' be specified quantitatively? This problem reminds

[1] Pun intended.
[2] Ditto.

doi:10.1088/978-0-7503-5157-7ch10

me of the state of affairs in the 1970s, when irregularly shaped soot aggregates were given the adjective 'aciniform' to mean shaped like a cluster of grapes. Aciniform soot aggregates. What's in a name? In this case, not much. But then came the 'fractal revolution' when Mandelbrot (1982) taught us that much of nature had a fractal geometry and Forrest and Witten (1979) applied this concept to soot aggregates. With these seminal insights, we came to learn, through an explosion of scientific work, which aggregates were fractal (or, to be precise, fractal-like), a discriminating name that implies a scale invariant structure with a non-integer scaling dimension. The adjective 'fractal' has physical substance. From this arose mechanisms for the creation of their structures, such as diffusion limited cluster aggregation (DLCA), and a quantitative ability to determine aggregate physical properties, such as mobility, aggregation kinetics, and light scattering and absorption properties, in terms of the scaling, i.e. the fractal dimension. Ironically, with all this, fractal aggregates became less irregular.

The problem of irregular shape needs a similar revolution to yield a universal, quantitative description. I do not offer one here. However, we will soon see that Q-space analysis can discriminate the irregularly shaped fractals from other irregular shapes. Thus, at minimum, we will have two types of irregularly shaped particles.

The problem of how irregularly shaped particles scatter and absorb light is an important one. The bulk of aerosol mass in the atmosphere, including that of entrained mineral dust, volcanic ash, and soot, consists of particles with irregularly shapes. Thus, the manner in which they scatter and absorb light has implications for the radiative forcing component of climate models and for aerosol optical remote sensing, especially satellite remote sensing. The angular scattering pattern, the phase function, is a necessary component in these calculations and data interpretations.

In this chapter, Q-space analysis will be applied to the angular distribution of scattered light for a wide range of irregularly shaped particles. The effects of absorption on scattering will be observed, but absorption by particles will not be studied *per se*. Experimental data and results obtained from theoretical calculations will be used. As in the chapter concerning spherical particle light scattering, we will not discuss the mathematical methods to compute the scattering.

10.1 Experimental data

10.1.1 The Amsterdam–Granada data set

The most extensive data set for light scattering by aerosolized dusts is from the Amsterdam–Granada group. The term 'dust' is general enough to encompass many things. Here 'dust' will refer to powders of solid particles with sizes ranging roughly from less than one to a few hundred microns. Most dusts have particles with irregular shapes. Examples include mineral dust particles from desert and agricultural regions, road dusts, plant fragments and other biologically derived particles, volcanic ash, and dusts that occur in the workplace like construction sites, smelters, refineries and textile industries.

The Amsterdam–Granada data set is for mineral dusts such as feldspar, red clay, quartz, volcanic ash and many others. The particles were collected in different years and times from different locations. Figure 10.1 shows an example of Saharan dust from Libya. The scattering data were obtained in the lab by scattering light of wavelengths 441.6 nm or 632.8 nm from the aerosolized dusts (Munoz *et al* 2000, 2001, 2002, 2004, 2006, 2007, 2012, Volten *et al* 2001, 2006, Laan *et al* 2009). Scattered light was collected with a movable detector at angles ranging from either 3° or 5° to 177°. Six elements of the scattering matrix where determined. The group has generously made the data available online at http://www.iaa.es/scattering/granada/site_media/download/MunozetalJQSRT2012.pdfH. Figure 10.1 shows an example of their dusts.

Figure 10.2 is an attempt to clearly demonstrate and emphasize the difference between plotting light scattering data in the usual manner of log intensity versus linear scattering angle θ (plot on left) and the method of Q-space analysis where log intensity is plotted versus log scattering wave vector magnitude q (plot on right)

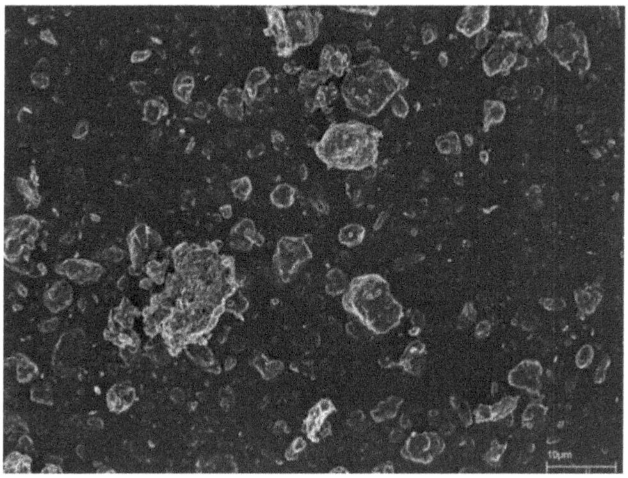

Figure 10.1. FESEM images of green clay particles directly collected from the aerosolized dust jet from which light was scattered. Adapted from Munoz *et al* (2012), Copyright (2012), with permission from Elsevier.

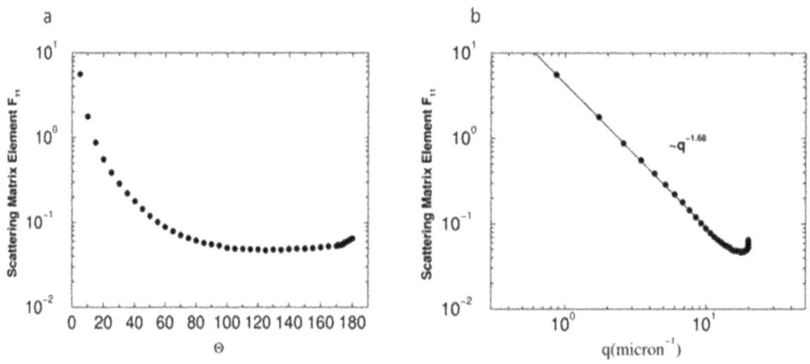

Figure 10.2. The F_{11} ($\equiv S_{11}$) matrix element (scattered intensity for unpolarized incident light) for light scattering from Saharan sand particles collected from the Libyan Desert. (a) Scattering plotted versus scattering angle θ, (b) scattering plotted versus the scattering wave vector magnitude, q. Adapted from Sorensen (2013), Copyright (2013), with permission from Elsevier.

(Sorensen 2013). Plotted are data for scattering of unpolarized light (hence proportional to S_{11}) from aerosolized Saharan sand particles collected from the Libyan Desert (Munoz *et al* 2007). Plotting versus linear angle (θ-space analysis) yields a nondescript curve; plotting versus log q (Q-space analysis) yields a straight line. The curve has no quantifier. The straight line is quantified by a slope to imply a power law functionality with an exponent of 1.68. Moreover, the straight line deviates at large q to emphasize a new functionality. Although the angle is dimensionless, the scattering wave vector has dimensions of inverse length. The Q-space plot shows no change from linearity, i.e. no Guinier regime, at small q down to $q \simeq 0.9\ \mu m^{-1}$ to imply that there are particles larger than $q^{-1} \simeq 1.1\ \mu m$ significantly affecting the scattering. Obviously, Q-space analysis shows a superior ability to describe the scattering.

In our study of the Amsterdam–Granada data set we (Heinson *et al* 2016) applied Q-space analysis to the entire set; a total of 43 aerosol data sets available on the website. Some of these results are shown in figure 10.3. In all cases Q-space analysis with its double logarithmic axes revealed power laws with q. It is uncertain to what

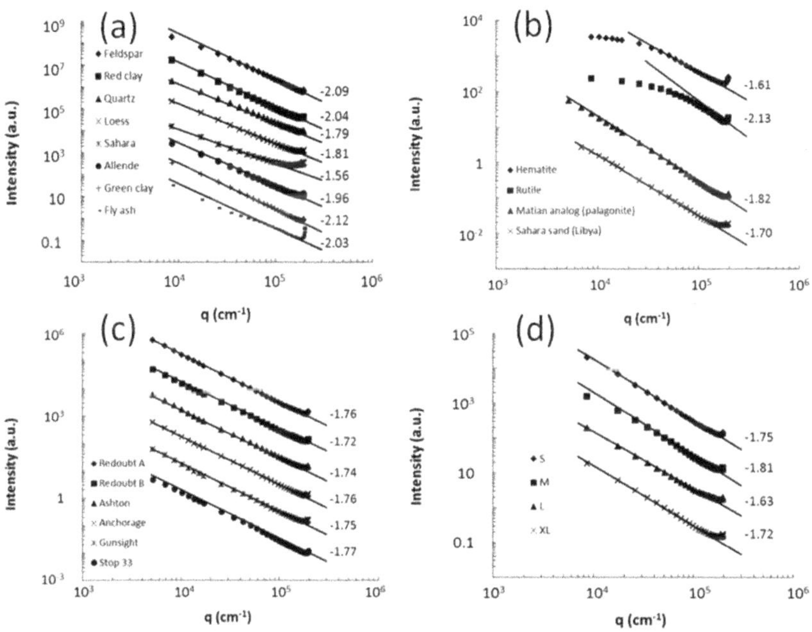

Figure 10.3. Q-space plots of the scattered intensity for unpolarized incident light (S_{11} matrix element) for light scattering from some of the Amsterdam–Granada data set (Munoz *et al* 2012) (a) Feldspar, Red clay, Quartz, Loess, Sahara, (Volten *et al* 2001) Allende (Munoz *et al* 2000), Green clay, Fly ash (Munoz *et al* 2001) measured at λ = 441.6 nm. (b) Hematite (Munoz *et al* 2006), Rutile (Munoz *et al* 2006), Martian analog (palagonite) (Laan *et al* 2009), Sahara sand (Libya) (Munoz *et al* 2007) measured at λ = 632.8 nm. (c) Volcanic ash (Redoubt A, Redoubt B, Spurr Ashton, Spurr Anchorage, Spurr Gunsight, Spurr Stop 33) (Munoz *et al* 2004) measured at λ = 632.8 nm. (d) Olivine S, Olivine M, Olivine L, Olivine XL (Munoz *et al* 2000) measured at λ = 441.6 nm. The lines are power law fits to the data; the number to the right of the plot is the power law exponent. Adapted from Heinson *et al* (2016) John Wiley & Sons. ©2015. American Geophysical Union. All rights reserved.

extent the shape of the dust particles and the aerosol polydispersity contribute to create these power laws. Very often enhanced back scattering was observed at q values larger than where the power law resided. Some of the data showed a declining slope with q decreasing, e.g. figure 10.3(b), to imply the onset of a crossover, Guinier regime. The exponents for the power law regime were obtained by fitting the data after the Guinier regime, when present, and before the enhanced back scattering.

The Q-space perspective indicates that the data are limited in that they do not show a constant forward scattering lobe and only some of them show a portion of a Guinier regime. This limitation is due to the minimum scattering angles of either 3° or 5° used by the Amsterdam–Granada group. Consequently, neither the true forward scattering regime nor the Guinier regime are captured in the scattering data for sizes (~radii) greater than $R \sim \lambda/2\theta$ which is approximately 5 microns for 3°. Said differently, the Guinier regime is centered near $qR \simeq 1$, which means that the size at the Guinier center is at $R \simeq q^{-1}$. The smallest q-value is $q = 5 \times 10^3$ cm^{-1}, hence $q^{-1} = 2 \times 10^{-4}$ cm = 2 microns, comparable to 5 microns. We conclude that most of the dusts studied had larger intensity weighted mean sizes. Note the term 'intensity weighted' and recall the weighting factor in the forward direction for large ρ' particles is R^4. This weighting will be discussed later in this chapter.

Now the obvious question arises: why are essentially all these data showing power laws? That question is as yet unanswered.

10.1.2 The Kansas state data set

It is with some trepidation that I name this section as seen above because the data set we have obtained here at Kansas State University is, by far, much less extensive than that obtained by the Amsterdam–Granada group. Nevertheless, the term is completely descriptive so I humbly proceed.

We have studied four dust systems with the multi-angle light scattering apparatus shown schematically in figure 10.4. This is an earlier version of the setup described in figure 8.18. The dusts are aerosolized and sprayed at the point 'aerosol' in figure 10.4 perpendicular to the diagram; the vertical direction in our lab. The apparatus was designed specifically to have access to very small angles to allow for comprehensive measurements of large particles. It also detects light scattered at all available angles simultaneously, hence there is no need to precisely control the time uniformity of the aerosol. A laser beam with $\lambda = 532$ nm passes through a linear polarizer and/or a wave plate toward the aerosol from which it is scattered. Linear polarizers and/or a wave plates can also be placed before the forward and side scattering detectors. The forward detector followed a design by Ferri (1997). It uses a lens, the Fourier lens, to create a Fourier transform of the scatterer at its focal plane. A beam stop blocks the light directly from the laser. This is another reason to use the Fourier lens because by focusing the incident beam on the beam stop, that stop can be narrower than without the Fourier lens, hence allowing detection of light at smaller angles. The scattered light Fourier transform is imaged by a second lens, the imaging lens, onto the forward detector. Scattered light is detected at sixteen angles in the range of $0.32° \leqslant \theta \leqslant 9.9°$ with angular widths of ± 0.3°. A second detector uses an elliptical mirror

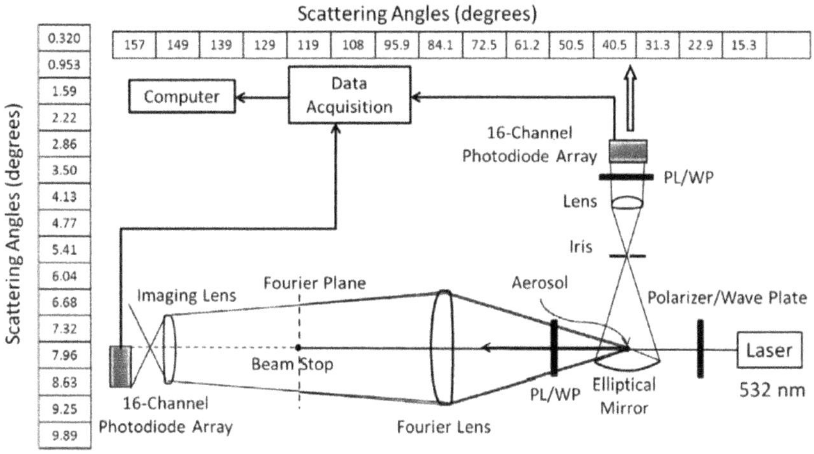

Figure 10.4. Schematic diagram of the Kansas State University multi-angle light scattering apparatus. Reprinted from Wang *et al* (2015), Copyright (2015), with permission from Elsevier.

to collect scattered light over a wide range of angles. The scattering volume in the aerosol is at one focal point of the elliptical mirror and an iris is placed at the other focal point. This scattered light is then collimated by a lens to direct the light to the side detector that detects light over the angular range of 15.3°–157°. One channel of the side detector is sacrificed as a monitor. Both detectors are photodiode arrays, and some manifestations of our device use a 512 channel array for the forward scattering, see e.g. figure 5.19. The device can be calibrated using single slit diffraction in the forward direction. Calibration of the two detectors relative to each other is achieved by redirecting the incident beam to get an overlap of scattering angles. In total light is detected at 31 angles in the range $0.32° \leqslant \theta \leqslant 157°$ for a q range at $\lambda = 532$ nm of $660\ \text{cm}^{-1} \leqslant q \leqslant 2.3 \times 10^4\ \text{cm}^{-1}$; a range of 2½ orders of magnitude. The small angles allow for a complete (i.e. including the Guinier regime and part of the forward scattering lobe) measurement of the scattering profile for particles as large as 20 μm in diameter. The use of photodiode arrays to detect light at many angles simultaneously allows for quick and efficient procurement of data, eliminates problems regarding aerosol stability, and makes detection at small angles easier.

10.1.2.1 Arizona Road Dust

Arizona Road Dust (AZRD) is a well-known dust often used as an aerosol standard. An optical microscope picture of AZRD is given in figure 10.5 (left). One can see relatively clear, irregularly shaped, and non-descript, particles on the order of microns in size. The refractive index of AZRD does not appear to have been measured but based on its composition is thought to be approximately $m = 1.54 + i0.0001$ in the visible region of the spectrum (Curtis *et al* 2008).

Our group conducted light scattering measurements on three AZRD samples (Wang *et al* 2015) for an angular range of $0.32° \leqslant \theta \leqslant 157°$ at a wave length of $\lambda = 532$ nm; the apparatus in figure 10.4. Figure 10.6 shows the Q-space analysis of

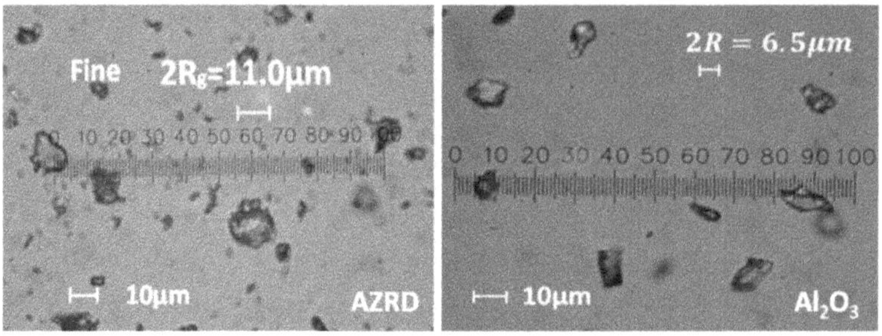

Figure 10.5. Optical microscope pictures of (left) Arizona Road Dust (AZRD) fine grade (Wang *et al* 2015) and (right) 800 grit size Al_2O_3 abrasive powders (Heinson *et al* 2016). Sizes, $2R_g$ and $2R$, were inferred from Guinier analysis of the scattered light.

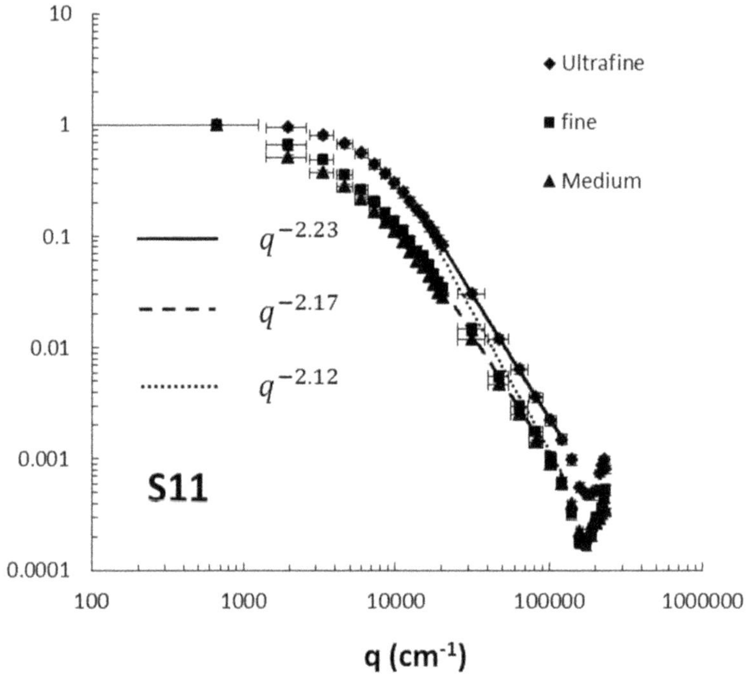

Figure 10.6. Forward normalized S_{11} matrix element (scattered intensity for unpolarized incident light) versus q normalized at the smallest q for light scattering from Arizona Road Dust. Adapted from Wang *et al* (2015), Copyright (2015), with permission from Elsevier.

the scattering by Ultrafine, Fine and Medium (as specified by the supplier) dust samples. The incident light was unpolarized. The scattering shows a hint of a constant forward scattering lobe followed by a Guinier regime, a power law, and at largest q, enhanced backscattering. As for the Amsterdam–Granada dusts, it is uncertain to what extent dust particle shape and polydispersity contribute to create these power laws.

Analysis of the Guinier regime yielded mean radii of gyrations of 2.7 μm, 5.5 μm, and 9.7 μm for the Ultrafine, Fine, and Medium dust samples, respectively. Recognize that these are intensity weight averages of the polydisperse size distribution. This means that when equal numbers of small and big particles are present in the scattering volume, the scattered light detected and will be heavily weighted by the larger particles. The power law slopes yield exponents of −2.23, −2.17, and −2.12 for the Ultrafine, Fine, and Medium dust samples, respectively, each with an error of ±0.05.

10.1.2.2 Al_2O_3 abrasive grits

Our group also studied light scattering by irregularly shaped Al_2O_3 abrasive powders of various grit sizes using the apparatus in figure 10.4 (Heinson *et al* 2016). While not a mineral dust, these abrasive grits are similar in composition and shape. They were chosen in an attempt, somewhat successful, to systematically vary the size while the material and shape remained the same. Figure 10.5 (right) shows an optical microscope picture of one of the grits. The index of refraction of Al_2O_3 is 1.764 with a small birefringence and very small or zero imaginary part at the sodium D line, $\lambda = 589$ nm (1976–77).

Figure 10.7 shows the scattered intensity versus the magnitude of the scattering wave vector q for all six abrasive dusts studied. The sizes range by about a factor of four, but the composition and probably the shape remain the same. The multi-angle light scattering apparatus shown schematically in figure 10.4 was used. The incident light was unpolarized and had a wavelength of 532 nm. The scattering shows a forward scattering regime tending to be constant as $q \to 0$, a Guinier regime, a power law regime with quantifiable exponents, and an enhanced backscattering regime. The exponents of the power laws decrease with increasing size.

At small q, notice how the Guinier regime disappears as the size increases. This happens because the size goes from smaller than the inverse q to bigger than inverse q for the smallest q. The smallest q is $q = 660$ cm^{-1} (corresponding to a scattering angle of 0.32°), thus $q^{-1} = 15$ μm. Thus, Guinier analysis for size could not be performed for the three largest grits, 600, 400 and 320. A rather convoluted analysis of the Guinier regime for the 1200, 1000, and 800 grits led to mean perimeter diameters of 5.2 μm, 8.4 μm, and 14.4 μm, respectively. These are about a factor of two larger than the sizes claimed by the manufacturer but consistent with figure 10.5. As for the AZRD, remember that these are intensity weight averages of the polydisperse size distribution.

The power law regime for the Al_2O_3 abrasives is not well defined as it is for AZRD and the Amsterdam–Granada data, and in many cases not a power law. Ignoring this fact and forcing power laws, solid lines in figure 10.7, leads to exponents decreasing with increasing internal coupling parameter ρ', which could be calculated approximately assuming a spherical shape from the size, wavelength and refractive index of Al_2O_3 which is $m = 1.764$. However, in some cases the Guinier regime bulges forth, and the largest three abrasives, for which $\rho' \gtrsim 100$, show a kink in the power law. The presence of this kink is near the crossover from

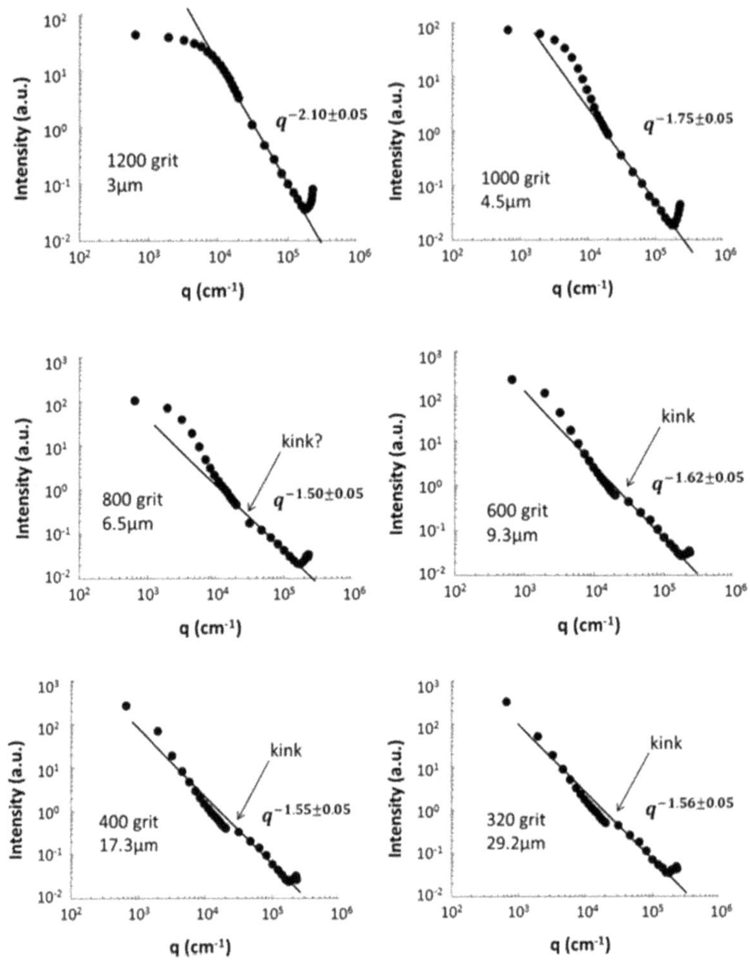

Figure 10.7. Scattered intensity (arbitrary units) versus the magnitude of the scattering wave vector q for six different sizes of Al_2O_3 abrasive grits. The incident light was unpolarized and had a wavelength of 532 nm. The manufacturer's size is labeled in each figure. Power law regimes are indicated by straight lines and the power law is labeled. Dashed lines are suggested power laws with slope −3. Adapted from Heinson *et al* (2016), Copyright (2016), with permission from Elsevier.

the forward detector to the side detector to suggest an instrumentation artifact, but this possibility was carefully checked to find that the kink is real. A tenuous explanation draws the dashed lines in figure 10.7 with slope −3 to imply a regime of 2d Fraunhofer diffraction. Then the kink is the onset of a mild hump in the Guinier regime. Such behavior occurs for spheres when $\rho' > 30$, see e.g. figure 8.9, and this implies that the abrasive particles have a sphere-like shape. This explanation is at best tentative. Nevertheless, we ask (but do not answer), are the abrasives more sphere-like than the AZRD and the great variety of dusts in the Amsterdam–Granada data set?

10.1.2.3 Molybdenum disulfide (MoS_2) particles

Micron sized, irregularly shaped, molybdenum disulfide (MoS_2) particles provide useful insights to scattering by irregular particles. In particular molybdenum disulfide has rather extreme values of the real and imaginary parts of the refractive index, hence expands the phase space of independent variables that control scattering that we have so far explored. The complex refractive index has been reported as $m = n + i\kappa = 5.24 + i1.16$ (Beal and Hughes 1979), $4.43 + i1.17$ (Yim *et al* 2014) and $4.41 + i0.63$ (Zhang *et al* 2015) for the wavelength $\lambda = 532$ nm. This averages roughly to $m = 5 + i$. We studied scattering in an angular range of $0.32° \leqslant \theta \leqslant 157°$ at a wavelength of 532 nm using our multi-angle apparatus described above in figure 10.4 (Gautam and Sorensen 2020). The incident light was polarized perpendicular to the scattering plane. We also sampled the aerosols with a commercially available aerodynamic particle sizer spectrometer (APS 3321 (TSI)) to determine the size distribution of the aerosolized MoS_2 particles from which we scattered light. Optical and TEM pictures are shown in figure 10.8.

The measured aerodynamic diameters were converted to the volume equivalent diameters. Figure 10.9 shows the volume equivalent particle number distribution for MoS_2 and their transformed mass and light scattering intensity weighted distribution. The number weighted distribution peaks around the diameter 1.13 ± 0.10 μm with geometric standard deviation spread of $\sigma = 1.6$. The size parameter corresponding to this diameter is $kR = 6.7$. This number size distribution was used for Mie spherical particle for calculations with $m = n + i\kappa = 5.24 + i1.16$ (Beal and Hughes 1979). Using other reported values of m did not significantly change the calculations. Notice how different the number distribution is to the mass and intensity distributions. The Guinier inferred size is consistent with the intensity-weighted distribution. Obviously, light scattering heavily weights the large size tail of the number distribution.

Figure 10.10 shows the scattering data in both Q-space and θ-space and a Mie calculation for spheres of the same APS determined size distribution for comparison.

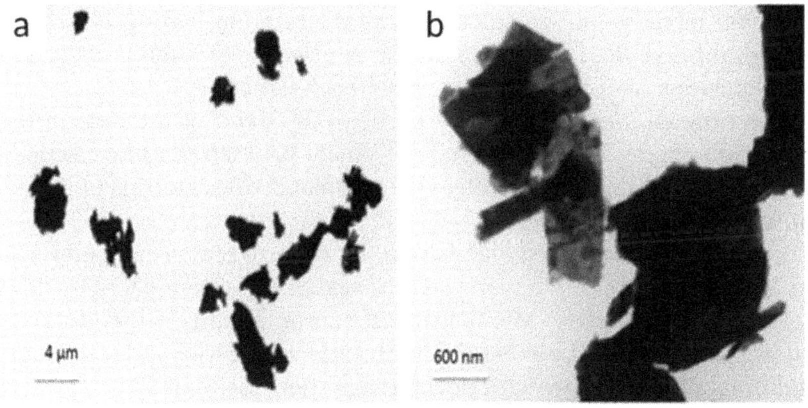

Figure 10.8. Transmission electron microscope (TEM) images of molybdenum disulfide (MoS_2) particles. Adapted from Gautam and Sorensen (2020), Copyright (2020), with permission from Elsevier.

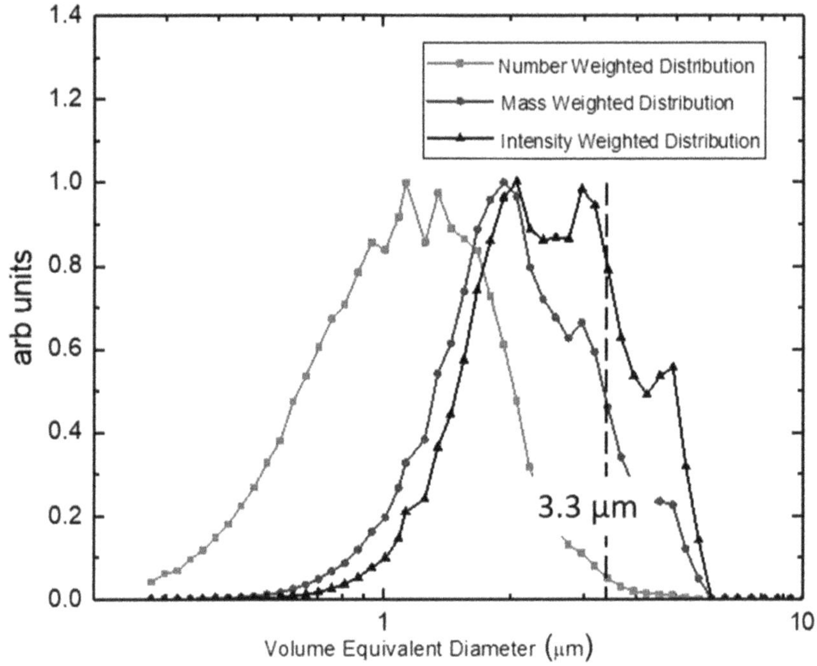

Figure 10.9. Particle number (grey), mass (brown) and intensity weighted (black) distributions for the molybdenum disulfide (MoS_2) particles. The number distribution was measured with an APS 3321 after converting aerodynamic diameter to the volume equivalent diameter. Transformations of this distribution to mass and intensity-weighted distributions were achieved with weighting factors of diameter to the third and fourth power, respectively. Dashed line marked $D = 3.3$ µm indicates the size determined by Guinier analysis of the light scattering data. Adapted from Gautam and Sorensen (2020), Copyright (2020), with permission from Elsevier.

The Q-space plot for the data shows a forward scattering lobe and a Guinier regime, but the subsequent power law regime is ill defined. This is in contrast to the distinct power laws seen in the Amsterdam–Granada data set and the AZRD data. Furthermore, there is no enhanced backscattering, in strong contrast to the AZRD and abrasive data and much of the Amsterdam–Granada data. Note that the Mie calculation also shows no enhanced backscattering.

The conventional plot more clearly portrays the trend of the data in the larger scattering angle range. There is general agreement between the Mie calculation and experiment, but the Mie result slightly underestimates the scattering at larger angles. Here, it is valuable to recognize that when the parameter κkR (section 7.2.1) is large, essentially all the light scattered energy is in the forward scattering regime (Sorensen *et al* 2017), see section 8.3. Furthermore, the scattering is dominated by diffraction, so in such a limit, who needs Mie theory? Of course, 2d diffraction has a −3 power law, but would a rotationally averaged 2d obstacle with a wide range of length scales yield a diffraction power law with power magnitude less than 3?

The lack of enhancement in the backscattering up to 157° can be explained with the value of the parameter κkR. For our MoS_2 particles we found above an intensity

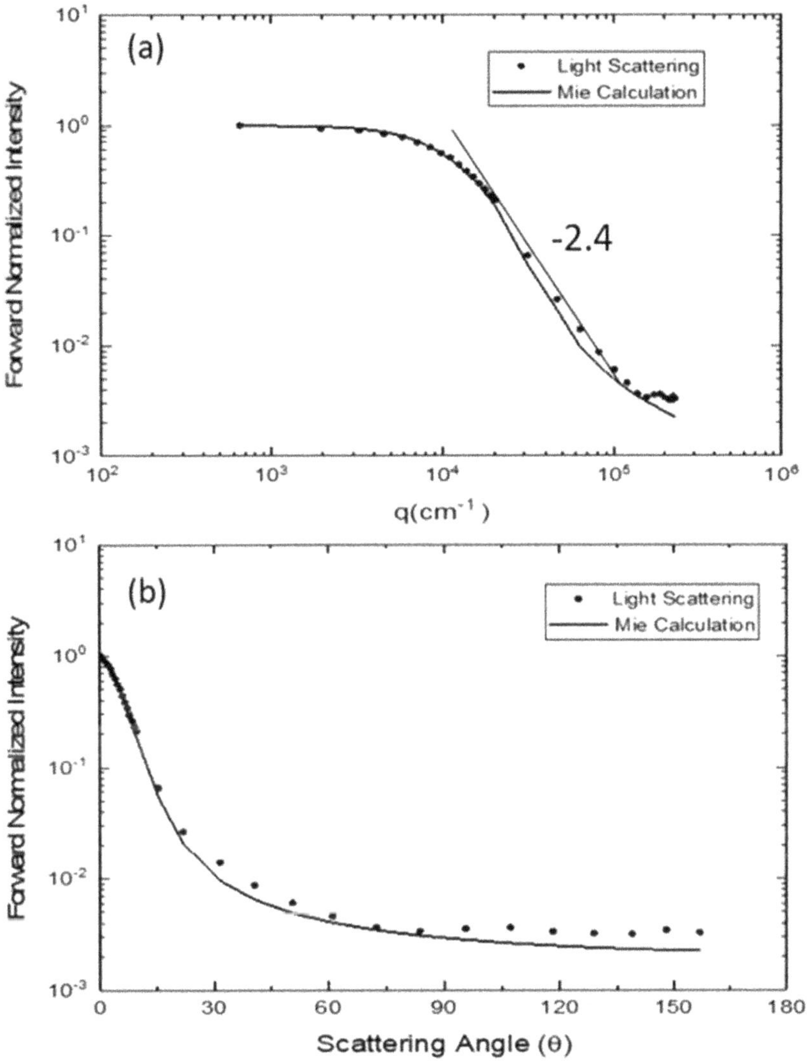

Figure 10.10. Comparison of the experimentally observed forward normalized light scattering data with Mie calculations for spheres with the APS observed molybdenum disulfide (MoS_2) particle size distributions, (a) plotted logarithmically versus q, the method of Q-space analysis. The line indicates a power law of $q^{-2.4}$. (b) Plotted linearly as a function of scattering angle (θ). Adapted from Gautam and Sorensen (2020), Copyright (2020), with permission from Elsevier.

weighted diameter of $D = 3.3$ μm, hence κkR in the range of 12.5 to 22.75, depending on which of the three reported values of κ are used. These are all very large values well in the saturated regime. Backscattering occurs in micron sized spheres due to internal rays that reflect at least once off the inner surface of the sphere; the $p \geqslant 2$ rays of the Debye series, see section 8.2.8. When $\kappa kR \geqslant 3$, these internal rays are completely extinguished to result in no enhanced backscattering.

Now, with the observations that Amsterdam–Granada, AZRD and abrasive particles essentially all show enhanced backscattering and yet molybdenum disulfide particles with large κkR do not, we must conclude that interior rays are responsible for the presence of, or lack of, enhanced backscattering in all these particles despite their irregular shapes.

Guinier analysis in Q-space shows the radius of gyration R_g to be 1.51 μm. By multiplying R_g by $\sqrt{5/3}$, the three-dimensional object equivalent radius is $R = 1.9$ μm. Thus, the light scattering determined diameter is $D = 3.8$ μm. Recall in section 7.2.4.3 it was shown that the Guinier inferred radius of gyration is as much as 50% larger than true radius of gyration when the $\rho > 1$ (Sorensen and Shi 2000). For our MoS_2 at 532 nm, $\rho' \approx 13$. At that value, $R_{g,G}$ is greater than R_g by around 12%. Reducing $D = 3.8$ μm by 12% leads to $D = 3.3$ μm. This value is included in figure 10.9 and seen not to agree with the APS number distribution, but it agrees fairly well with the intensity-weighted distribution.

It is useful to look at these Guinier result from the perspective that the Guinier equation was derived for x-ray scattering, and x-rays have a refractive index of essentially one. Indeed, x-ray scattering is simply wave diffraction, the electromagnetic character is insignificant. Nevertheless, our results here show that the Guinier equation applied to light scattering from micron size, highly refractive particles still yields a valuable measure of the size. In addition, look again at the distributions in figure 10.9. The weighting by mass and scattering magnifies a very small faction (by number) of the distribution. Moreover, the random error in the APS measurement is relatively large at the large end of the distribution and this large relative error remains with the weighting. All these issues must be remembered when using Guinier analysis on light scattering, but they should not deter us from making such measurements and using them with due caution.

10.1.2.4 Hematite (α-Fe_2O_3) particles

Subsequent to the study of molybdenum disulfide, we studied scattering by hematite (α-Fe_2O_3) particles (Gautam *et al* 2020). Here, we follow that paper closely. Like molybdenum disulfide, hematite has high values of the real and imaginary parts of the refractive index. On Earth, hematite is known as an iron ore. On Mars, it lends its orange color to that of the planet's. On both planets, it can be windblown, and on Mars it is a component of the great global Martian sandstorms. It is a birefringent material with significant dispersion. At a wavelength of $\lambda = 532$ nm, the real and imaginary values of the refractive index for extraordinary ray is $n = 2.8$ and $\kappa = 0.5$ and for the ordinary ray is $n = 3.2$ and $\kappa = 0.5$ for 532 nm (Ordal *et al* 1985, Sokolik and Toon 1999). A useful average is $m = n + i\kappa = 3 + i0.5$.

Scanning electron microscopy (SEM) indicated that the particles we used had a compact, aggregate structure, as shown in figure 10.11. Ultra small angle x-ray scattering (USAXS) data obtained at the Advanced Photon Source, a U.S. Department of Energy user facility at Argonne National Laboratory, showed Porod regimes with q^{-4} power laws. This indicated that the aggregates were not fractals and have mass and scaling dimensions of 3 and 2, respectively (see section 5.3). The largest

Figure 10.11. (a)–(c) Scanning electron microscope (SEM) images; (d) an optical microscope image of the hematite particles used in our light scattering study. Reprinted from Gautam *et al* (2020), Copyright (2020), with permission from Elsevier.

particles were ~10 μm, and composed of small grains, monomers, with a flat, disk shape and mean size of approximately $2a \simeq 200$ nm.

We studied scattering in the angular range of $0.32° \leqslant \theta \leqslant 157°$ at a wavelength of 532 nm using our multi-angle apparatus described above in figure 10.4. The incident light was polarized perpendicular to the scattering plane. Figure 10.12 (left) shows Q-space analysis of the scattered intensity. The scattering data are normalized to 1 at the smallest measuring angle of 0.3° ('forward normalized'). At smallest q, the data appear to be leveling off to a forward scattering lobe with no q, hence angular, dependence. Then, at larger q, an extended and somewhat unusual Guinier regime follows, and after that a power law of slope −3. None of these features were discernable in the θ-space analysis. Finally, enhanced backscattering is seen at large q corresponding to angles of $\theta = 129°$, 139°, 149° and 157°.

Figure 10.12 also shows spherical particle scattering for hematite refractive index of $m = 3 + i0.5$ for comparison on the right side. Spheres with a radius of $R = 4.6$ μm, similar to the larger hematite particles, show a forward scattering lobe, a clean Guinier regime to indicate a single size, a ripple regime with a power law envelope of q^{-3}, followed by a tail with no enhanced backscattering. Spheres with the same refractive index but a size comparable to the monomer grains, $R = 100$ nm, are also included on the right side of figure 10.12. These small spheres have uniform scattering until large q where there is Guinier dip followed by enhanced backscattering.

Chapters 7 and 8 have armed us to understand the spherical particle scattering in figure 10.12. For the larger sphere the skin depth parameter $\kappa kR = 27.2$. With this

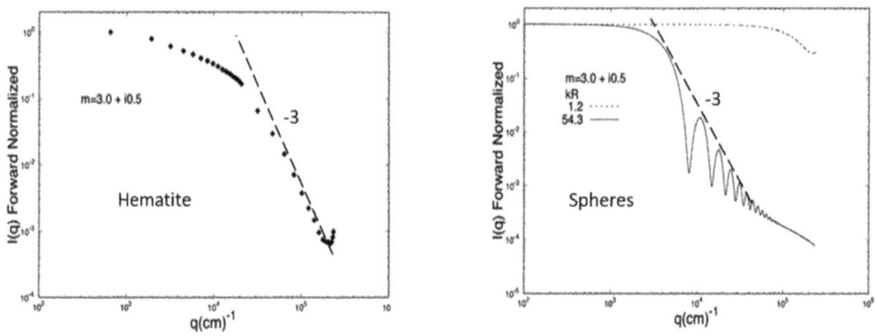

Figure 10.12. (Left) Forward normalized Q-space analysis, scattered light intensity versus q plotted double logarithmically, of the light scattered by hematite particles. (Right) Mie scattered intensity for spheres with the same refractive index as hematite with a radius of $R = 4.6$ μm, similar to the larger hematite particles, solid curve, and a spherical hematite grain with radius 100 nm, dash-dot curve. In each half, dashed lines with slope −3 are shown. Adapted from Gautam *et al* (2020), Copyright (2020), with permission from Elsevier.

very large value, the sphere is deep in the regime in which the incident light leads to only surface electric polarization. The sphere is essentially an opaque object with a circular, geometric cross section, so that the scattering is mostly equivalent to 2d Fraunhofer diffraction. This explains the ripples encased under a q^{-3} power law. Furthermore, with this large κkR value, there are no internal waves to lead to enhanced backscattering. For the smaller sphere the parameter $\kappa kR = 0.6$. This finite but small value will attenuate but not eliminate interior waves. Thus, the small sphere will not act like a 2d, opaque object. Furthermore, the interior waves, although somewhat attenuated, will lead to some enhanced backscattering.

With these spherical particle perspectives, we now give a plausible explanation of the scattering by the hematite aggregates. The low power, wide field, optical microscope of view picture in figure 10.11(d) indicates sub-micron particles dominate in number. Unfortunately, a quantitative size distribution measurement could not be done with our APS because the hematite aerosol would clog and ruin the device. Nevertheless, we were able to show, see publication, that the extended Guinier regime seen in figure 10.12 (left) and the subsequent q^{-3} power law could be explained with a bimodal size distribution of particles with modes of $R = 0.5$ and 10 μm with large κkR. This bimodality is consistent with the picture in figure 10.11.

The fact that same sized, large hematite aggregates and spheres show different backscattering implies that the aggregate structure is the origin of the enhance backscattering. To explore this implication we created a computer-generated spherical aggregate. The overall sphere radius was $R = 1.8$ μm, and it was filled with N spherical monomers with radii of $a = 100$ nm. The scattering for $\lambda = 532$ nm light was calculated using the T-matrix method (Mackowski and Mishchenko 2011). Results are shown in figure 10.13.

Figure 10.13 (top) shows calculations for a refractive index of $m = 3 + i1.0$, which has twice the imaginary part κ as the hematite. This κ was used to eliminate the possibility that the small amount of backscattering from the monomer when $m = 3 + i0.5$ was not being enhanced by the aggregate structure, i.e. we quenched the

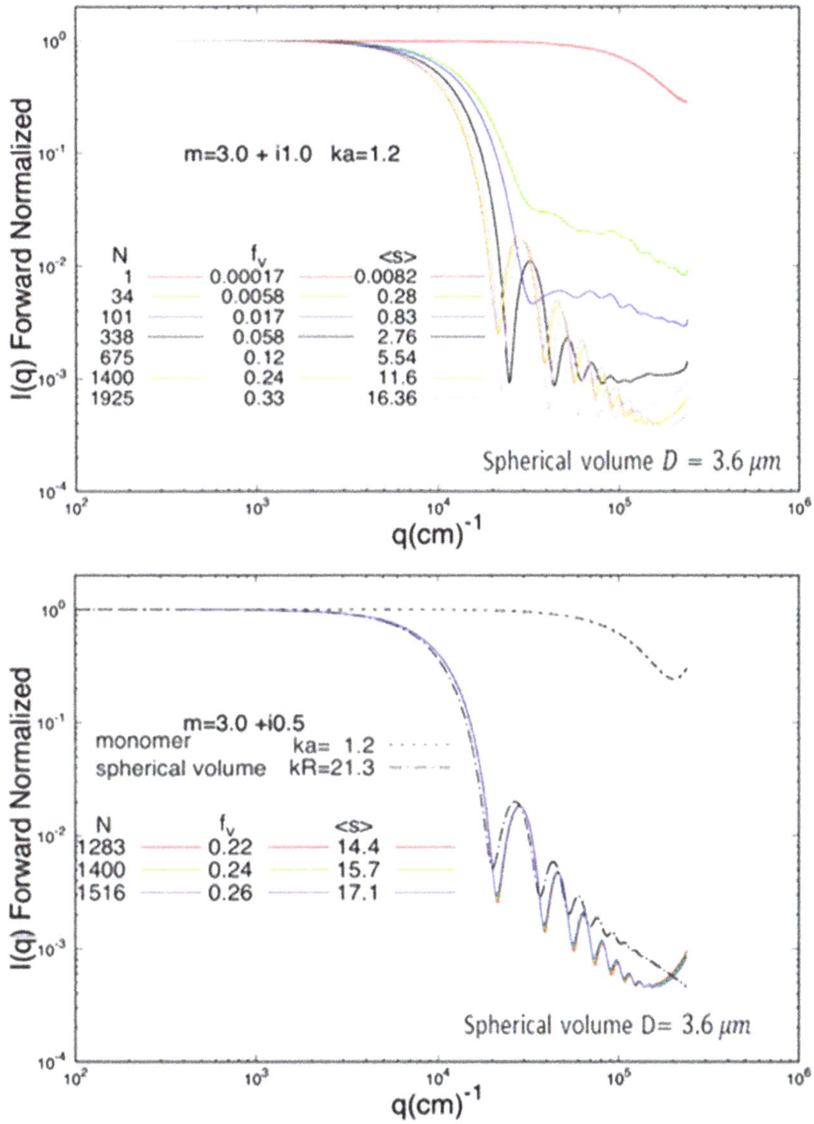

Figure 10.13. Calculated scattering for spherical aggregates of overall radius R = 1.8 μm filled with N monomers of radius a = 100 nm. The volume fraction of monomers is f_V, and $\langle s \rangle$ is the average number of photon scattering events within the aggregate. (Top) For a monomer refractive index of $m = 3 + i1.0$; (bottom) for $m = 3 + i0.5$, the refractive index of hematite at $\lambda = 532$ nm. Also included are Mie calculations for uniform spheres with the monomer radius and the aggregated radius.

backscattering of the monomer. Figure 10.13 (top) shows no enhanced backscattering for simulated aggregates with small numbers of monomers, N, hence the small volume fraction, f_V. However, when $f_V \geqslant 0.058$, enhanced backscattering begins to appear.

The bottom half of figure 10.13 shows calculations for the aggregate with the refractive index of hematite in the range of volume fractions measured for the studied aggregates. Calculations show significant enhanced backscattering comparable to that observed in the experiment, figure 10.12. In stark contrast, the same sized uniform sphere shows no enhanced backscattering. We conclude the enhanced backscattering is due to the porous aggregate structure.

The interfaces of the monomers with the voids inside the aggregate present discontinuities of refractive index within the aggregate that are not present when the spherical volume is uniform, i.e. when the monomer volume fraction is unity. This will lead to internal multiple scattering between monomers[3]. A photon picture of multiple scattering involves the photon mean free path (mfp) (Mokhtari *et al* 2005), as described in section 2.6. This is the average distance travelled by the photon between successive scattering events within the scattering volume. The photon mean free path (mfp) is given by equation (2.22)

$$\text{mfp} = (n\, C_{\text{sca,mon}})^{-1}, \tag{10.1}$$

where $C_{\text{sca,mon}}$ is the total scattering cross-section of a monomer, and n is the number density of the monomers. The average number of scattering events $\langle s\rangle$ is the ratio of the path length of the light beam in the scattering volume to the mfp. The path length in the spherical volume ranges from 0 to D, the diameter. Thus, to a good approximation, we take the radius R as the path length and follow equation (2.21) to find

$$\langle s\rangle = \frac{R}{\text{mfp}}, \tag{10.2}$$

The probability for a given number of scattering events s for a photon is determined by the Poisson distribution, equation (2.17).

$$P(s) = \frac{\langle s\rangle^s}{s!} e^{-\langle s\rangle} \tag{10.3}$$

Multiple scattering plays a significant role for non-absorptive particles when the average number of scattering events $\langle s\rangle$ becomes large.

We have calculated the average number of scattering events $\langle s\rangle$ for the spherical aggregate with radius R = 1.8 μm. The enhancement in the backscattering for different volume fractions was quantified by calculating the ratio of scattered intensity at 180° to the minimum intensity just before 180°, i.e. (I(180)/I(Min)) This ratio is plotted versus the average number of scattering events in figure 10.14.

Figure 10.14 shows that when the average number of scattering events in the aggregate is very small, there is no enhanced backscattering. With increasing scattering events, there is an increase in the backscattering enhancement, especially after the average number of scattering events $\langle s\rangle$ exceed 1.0. This strongly supports the hypothesis that the enhanced backscattering is due to multiple scattering within

[3] In contrast to internal multiple scattering between different volume elements of a uniform particle.

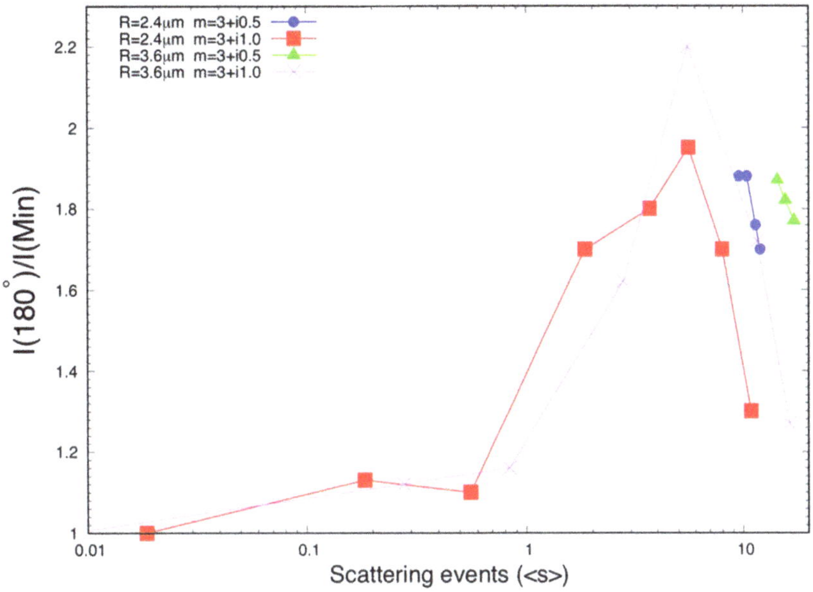

Figure 10.14. Plot showing the enhancement in the backscattering $I(180°)/I(\mathrm{Min})$ versus the average number of scattering events in the aggregate, $\langle s \rangle$.

the hematite aggregate. The enhancement peaks around $\langle s \rangle \approx 6$, and with further increase in the scattering events, the backscattering enhancement decreases. This decrease may be due to two possible reasons. First is that with increasing number of monomers, the spherical volume starts to act like a homogenous particle. Secondly, since the monomers are highly absorptive in nature, the absorption starts to dominate, causing a decrease in the backscattering enhancement. Regardless, the conclusion is that the aggregate structure of the hematite leads to internal multiple scattering, which causes enhanced backscattering that does not occur for spheres with similar large imaginary part to the refractive index.

10.1.2.5 Summary for mineral dusts

In summary, the three data sets above all concern dusts and all show definite q-space power laws in the scattering except the three largest abrasives. All these dust samples are polydisperse, some extremely so. Polydispersity will smooth out the angular dependence and perhaps such smoothing will remove bumps and dips in the phase function to lead to smooth power laws. However, unpublished attempts on our part have so far not been able to show the viability of this proposal. For now, we conclude that the q-space view of the angular scattering pattern, the phase function, for ensembles of randomly oriented dust particles, with the shape yet to be quantitatively described, includes a constant intensity at smallest q followed by a Guinier regime leading to power laws after the Guinier regime and often a brief enhanced backscattering. Furthermore, the relative skin depth parameter, κkR, is effective for describing scattering by irregularly shaped particles.

10.2 The general Rayleigh method for a particle of arbitrary shape

10.2.1 Rayleigh scattering by an arbitrary shape

The spherical particle Rayleigh differential scattering cross section was given in chapter 6, equation (6.5), as

$$dC_{\mathrm{sca}}/d\Omega = k^4R^6 \left| \frac{m^2 - 1}{m^2 + 2} \right|^2 \sin^2\psi \tag{10.4}$$

This equation includes the Lorenz–Lorentz term, $|(m^2 - 1)/(m^2 + 2)|$, the functional form of which is solely for spheres. This same spherical shape term is found in our formulation of the internal coupling parameter, equation (7.17). The Lorenz–Lorentz term is the time varying generalization of the Claussius–Mossotti relation which describes, among other things, the polarization of a spherical dielectric in a static applied electric field (Jackson 1999). Other shapes will have other functionalities on the refractive index. In this section, methods will be described to determine the Lorenz–Lorentz term and the concomitant Rayleigh scattering cross section and internal coupling parameter for other shapes.

van de Hulst (1957) showed that the Rayleigh differential cross section for a particle of arbitrary shape is

$$\frac{dC_{\mathrm{sca,\ Ray,\ arb}}}{d\Omega} = k^4V^2 \,|\, \alpha(m) \,|^2 \tag{10.5}$$

where V is the volume of the particle and $\alpha(m)$ is the average volume polarizability with a shape dependent functionality on the refractive index m. For ellipsoids with volume $V = (4/3)\pi abc$ an analytic form is available as

$$| \alpha(m) |^2_{\mathrm{ell}} = \frac{3}{(4\pi)^2}\left[\left| \frac{(m^2 - 1)}{(3+3L_1(m^2 - 1))} \right|^2 + \left| \frac{(m^2 - 1)}{(3+3L_2(m^2 - 1))} \right|^2 + \left| \frac{(m^2 - 1)}{(3+3L_3(m^2 - 1))} \right|^2\right] \tag{10.6}$$

where a, b, and c are the semi-principle axes, the subscript 'ell' means 'ellipsoid', and L_1, L_2,and L_3 are geometric parameters given by the integrals

$$L_x = \frac{abc}{2}\int_0^\infty \frac{dq}{(q + x^2)\sqrt{(q + a^2)(q + b^2)(q + c^2)}} \tag{10.7}$$

In (10.7) x is equal to a, b or c for L_1, L_2 and L_3, respectively. For spheroids, with $a \neq b = c$, $L_2 = L_3$, equation (10.6) leads to

$$| \alpha(m) |^2_{\mathrm{spheroid}} = \frac{3}{(4\pi)^2}\left[\left| \frac{(m^2 - 1)}{\left(3+3L_1(m^2 - 1)\right)} \right|^2 + 2\left| \frac{(m^2 - 1)}{\left(3+3L_3(m^2 - 1)\right)} \right|^2\right] \tag{10.8}$$

For the special case of spheres, $L_1 = L_2 = L_3 = 1/3$ one finds

$$|\alpha(m)|^2_{\text{sph}} = \frac{9}{(4\pi)^2}\left|\frac{(m^2-1)}{(m^2+2)}\right|^2. \tag{10.9}$$

Equation (10.9) regains the Claussius–Mossotti, Lorenz–Lorentz form for the spherical polarizability. Substitution of equation (10.9) into equation (10.5) yields the well-known equation for the Rayleigh differential cross section for a sphere, equation (10.4).

There are no known analytical results for $|\alpha(m)|^2$ for shapes more complex than generalized ellipsoids. Therefore, for arbitrary shapes we used computations to determine $|\alpha(m)|^2$ numerically. To do this the differential cross section for a given particle shape was calculated using an ADDA code (Yurkin and Hoekstra 2011) for a series of volumes within the Rayleigh limit that the wavelength of light is much greater than the largest linear size of the particle (Maughan *et al* 2017). The differential cross section in the forward direction (i.e. $\theta=0$) divided by k^4 was then plotted versus V^2, the slope of which yields $|\alpha(m)|^2$ by equation (10.5). Once $|\alpha(m)|^2$ is known for a given shape, equation (10.5) can be used to calculate the Rayleigh scattering cross section for that shape. We call this procedure the *general Rayleigh method (GRM)*.

10.2.2 Internal coupling parameter for arbitrary shapes

As with spheres, the internal coupling parameter for an arbitrary shape is derived by taking the ratio of the forward scatterings of the particle in the 3d and 2d diffraction limits (section 7.1.3.2). The former of these is the Rayleigh differential cross section, as defined by equation (10.4), and the calculated average volume polarizability $|\alpha(m)|^2$ obtained by the GRM. The latter is Fraunhofer diffraction from a two-dimensional obstruction of arbitrary shape given by (Born and Wolf 1975, Bohren and Huffman 1983, Bohren and Koh 1985)

$$\frac{dC_{\text{sca, arb}}}{d\Omega}(0) = \frac{k^2 A^2_{\text{proj}}}{4\pi^2} \tag{10.10}$$

where A_{proj} is the rotationally averaged projected area of the three-dimensional shape in the direction of the incident beam. Taking the square root of the ratio of these two limits results in

$$\rho'_{\text{arb}} = 2\pi k \frac{V}{A_{\text{proj}}} |\alpha(m)| \tag{10.11}$$

Equation (10.11) is the general definition for the internal coupling parameter for any arbitrary shape, of which equation (7.17) in section 7.1.3.2 is a specific case for spheres.

The rotationally averaged projected area is equal to $S/4$, where S is the total surface area of the particle (Slepian 2012). This applies when the particle has no concavities on the surface. This condition is true for spheres, ellipsoids and a variety of elongated or flattened shapes like cylinders and prisms. One can show that as the aspect ratio of the shape becomes more different from unity, the ratio V/A_{proj} in equation (10.11) becomes dominated by the smaller dimension and consequently smaller than the volume equivalent sphere radius. Thus, ρ'_{arb} in equation (10.11) can be quite different from its volume equivalent sphere counterpart.

Figure 10.15 demonstrates the efficacy of the GRM by comparing the square of the average volume polarizability $|\alpha(m)|^2$ for spheroids with $m = 1.5$ calculated using the exact analytic equation (10.8) to the GRM. It can be seen that the GRM agrees well with the analytical solution, deviating only slightly as the aspect ratio a/b becomes extreme. We believe this deviation is due to problems with the ADDA code. Figure 9.7 also shows the results of the application of the GRM to cylinders with $m = 1.5$, and a comparison can be made to spheroids. When the ratio $L/D = 1$, where D is the diameter and L is the length of the cylinder, the $|\alpha(m)|^2$ term for cylinders sits slightly above the spheroids, and then still follows the same general trends of spheroids as the shape is elongated or flattened. It seems reasonable to expect the same quantitative trend for cylinders with geometric cross sections other than circular. A general conclusion is that the Rayleigh scattering increases as one deforms a given volume of material into non-spherical shapes. However, note that the variation in $|\alpha(m)|^2$ is rather small, 25% or less as the aspect ratio varies an order

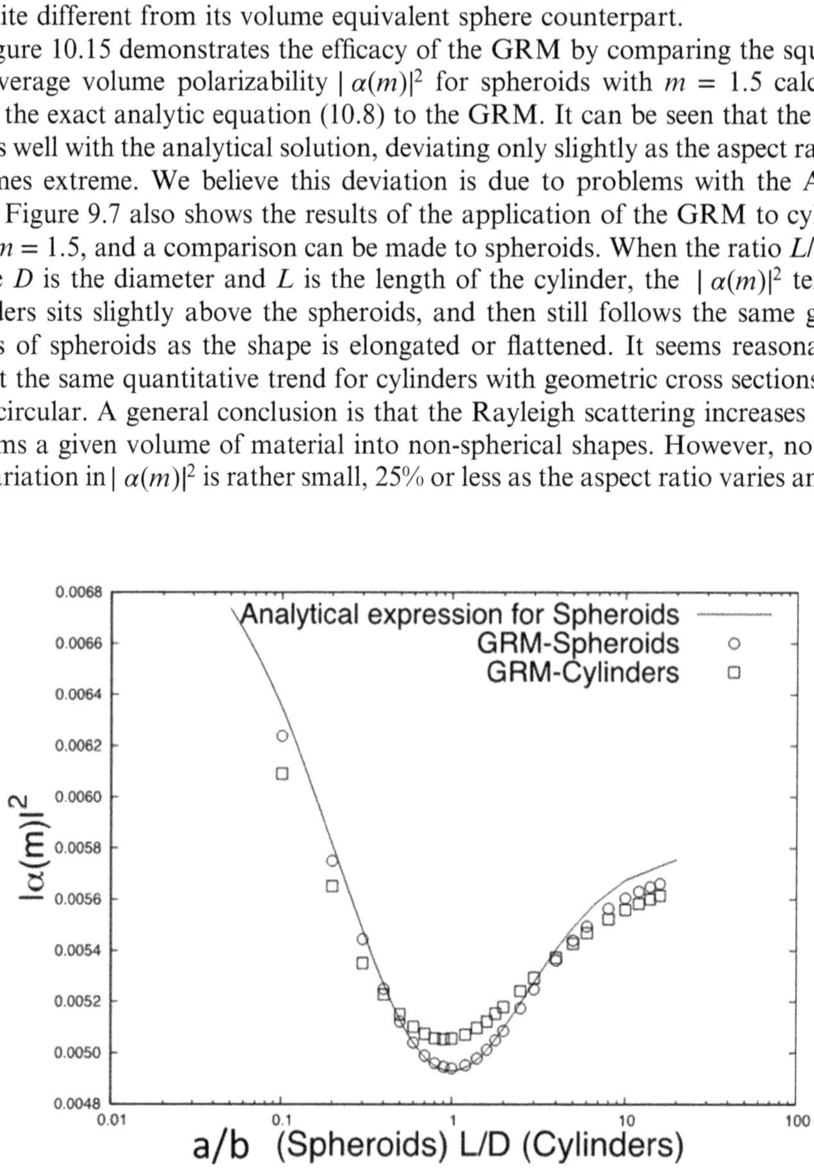

Figure 10.15. A comparison of the average volume polarizability squared $|\alpha(m)|^2$ for spheroids and cylinders with $m = 1.5$ calculated using the general Rayleigh method (GRM) versus the aspect ratio a/b for spheroids (open circles) or L/D for cylinders (open squares). The solid line shows the analytical solution for spheroids calculated with equation (10.8). The minimum for spheroids near $|\alpha(m)|^2 \simeq 0.0049$ is in agreement with equation (10.9) for spheres. Reprinted from Maughan *et al* (2017), Copyright (2017), with permission from Elsevier.

of magnitude in either direction. This corresponds to variations in ρ'_{arb} of ~12% or less.

Comparing the effects of non-sphericity on scattering and the internal coupling parameter for a constant volume V shows that scattering, described by equation (10.5), is only affected via the changes in polarizability α(m), which are rather modest. On the other hand, the internal coupling parameter, described by equation (10.11), depends on the projected area A_{proj} as well as the polarizability, and A_{proj} can change significantly with shape for a fixed volume.

With the generalized polarizability, the generalized internal coupling parameter ρ'_{arb} can be calculated for any arbitrary shape via equation (10.11). Figure 10.16 demonstrates the efficacy of this generalized internal coupling parameter. To set the stage for this demonstration recall figure 7.6 that showed that ρ' could provide a universal parameter for forward scattering by spheres of any size or refractive index. Figure 10.16 is a similar attempt to find a universal parameter in ρ'_{arb} for particles of arbitrary shape. Plotted there is the forward scattering differential cross section multiplied by the square of the internal coupling parameter ρ' versus ρ' for cylinders of various aspect ratios L/D. Recall for spheres that such a plot yields a quasi-universal description of the forward scattering dependent only upon ρ' regardless of the values of the size parameter and the refractive index that determine ρ'. Furthermore, for spheres such plots approach unity as $\rho' \to \infty$; see figure 7.6(d). The left hand side of figure 10.16 shows the scattering for cylinders analyzed by application of ρ'_{arb} for spheres, whereas the right hand side shows the analysis by application of the generalized (for arbitrary shapes) ρ'_{arb} for the cylindrical shape. One can see that the use of the generalized ρ'_{arb} for cylinders yields a better unification of the plots and the asymptotic approach to unity. This indicates the need for the generalization.

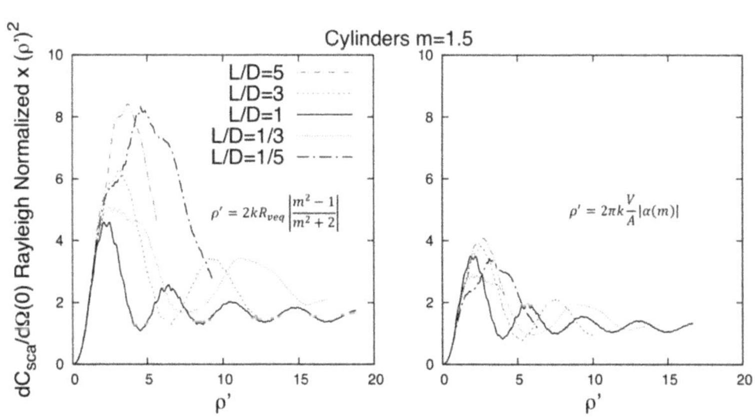

Figure 10.16. (Left) Rayleigh normalized differential cross section for cylinders with refractive index $m = 1.5$ with different aspect ratios L/D multiplied by $\left(\rho'_{\mathrm{sph}}\right)^2$ versus ρ'_{sph} using the volume equivalent radius. (Right) Rayleigh normalized differential cross section of cylinders multiplied by $\left(\rho'_{\mathrm{arb}}\right)^2$ versus ρ'_{arb}. Reprinted from Maughan *et al* (2017), Copyright (2017), with permission from Elsevier.

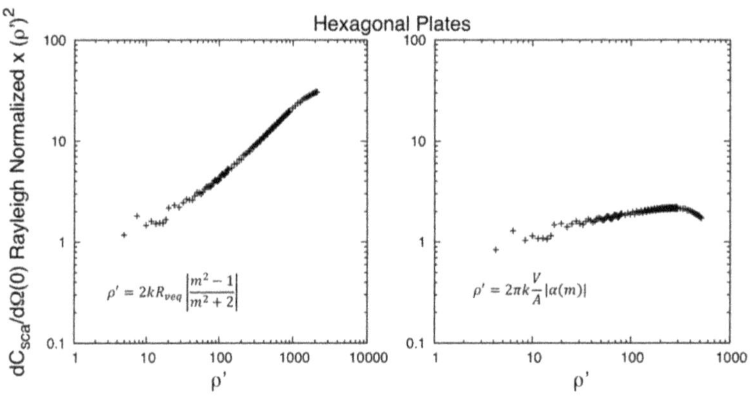

Figure 10.17. (Left) Rayleigh normalized differential cross section of hexagonal plates with refractive index $m = 1.31$ and a variety of aspect ratios multiplied by $(\rho'_{\rm sph})^2$ versus $\rho'_{\rm sph}$ using the volume equivalent radius of the plate. (Right) GRM Rayleigh normalized differential cross section of hexagonal plates multiplied by $(\rho'_{\rm arb})^2$ versus $\rho'_{\rm arb}$. Reprinted from Maughan *et al* (2017), Copyright (2017), with permission from Elsevier.

Yet another example is given in figure 10.17. As can be seen on the left of figure 10.17, when the sphere Rayleigh differential cross section and $\rho'_{\rm sph}$ are used to study the forward scattering of ice crystal plates, the Rayleigh normalized forward scattering varies over a range of ~30. On the other hand, figure 10.17 shows on the right that when the GRM is used for Rayleigh normalization and $\rho'_{\rm arb}$, the data is more tightly grouped over a range of ~2 as $\rho'_{\rm arb}$ increases. Note that $\rho'_{\rm arb}$ and $\rho'_{\rm sph}$ differ by about a factor of four, due mostly to the different values of $V/A_{\rm proj}$.

The improvements that the GRM $\rho'_{\rm arb}$ yields for description of scattering for non-spherical shapes over $\rho'_{\rm sph}$ of equation (7.17) is significant. However, $\rho'_{\rm arb}$ fails to be a completely unifying parameter. Nevertheless, this is better than a poke in the eye with a sharp stick, and within this caveat, we conclude that for non-spherical shapes the forward scattering is better described by using the generalized expressions for the Rayleigh scattering and the internal coupling parameter.

Given the success of $\rho'_{\rm arb}$ as defined in equation (10.11) and the fact that this expression includes the result for a sphere, equation (7.17), we hereafter drop subscript such as 'arb' and 'sph' and simply write ρ' and rely on context to indicate if accounting for non-spherical shape is necessary.

10.3 Theoretical calculations of scattering by various shapes

10.3.1 Solid geometric shapes orientationally averaged

Most experimental and field conditions will encounter light scattered by particles with random orientations. Thus, it is very worthwhile to study light scattered from particles averaged over random orientations. Here, we present our studies of solid, geometric shapes: spheroids and right cylinders with circular, square and hexagonal cross sections. Diffraction for these shapes was studied in section 5.4. A major theme of this book is that scattering evolves from diffraction as the refractive index increases from unity, and that theme will be amplified here.

In section 5.4, it was demonstrated that the shape of the diffraction pattern was fundamentally related to the aspect ratio of the particle. The aspect ratio was defined by equation (5.24), which states that the aspect ratio is $\varepsilon = a/b$ for spheroids or $\varepsilon = L/2R$ for the cylinders, where L is the cylinder length and $2R$ is a diameter, a square side or a hexagon corner to opposite corner distance for circular, square or hexagonal geometric cross sections, respectively.

The differential light scattering cross section was calculated for each of the four shapes for a variety of real refractive indices and an optical wavelength of $\lambda = 628$ nm. The method of calculation was the IITM code (Bi *et al* 2013). These cross sections were normalized by the Rayleigh cross section calculated by the GRM embodied in equation (10.5), above. Refractive indices and sizes were selected to generate a set of internal coupling parameters, $\rho' = 3$, 10 and 30, for arbitrary shapes using the GRM equation (10.11). The results for three different aspect ratios $\varepsilon = 1/3$, 1 and 3 are shown in figures 10.18, 10.19 and 10.20, respectively. Included in these figures are the diffraction scattering patterns, the $\rho' \to 0$ limit. These were described in section 5.4. Note that unlike the results there, here the aspect ratio is too small to show two Guinier regimes.

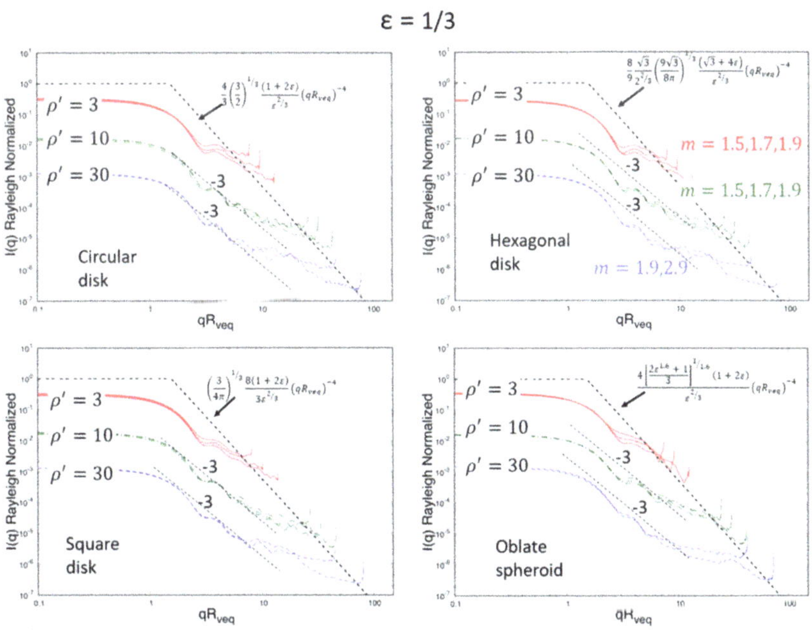

Figure 10.18. Rayleigh normalized scattered intensity versus qR_{veq} (R_{veq} is the volume equivalent radius) for rotationally averaged circular, hexagonal and square disks and an oblate spheroid all with an aspect ratio of $\varepsilon = 1/3$. Plots with three different internal coupling parameters ρ' are given. These internal coupling parameters were obtained from a variety of real refractive indices, as marked in the upper right, and sizes (not marked). Dashed line with a bend is the diffraction pattern, i.e. the structure factor, of the particle shape averaged to remove the ripple structure; see section 5.4. This is also the averaged optical scattering in the $\rho' \to 0$ limit. Other dashed lines indicate -3 power laws.

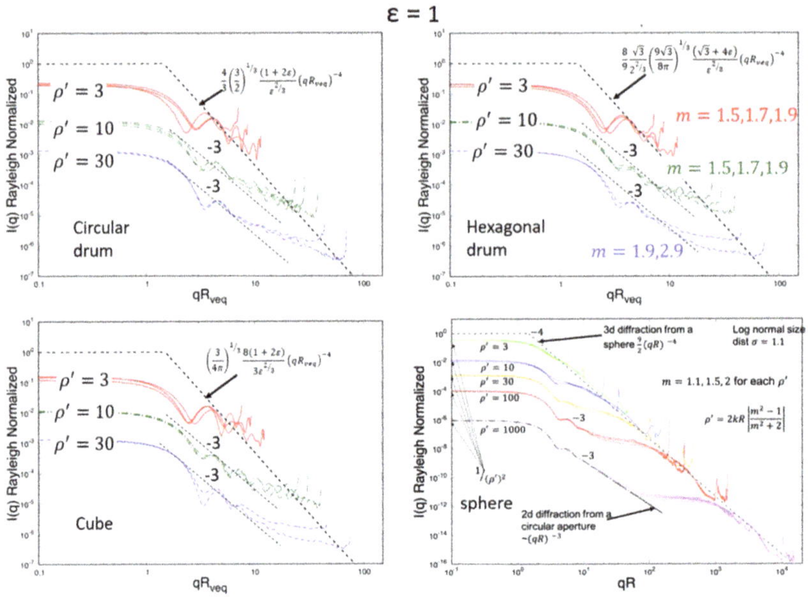

Figure 10.19. Same as figure 10.18, viz. Rayleigh normalized scattered intensity versus qR_{veq} except for circular, hexagonal and square drums and a sphere all with an aspect ratio of $\varepsilon = 1$. Plots for the sphere were obtained with the Mie equations and averaged over a slight size distribution to eliminate the ripples.

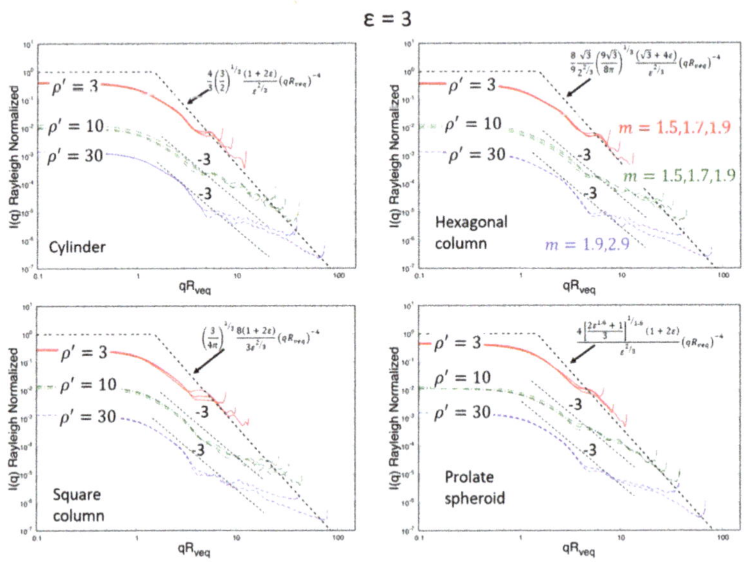

Figure 10.20. Same as figure 10.18, viz. Rayleigh normalized scattered intensity versus qR_{veq} except for circular, hexagonal and square columns and a prolate spheroid all with an aspect ratio of $\varepsilon = 3$.

Inspection of figures 10.18, 10.19 and 10.20 shows the systematic evolution of the scattering away from the diffraction limit with increasing ρ'. The ρ' provides a quasi-universal description of this evolution consolidating the size parameter and the refractive index. For all these shapes one can conclude that if the ρ' is the same the scattering will be very similar. Given this, one must acknowledge, as I have done before, that these plots are on logarithmic axes and differences exist for a given ρ'. Nevertheless, the differences for a given ρ' are orders of magnitude less than the differences between different ρ'.

Other features to stress are the q, hence scattering angle, independent forward scattering regime. The scattering in this forward regime is well described by the GRM Rayleigh scattering divided by ρ'^2. There is a Guinier regime near $qR_{\mathrm{veq}} \simeq 1$. Two-dimensional diffraction ensues after this regime for larger ρ' as indicated by the -3 power laws. However, at yet larger q, the hump feature present in spherical particle scattering is diminished if not gone. At largest q, the scattering joins the scattering for the diffraction limit and then ends by surpassing that with an enhanced backscattering.

10.3.2 Gaussian random spheres

A Gaussian random sphere (GRS) is based on smooth, random surface deviations relative to an underlying spherical shape. The GRS shape has been used to model a wide range of irregularly shaped particles such as desert dust particles. Figure 10.21 shows an example of a GRS characterized by three parameters: (1) σ, the relative standard deviation of the distribution of deviations from a perfect sphere in the radial direction; (2) ν, the power law index of the covariance function which controls the number of bumps and dips in the tangential direction; and (3) R_{mean}, the mean radius from which deviations occur and that sets the overall size of the particle (Muinonen *et al* 2007).

Our study of GRSs (Maughan *et al* 2016) used DDA methods to calculate scattering averaged over 40 random orientations. We found that once $\sigma \geqslant 0.2$, the ripple structure and enhanced backscattering of the $\sigma = 0$ sphere where essentially wiped out. We also found that there was very little difference in the scattering for $\nu = 2$, 3 or 4. Given these results, we studied in detail the $\sigma = 0.2$, $\nu = 3$ situation. We found the characteristic features seen so often by now in this book, viz. a forward scattering lobe, a Guinier regime, a power law regime, and as ρ increases, enhanced backscattering.

Figure 10.22 shows studies over a broad range of randomization parameters, σ and ν, size parameters, kR_{mean}, and real refractive indices. The scattering properties were calculated using the SIRIS 2 program (Muinonen *et al* 2009) involving averaging the ray tracings of a 1000 different GRSs with the same R_{mean}, σ and ν parameters at fixed orientations.

Figure 10.22 demonstrates quasi-universality with the generalized internal coupling parameter, ρ', i.e. plots are similar for different size parameters and refractive indices as long as ρ' is the same. Scattering features include a forward

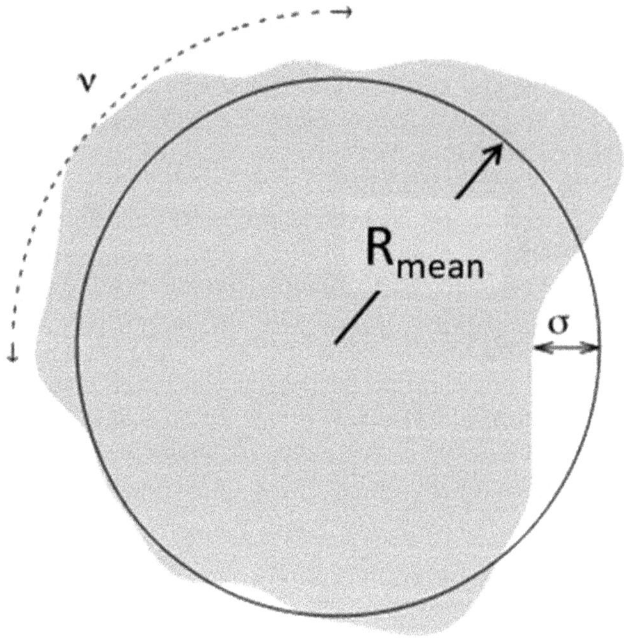

Figure 10.21. Geometric cross section of a Gaussian random sphere (GRS) with mean radius R_{mean}, random radial deviations measured by σ, with random tangential extent measured by ν. For this example GRS $\sigma = 2$ and $\nu = 3$. Reprinted from Maughan *et al* (2016), Copyright (2016), with permission from Elsevier.

scattering lobe with intensity equal to the Rayleigh intensity divided by ρ'^2, a Guinier regime near $qR_{\mathrm{mean}} \simeq 1$, a q^{-3} power law regime, especially for the largest size, due to 2d diffraction, the hump near $qR_{\mathrm{mean}} \simeq \rho'$, an equal tangent q^{-2} power law between the Guinier and hump regimes. Also in figure 10.23 we find that near the maximum $qR_{\mathrm{mean}} = 2kR_{\mathrm{mean}}$, the scattering magnitude becomes comparable to the 3d sphere diffraction limit, and there are hints at enhanced backscattering. In fact, all of these features are those of unmodified spheres with scattering described by the Mie equations except the backscattering is suppressed for GRSs. Apparently, GRSs do not have rainbows and glories, which is not surprising because these phenomena rely on smooth and regular surfaces not found with GRSs.

One can see in figure 10.22 with a little mental math that the forward scattering lobe is decreasing relative to the Rayleigh limit proportional ρ'^2, a characteristic apparently common to all shapes. The figure also shows that GRS scattering evolves from the diffraction limit with increasing ρ'.

10.3.3 Irregular spheres

Irregular spheres can be created with an algorithm that starts with a sphere and then messes with it in one manner or another to produce ensembles of stochastically different particles (Zubko *et al* 2006, 2008, 2009). Four different types of irregular

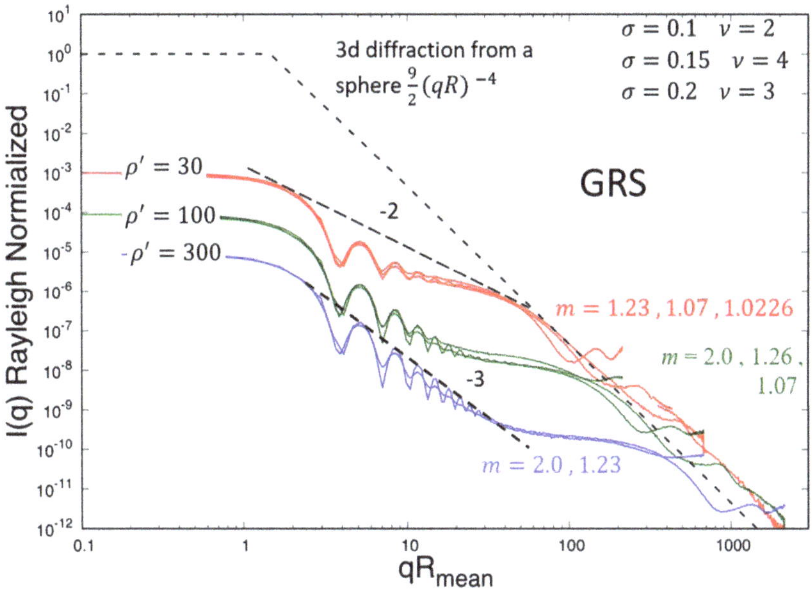

Figure 10.22. Q-space analysis of the Rayleigh normalized scattered intensity from GRSs for $\rho' = 30, 100, 300$. Scattering for three sets of generalized internal coupling parameters, ρ' are plotted. Three sets of fluctuation parameters σ and ν are used. Size parameters are $kR_{\text{mean}} = 100$, 300 and 1000 (note that the maximum qR_{mean} is $2kR_{\text{mean}}$). The eight plots are $\rho' = 30$, $kR_{\text{mean}} = 100$, $m = 1.23$, $\sigma = 0.15$, $\nu = 4$; $kR_{\text{mean}} = 300$, $m = 1.07$, $\sigma = 0.2$, $\nu = 3$; $kR_{\text{mean}} = 1000$, $m = 1.0226$, $\sigma = 0.1$, $\nu = 2$. $\rho' = 100$, $kR_{\text{mean}} = 100$, $m = 2.0$, $\sigma = 0.15$, $\nu = 4$; $kR_{\text{mean}} = 300$, $m = 1.26$, $\sigma = 0.2$, $\nu = 3$; $kR_{\text{mean}} = 1000$, $m = 1.07$, $\sigma = 0.1$, $\nu = 2$. $\rho' = 300$, $kR_{\text{mean}} = 300$, $m = 2.0$, $\sigma = 0.2$, $\nu = 2$; $kR_{\text{mean}} = 1000$, $m = 1.23$, $\sigma = 0.1$, $\nu = 2$.

spheres have been made: strongly damaged spheres, rough surface spheres, pocked spheres, and agglomerated debris particles, and their family portrait is in figure 10.23.

The light scattering properties of these irregular spheres were calculated using DDA. For each particle type three refractive indices were used: $m = 1.313$, $1.5 + 0.1i$ and $1.6 + 0.0005i$. The size parameter $x = kR$, where R is the radius of the smallest circumscribing sphere of the perturbed particle, ranged from 2 to 14 with steps of 2. For a wavelength of 532 nm, these values correspond to radii of $R = 0.17$ to 1.2 μm. Also note that all the internal coupling parameters are $\rho' < 10$. The Q-space analysis of the scattering was presented in Sorensen *et al* (2014).

Figure 10.24 shows a Q-space analysis of the scattering by strongly damaged spheres as an example. The upper left plot, figure 10.24(a), is the structure factor of the particle. There is a forward scattering lobe, a Guinier regime, and a power law regime with slope −4. The apparent enhanced backscattering is consequence of the spacing of the cubic lattice that the real space particle was represented on and hence an artifact of no consequence. The −4 slope in the power law region indicates the particles have mass and surface scaling dimensions of $D_m = 3$ and $D_s = 2$, respectively. Thus, the perturbations did not change these scaling dimensions, and the particles are not fractals. This is true for the three other types in figure 10.23.

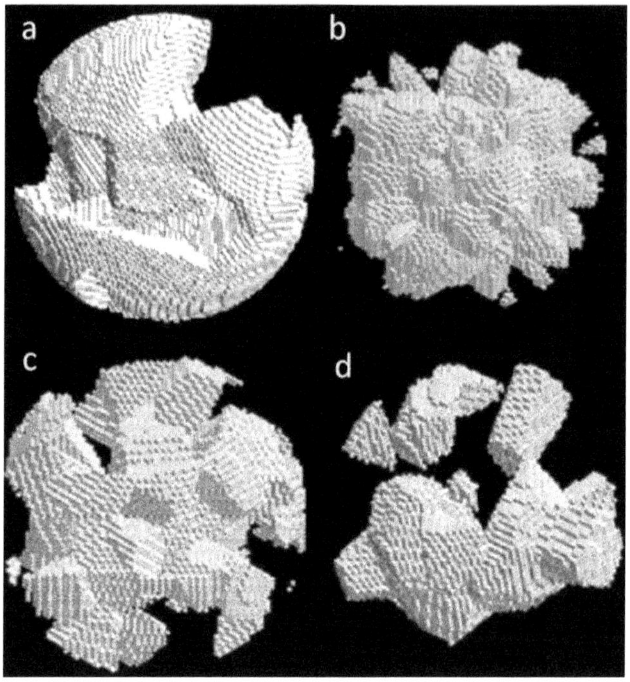

Figure 10.23. Irregular spheres: (a) strongly damaged, (b) rough surface, (c) pocked spheres, and (d) agglomerated debris particles. Adapted from Zubko *et al* (2006), Copyright (2006), with permission from Elsevier.

Figures 10.24(b)–(d) show the scattered light intensity for three different refractive indices. These finite refractive indices change the $\rho' = 0$ structure factor in the power law regime. The power law remains for about one order of magnitude in qR, but the slope decreases with increasing size parameter kR. This occurs as the internal coupling parameter ρ' increasing from zero to nearly 10 (given the relative symmetry of the particles, we expect $\rho'_{\text{arb}} \simeq \rho'_{\text{sph}}$), and we infer that the increased internal coupling is the cause of power law regime change. Note that the refractive index for the situation in figure 10.24(c) has an imaginary part of $\kappa = 0.1$. This leads to κkR as large as 1.4, which would certainly affect the scattering, see section 8.2.7. Similar results were obtained for pocked spheres and agglomerated debris particles, but the power law regime for the rough spheres had strong ripples that masked any possible power law. Another feature is the occurrence of some enhanced back-scattering with increasing ρ'. Finally, note the dip near $qR \simeq 3.5$.

The behavior of the power law exponent magnitude in these plots is somewhat enigmatic. This decrease starts near $\rho' = 0$ in an expected manner with the exponent magnitude equal to 4, consistent with 3d diffraction from a $D_m = 3$, $D_s = 2$ object. With increasing ρ', the exponent magnitude passes through 3 consistent with 2d diffraction from such an object. However, it continues, enigmatically, to smaller values with increasing ρ'. Given that $\rho' < 10$, one can propose that the results are in a transition state, and if ρ' where larger, 2d diffraction would eventually appear with

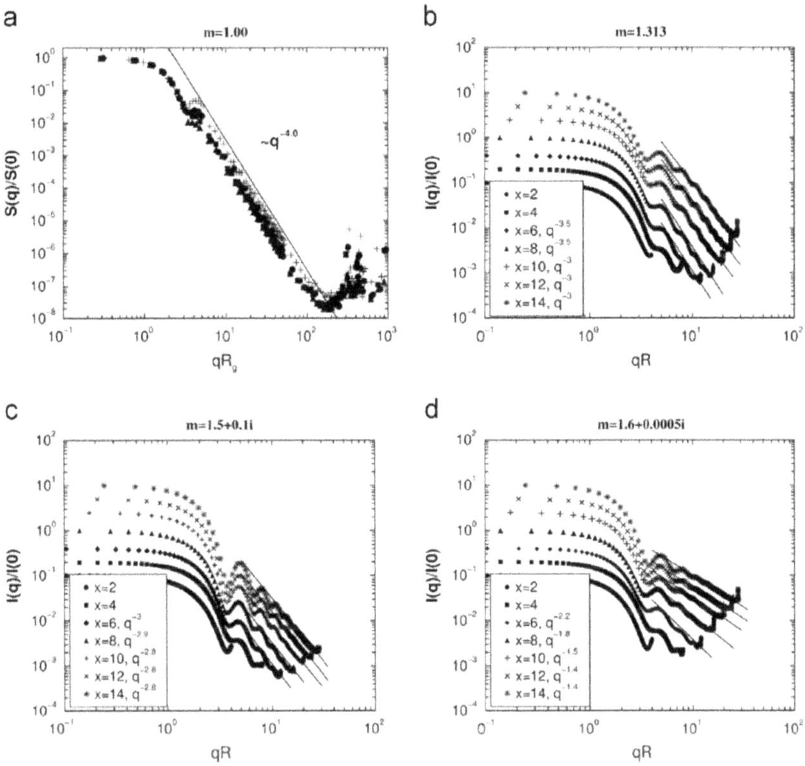

Figure 10.24. Q-space analysis of the scattering by strongly damaged spheres of different size parameters, $x = kR$, and refractive indices, m. (a) The structure factor ($m \to 1.00$). (b)–(d) With refractive indices as marked. Thin straight lines represent power laws with q as described in the legend. Plots are shifted along the intensity axis to avoid overlap. Reprinted from Sorensen *et al* (2014), Copyright (2014), with permission from Elsevier.

exponent magnitude 3. Testing this hypothesis requires future work with faster computers.

Figure 10.25 shows the behavior of the forward scattering for all the shapes and refractive indices as a function of the spherical particle ρ' using the volume equivalent sphere radius, which is smaller than the smallest circumscribing radius. A trend universal to all the data and quite similar to that for perfect spheres, see figure 7.6, is found.

10.3.4 Ice crystals

Ice crystals present a variety of very interesting shapes. The shapes are geometric in that the crystals have faces, edges and vertices situated, for the most part, in an orderly fashion with respect to each other. Their scattering and absorption properties are a subject of substantial study because these properties are of fundamental importance to the assessment of the radiative forcing of cirrus clouds that play an important role in the Earth's climate system.

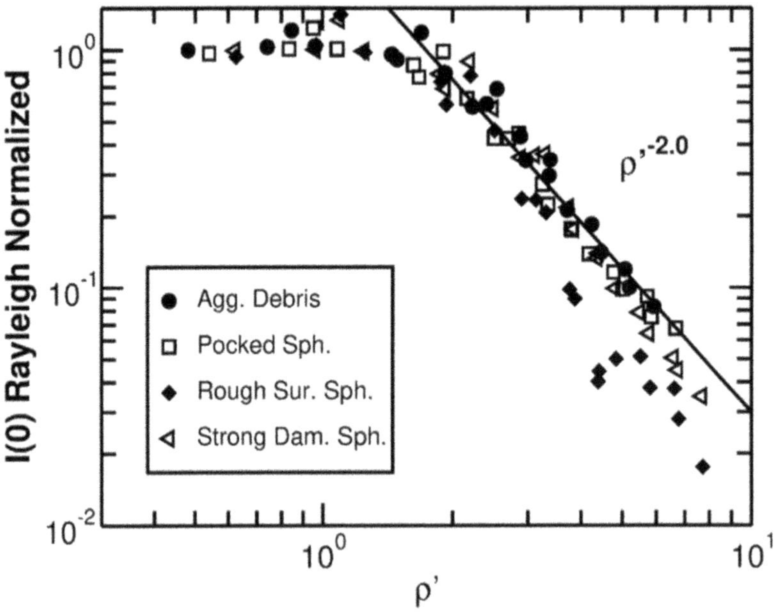

Figure 10.25. Rayleigh normalized forward scattering versus the internal coupling parameter ρ' for all four types of irregular spheres and three different refractive indices, $m = 1.313$, $1.5 + 0.1i$ and $1.6 + 0.0005i$ (hence a total of 12 sets of data).

We applied Q-space analysis to calculations of the ice crystal phase function, i.e. the angular scattering pattern (Heinson *et al* 2016) as calculated by Yang *et al* (2013). The volume equivalent radii, R_{veq}, were in the range 0.1 to 200 microns and the real part of the index of refraction was that of ice at optical wavelengths, 1.31. The differential cross section was calculated with a combination of the discrete dipole approximation (DDA) method (Purcell and Pennypacker 1973, Yurkin *et al* 2007, Yurkin and Hoekstra 2011) and an improved geometric optics method (IGOM) (Yang and Liou 1996). The scattering was averaged over random orientations of the crystals. The study included nine different ice crystal shapes.

Before presenting the results of Q-space analysis, it is of heuristic value to plot the phase functions for some of the crystals in the usual, θ-space manner, log scattered intensity versus linear scattering angle. Figure 10.26 shows these plots. There one sees an irregular pattern with an extreme increase in the scattering at forward directions. Increasing the surface roughness, σ, smooths the curves relative to the incident direction.

To perform the Q-space analysis Rayleigh normalization was applied using the methods of the GRM describe above in sections 10.2.1 and 10.2.2 to obtain the average volume polarizability $\alpha(m)$ from which the shape generalized Rayleigh differential cross section and the shape generalized internal coupling parameter were obtained via equations (10.5) and (10.11), respectively. Figure 10.27 displays the results in Q-space for the phase function for all nine shapes. The data set was large so some representative values of ρ' were chosen for the plots.

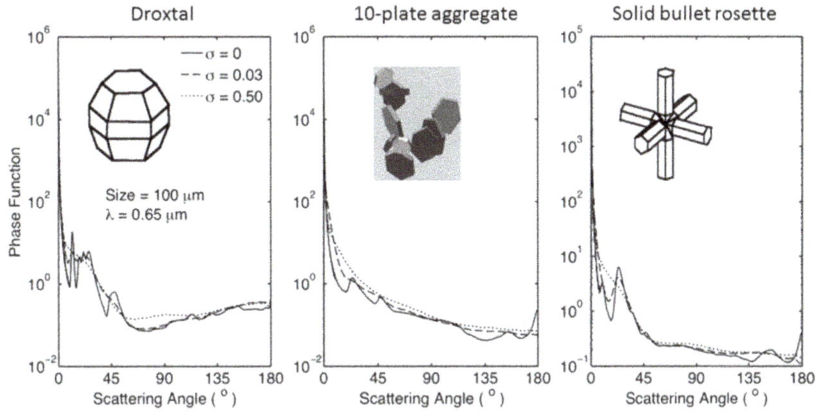

Figure 10.26. Scattered intensity versus scattering angle for three different ice crystal shapes. The symbol σ quantifies surface roughness. Adapted from Yang *et al* (2013), with permission from the American Meteorological Society.

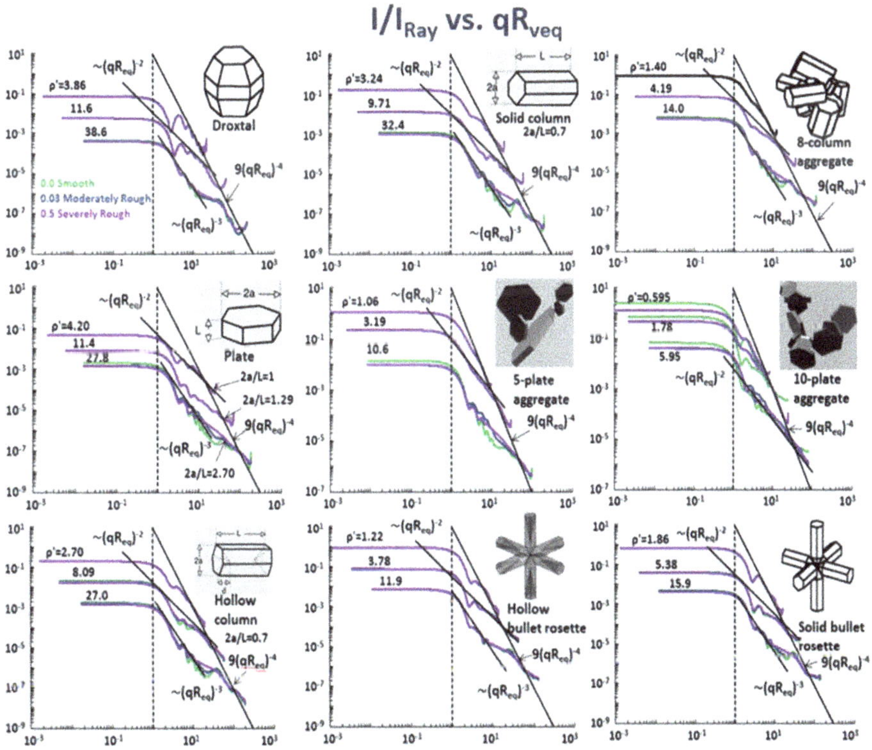

Figure 10.27. Rayleigh normalized scattering intensity, I/I_{Ray}, versus qR_{veq} for nine different ice crystal shapes. For each shape scattering for three different values of the internal coupling parameter ρ' as labeled. There are also three values of roughness, 0.0 (smooth), 0.03 (moderately rough) and 0.5 (severely rough), which mostly overlap. Lines for various power laws are marked. Included are pictures of the shapes. Adapted from Heinson *et al* (2016), Copyright (2016), with permission from Elsevier.

Figure 10.27 is distinctly different to figure 10.26. With Q-space analysis, features are uncovered that are invisible with normal plotting of scattered intensity versus linear scattering angle. These features will be summarized below, but before they are, the analysis is continued by examining the behavior of the forward scattered intensity.

Figure 10.28 illustrates the behavior of the Rayleigh normalized forward scattering for each individual shape. Once again the characteristic behavior of the forward scattering is that the intensity decrease proportional to the square of the internal coupling parameter, i.e. $I(0) \sim \rho'^{-2}$, once $\rho' > 1$.

Figure 10.29 illustrates the effects of significant absorption, as quantified by $\kappa k R_{\text{veq}}$, on the scattering. It also shows the effects of large anisotropy and makes a direct comparison to scattering by spheres.

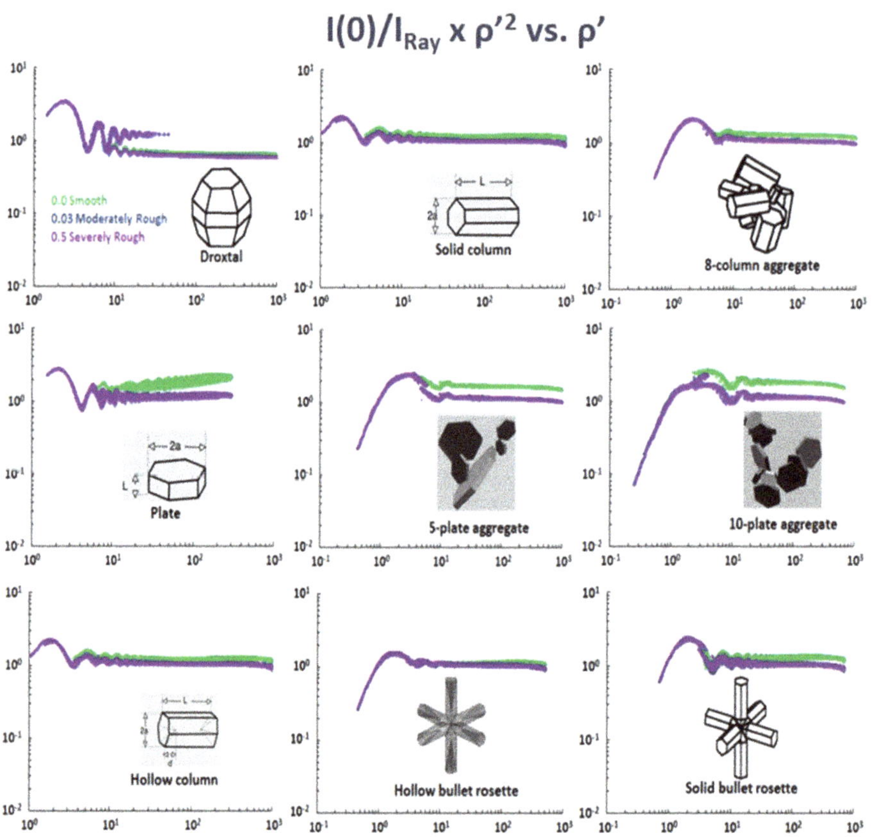

Figure 10.28. GRM Rayleigh normalized forward scattering intensity, $I(0)/I_{\text{Ray}}$, multiplied by the square of the GRM internal coupling parameter, ρ'^2, versus ρ' for nine different ice crystal shapes. Scattering for three different values of roughness, 0.0 (smooth), 0.03 (moderately rough) and 0.5 (severely rough) are shown which mostly overlap. Included are pictures of the shapes. These plots imply that $I(0) \sim \rho'^{-2}$, once $\rho' > 1$ for all these ice crystal shapes. These plot are very similar to those for spheres, figure 7.6(d). Adapted from Heinson *et al* (2016), Copyright (2016), with permission from Elsevier.

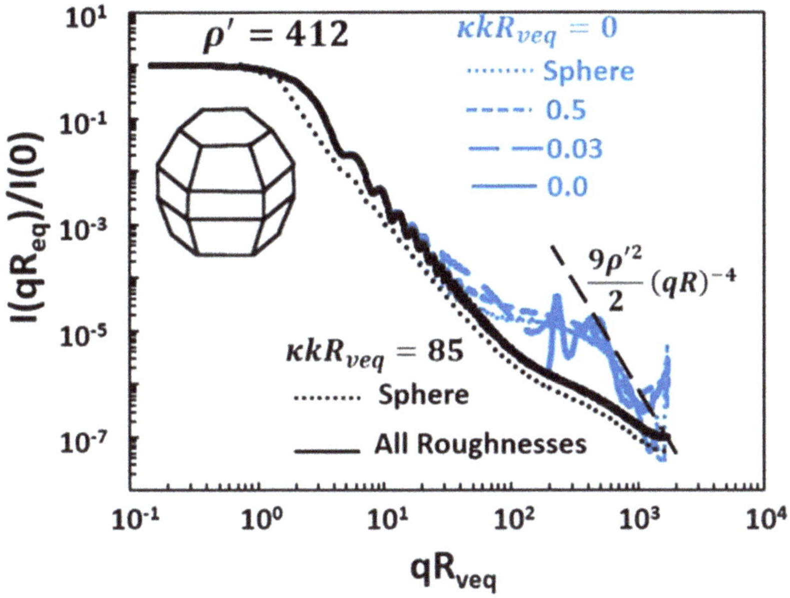

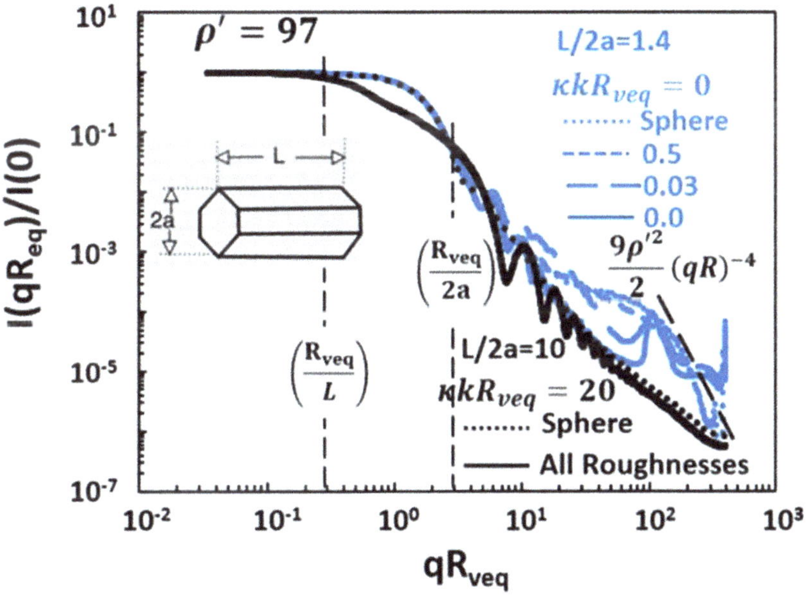

Figure 10.29. (Top) Forward normalized scattering versus qR_{veq} for a droxtal ice crystal, solid and dashed lines, with marked roughness, and a volume equivalent sphere, dotted lines, of the same internal coupling parameter, $\rho' = 412$. Blue plots are for $\kappa k R_{\mathrm{veq}} = 0$, black are for $\kappa k R_{\mathrm{veq}} = 85$. (Bottom) Forward normalized scattering versus qR_{veq} for two solid, hexagonal column structures with different aspect ratios, solid and dashed lines, with marked roughness, and a volume equivalent sphere, dotted lines, of the same internal coupling parameter, $\rho' = 97$. Blue plots are for $\kappa k R_{\mathrm{veq}} = 0$, black are for $\kappa k R_{\mathrm{veq}} = 20$. All the ice crystals and the spheres with significant absorption, as indicated by $\kappa k R_{\mathrm{veq}} > 1$, lose the hump and the enhanced backscattering. Note also the two Guinier regimes for the solid columns with $\kappa k R_{\mathrm{veq}} = 20$ match $qR_{\mathrm{veq}} = \left(L/R_{\mathrm{veq}}\right)^{-1}$ and $qR_{\mathrm{veq}} = \left(2a/R_{\mathrm{eq}}\right)^{-1}$. Adapted from Heinson *et al* (2016), Copyright (2016), with permission from Elsevier.

Figures 10.27–10.29 show that there are many common features between scattering by spheres and the ice crystals that evolve with the internal coupling parameter ρ' in a similar manner. These features are:

1. A forward scattering lobe when $qR_{\text{veq}} < 1$ (figures 10.27 and 10.29) the intensity of which is the GRM Rayleigh scattered intensity when the internal coupling parameter is ρ'<1 and the Rayleigh scattering divided by ρ'^2 when $\rho' > 10$, figure 10.28. This behavior is identical to that of spheres.
2. A Guinier regime near $qR_{\text{veq}} \simeq 1$ that can quantitatively indicate size scales of the particle, figures 10.27 and 10.29. In particular, notice how figure 10.29 has two Guinier regimes because the column has two length scales that are approximately an order of magnitude apart. This behavior is the same as seen in diffraction and described in sections 5.3 and 5.4.
3. A complex power law regime beyond the Guinier regime when $1 \lesssim qR_{\text{veq}} < 2kR_{\text{veq}}$, figures 10.27 and 10.29. For non-aggregate crystals in this regime a $(qR_{\text{veq}})^{-3}$ functionality starts to occur with increasing ρ' which can be ascribed to 2d diffraction from the projected crystal shape. A hump structure appears centered near ρ'. At larger qR_{veq}, there is a tendency to approach the spherical diffraction limit (RDG) of $9\left(qR_{\text{veq}}\right)^{-4}$ functionality. Aggregate ice crystals tend to have a more uniform power law regime similar to fractal aggregates.
4. Significant absorption, as quantified by $\kappa k R_{\text{veq}}$, removes the hump near $qR_{\text{veq}} \simeq \rho'$, figure 10.29, in a manner similar for spheres.
5. In many cases the ice crystals have enhanced backscattering near $qR_{\text{veq}} \simeq 2kR_{\text{veq}}$, which is equivalent to a scattering angle of $\theta = 180°$. However, no systematic trends were discerned.

Surface roughness plays a minor role in these features except in the hump region where smooth surfaces give a wavy structure to the hump.

A key point that has been stressed throughout this book is that the light scattering evolves with increasing ρ' from the diffraction limit where $\rho' \to 0$. Then the scattering is the structure factor, $S(q)$. Figure 10.30 shows the structure factor for two of the ice crystals, a rosette and an aggregate of ten hexagonal plates. The most important aspect of these plots is the power law regime displays a $(qR_{\text{veq}})^{-4}$

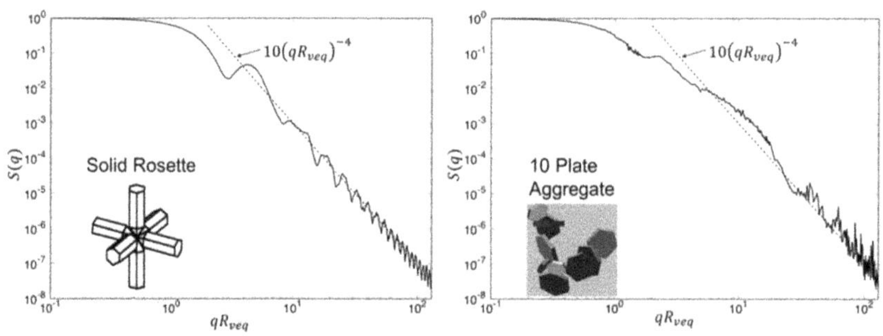

Figure 10.30. The structure factor of two ice crystal shapes.

functionality with exponent magnitude 4. This fact is true for all nine of the shapes studied. The exponent of magnitude 4 results because the ice crystals have mass and surface scaling dimensions of $D_m = 3$ and $D_s = 2$, respectively, and via the results of section 5.3, the exponent for diffraction is $2D_m - D_s$. This fact implies that none of the shapes is fractal, a fact particularly noteworthy for the aggregate. Although 10 monomers is rather small for a fractal-like aggregate, such a small aggregate would show the fractal dimension in the Porod regime, a dimension typically on the order of $D_f = 1.8$. Here, we conclude that there are at least two types of irregular shapes, fractal and non-fractal and they can be distinguished by their diffraction limit structure factor.

For both of the examples in figure 10.30, and indeed for all of the ice crystal shapes, the light scattering, for a given size and wavelength, would evolve from a structure factor with −4 power law power law regime into the scattering patterns seen in figures 10.27 and 10.29 with increasing refractive index (thus increasing ρ'). This is the same evolution as displayed by sphere.

10.3.5 Thickened percolation clusters

We are fortunate to have some aggregates which are quantitatively described by a fractal morphology because with quantification comes specification of the morphology. Scattering and absorption by fractal aggregates are the subjects of chapter 9. Many aggregates, however, are not as straightforward as fractal aggregates. For example, while freshly emitted soot particles are typically fractal-like, soot can undergo various aging processes resulting in changes in its morphology.

To simulate 'dust' agglomerate particles we used a classic site percolation theory as a backbone and then 'thickened' it (Sorensen *et al* 2017). The percolation clusters were made on a finite cubic lattice that had sites randomly filled. When filled sites were neighbors, they touched and were considered joined into a cluster. As more sites were filled, the cluster eventually percolated across the finite volume of the finite cubic lattice. This cluster had a fractal dimension of 2.5. All filled lattice sites not associated with the percolated cluster were set to empty. Addition of layers of material thickened the cluster and eliminated the fractal structure. A layer was defined as setting all neighboring unoccupied lattice sites to be filled. The final 'dust' particle had three layers of thickening. Figure 10.31 shows an example of a thickened percolation cluster.

A DDA algorithm (Purcell and Pennypacker 1973, Draine and Flatau 1994, Kalashnikova and Sokolik 2004) was used to calculate the rotationally averaged scattered intensity. Size parameters ranged from 5.2 to 17.3, which for a wavelength of 532 nm corresponds to volume equivalent radii of 0.44 to 1.5 μm. The radii of gyration of these very non-spherical particles is about twice the volume equivalent radii. Refractive indices used where real as $m = 1.05$, 1.33, 1.5 and 1.6. Figure 10.32 shows the results of these calculations plotted in Q-space. One finds a constant forward scattering lobe, the Guinier regime, a minor dip near $qR \simeq 3$, followed by power law regimes. There is no enhanced backscattering, but note that $\rho' < 12$, which is rather small for backscattering.

Figure 10.31. A thickened percolation cluster and the box it in which it was created.

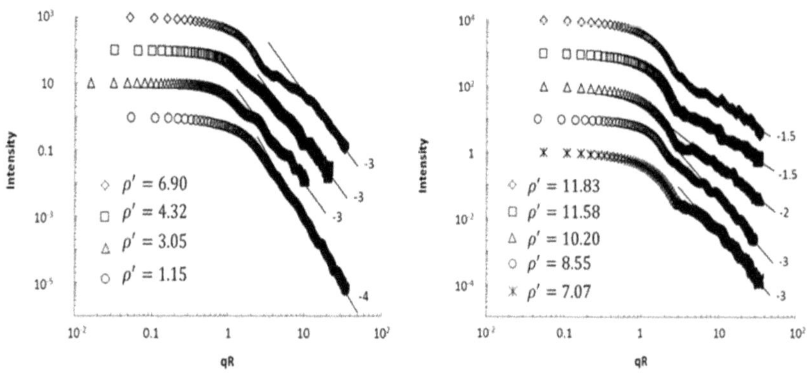

Figure 10.32. Scattered intensity, arbitrary units, versus qR_{veq} for thickened percolation clusters. Lines are the power law fits and the numbers to the right of the plots are the exponents of the power law. A different multiplication factor is applied to the intensity for each plot for clarity. Reprinted from Sorensen *et al* (2017) with permission from MDPI.

As for the irregular spheres of section 10.3.3, the magnitude of the exponents decrease with increasing ρ'. Also, similar to the irregular spheres,

These results are similar to those for irregular spheres discussed in section 10.3.3. Again, as for the irregular spheres, ρ' is limited to small values so the complete evolution of the scattering is not in these plots. I invite the reader to compare figures 10.24 and 10.32 for the scattering and then compare figures 10.23 and 10.31 for the real space pictures of the particles.

10.4 Summary and conclusions

I begin our conclusions with figure 10.33 which plots the Rayleigh normalized scattered intensity versus q times the volume equivalent radius for a compilation of the disks, columns, spheroids, Gaussian random spheres and spheres that have been studied above. All of these particles have in common a solid structure with a well-defined surface, although their shapes vary widely. Note that irregular spheres, ice crystals and thickened clusters are not included in figure 10.33. Ice crystals are not

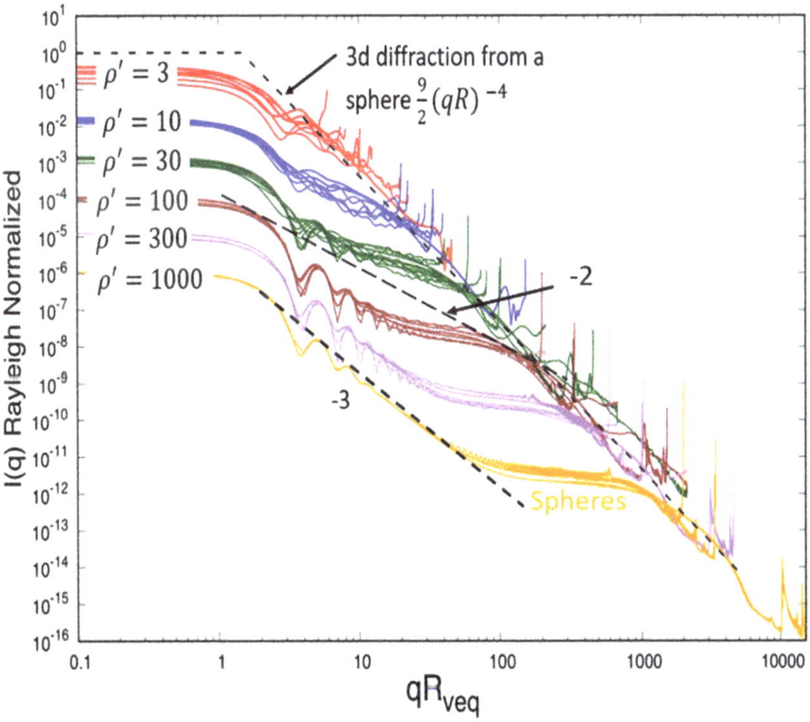

Figure 10.33. Rayleigh normalized scattered intensity versus $qR_{\rm veq}$ (volume equivalent radius). Plots where $\rho' \leqslant 300$ include a variety of disks, columns, and spheroids of various aspect ratios, Gaussian random spheres and spheres. Incident light had a polarization perpendicular to the scattering plane. All of these are also plotted in the figures above. Plots where $\rho' = 1000$ are spheres with a small size distribution to eliminate the ripples. Other shapes are not specifically identified because the intent of this figure is to demonstrate the global similarities of the scattering for all these shapes.

plotted on figure 10.33, but inspection of figures 10.27 and 10.29 suggests they would fit in well. On the grand scale of figure 10.33, where an order of magnitude is only a quarter of an inch, all of these objects scatter light the same. Moreover, the sameness is governed by the internal coupling parameter in its most general, GRM form. All of these particles have aspect ratios small enough to yield a single Guinier regime. All of these particles have mass scaling dimensions of $D_m = 3$ and surface scaling dimensions of $D_s = 2$; hence they all evolve from essentially the same diffraction limit.

Beyond solid particles with well-defined surfaces are the irregular spheres and thickened percolation clusters from this chapter and fractal aggregates from chapter 9. All of these are in some manner porous with ill-defined shapes and surfaces. The former two have $D_m = 3$ and $D_s = 2$, whereas fractal aggregates have $D_m = D_s < d$, where $d = 3$ is the spatial dimension. Scattering for these shapes differs from those in figure 10.33 in that the power law regime is essentially featureless. Note, however, that the irregular spheres and thickened percolation clusters studied here have small internal coupling parameters, and this results in a fairly featureless power law regime for solid objects with well-defined surfaces as well. More work is needed for these shapes with larger internal coupling parameters.

Experimental scattering measurements with dusts fit the description found for objects that are porous with ill-defined shapes and surfaces. But the dusts are, for the most part, not porous. Furthermore, their surfaces can range from smooth to irregular. Hence they are hard to classify. Future attempts to do so and experiments in a controlled manner would likely yield valuable results. Finally, the experiments studied here have all been with polydisperse systems (in some cases very polydisperse), which will smooth scattering profiles and hence could obscure important details.

I now proceed to summarize: *all shapes* have the following same features:

1. A forward scattering lobe of constant intensity with q and θ. This lobe appears when $qR < 1$. This condition is equivalent to $\theta < \lambda/2\pi R_{\mathrm{veq}}$. The magnitude of the forward scattering is that of its generalized Rayleigh scattering, I_{Ray}, when $\rho' \lesssim 1$, and I_{Ray}/ρ'^2 when $\rho' \gtrsim 10$. Moreover, it is reasonable to expect that when $\kappa k R_{\mathrm{veq}} < 0.1$, approximately half of the total scattered light energy occurs when $qR_{\mathrm{veq}} < \pi$ ($\theta < \lambda/2R_{\mathrm{veq}}$) (Sorensen *et al* 2017, Brillouin 1949, Bohren and Huffman 1983) and when $\kappa k R > 3$, nearly all the scattered light appears in this forward lobe.
2. A Guinier regime near $qR_{\mathrm{veq}} \simeq 1$ ($\theta \simeq \lambda/2\pi R_{\mathrm{veq}}$). This affords the ability to make size measurements although with some rather small uncertainty.
3. In some cases a minor dip near $qR_{\mathrm{veq}} \simeq 3$.
4. A power law regime when $1 \lesssim qR_{\mathrm{veq}} \lesssim 1.8kR_{\mathrm{veq}}$ ($\lambda/2\pi R_{\mathrm{veq}} \lesssim \theta \lesssim 130°$). This power law regime is a challenge to describe, and divides into two situations.
    a. For spheres, spheroids, geometric shapes, Gaussian random spheres and ice crystals, when $\rho' \lesssim 20$, a single power law appears with an exponent of about -2. With larger ρ', this evolves to an exponent of -3 immediately after the Guinier regime indicating the appearance of the 2d Fraunhofer diffraction. By the time $\rho' \geqslant 100$, the hump appears. Then there appears the enigmatic -2 equal tangent power law spanning

from the Guinier regime to the hump. The hump ebbs away with increasing $\kappa k R_{\mathrm{veq}}$.

b. On the other hand, for many of the experimental polydisperse dusts, fractal aggregates, and calculations for irregular spheres and thickened clusters, a single power law appears.

5. In many cases enhanced backscattering occurs for particles described in 4a near $qR_{\mathrm{veq}} \simeq 2kR_{\mathrm{veq}}$ (equivalent to a scattering angle of $\theta \simeq 180^\circ$) with larger ρ'.

These features are controlled by the internal coupling parameter ρ'. The small aspect ratio spheroids, Gaussian random spheres, irregular spheres and the ice crystal in the RDG, $\rho' \to 0$ limit all showed uniform power laws with exponent of −4. This diffraction limit obeys equation (5.11)

$$I(q) \sim q^{-(2D_m - D_s)} \tag{10.12}$$

where D_m is the particle mass scaling dimension and D_s is the particle surface scaling dimension. Thus we conclude that $D_m = 3$ and $D_s = 2$ for these particles, and, given the non-aggregate nature of all the other particles studied here, we conclude the same for them. This is an important distinction from fractal aggregates for which the fractal dimension $D_f = D_m = D_s$ and is less than the spatial dimension of $d = 3$. Thus we now have two types of irregularly shaped particles: fractal with $D_f < d$, and non-fractal with $D_m = 3$ and $D_s = 2$.

Nevertheless, in some cases scattering by fractals and non-fractals can look the same. This is especially true for dusts, which can have very uniform scattering power laws when $qR > 1$ just like fractal aggregates. For example, compare figures 10.2 and 10.3 to figures 9.12, 9.19 and 9.21.

The power law regime is quite diverse among the different shapes. This diversity broadens when one includes spheres and fractal aggregates, and it deepens with increasing ρ'. The remarkable linearity of the double-log plots for the dusts in the Amsterdam–Granada data set (see Heinson *et al* 2016 for more examples) and the Arizona road dusts is both stunning and in contrast to the other shapes which might have dips, kinks, humps or complexities similar to spheres. One important difference is that the Amsterdam–Granada and Arizona road dusts were polydisperse whereas the abrasive were very likely less so, and the theoretically calculated systems were monodisperse or nearly so. The dips, kinks, humps and complexities could well be averaged away with a size distribution. This is a hypothesis that should be tested.

Dips, kinks and polydispersity notwithstanding a trend for the power law exponent with internal coupling parameter is apparent. From the diffraction limit, where the power law is uniform with exponent −4, the magnitude of the power law exponent decreases with increasing ρ'. This empirical trend is plotted in figure 10.34 for all the shapes above except the ice crystals that are not well described by a single power law. The magnitude of the exponents starts from 4, and then beginning near $\rho' \simeq 1$, decreases with increasing ρ' until the exponents reach a constant 1.75 ± 0.25 when $\rho' \gtrsim 10$. This trend is universal among the various shapes and systems represented in the figure. Note, however, that the fractal aggregate does not fall on the trend in

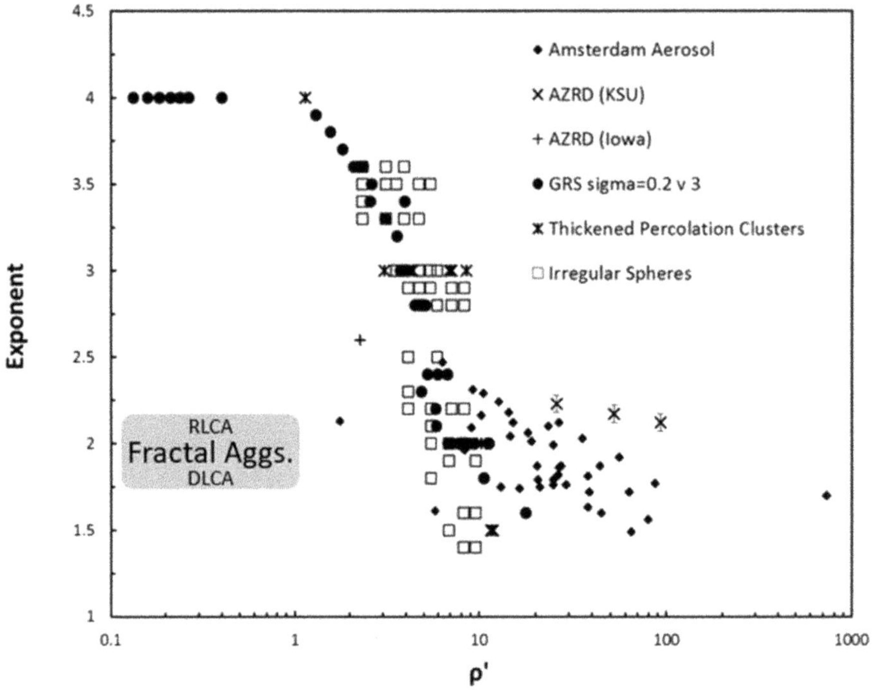

Figure 10.34. The exponents of the power laws versus the internal coupling parameter ρ' for experimental polydisperse dusts, fractal aggregates, and calculations for irregular spheres and thickened clusters, Included are results from four, polydisperse data sets and two theoretical calculations. Adapted from Heinson *et al* (2016) John Wiley & Sons. © 2015. American Geophysical Union. All rights reserved.

figure 10.34. In section 9.4.4 we showed that for fractal aggregates $\rho' \simeq \rho'_{\text{mono}}$ with a very weak dependence on the number of monomers in the aggregate, N. Thus, a fractal aggregate with small monomers, the typical situation, will have $\rho' < 1$ regardless of size, N. Hence, it will have a power law exponent equal to the fractal dimension, which is much less than 4, a result not on the trend in figure 10.34.

So, recognizing that the last few pages are not particularly definitive, and also a bit repetitive, as I try to describe phenomena that I really cannot yet describe, I'll use that as my cue to end this book. In truth, it's time to finish. I will end on a positive thought that uncertainty at the end, points to new directions in the future.

References

Beal A R and Hughes H P 1979 Kramers–Kronig analysis of the reflectivity spectra of 2H-MoS_2, 2H-$MoSe_2$ and 2H-$MoTe_2$ *J. Phys. C: Solid State Phys.* **12** 881–90

Bi L, Yang P, Kattawar G W and Mishchenko M I 2013 Efficient implementation of the invariant imbedding T-matrix method and the separation of variables method applied to large nonspherical inhomogeneous particles *J. Quant. Spectrosc. Radiat. Transf.* **116** 169–83

Bohren C F and Huffman D R 1983 *Absorption and Scattering of Light by Small Particles.* (New York: Wiley)

Bohren C F and Koh G 1985 Forward-scattering corrected extinction by nonspherical particles *Appl. Opt.* **24** 1023–9
Born M and Wolf E 1975 *Principles of Optics: Electromagnetic Theory of Propagation, Interference, and Diffraction of Light* (Oxford: Pergamon)
Brillouin L 1949 The scattering cross section of spheres for electromagnetic waves *J. Appl. Phys.* **20** 1110–25
Curtis D B, Meland B, Aycibin M, Arnold N P, Grassian V H, Young M A and Kleiber P D 2008 A laboratory investigation of light scattering from representative components of mineral dust aerosol at a wavelength of 550 nm *J. Geophys. Res. Atmos.* **113** D01820
Draine B T and Flatau P J 1994 Discrete-dipole approximation for scattering calculations *J. Opt. Soc. Am.* A **11** 1491–9
Ferri F 1997 Use of a charge coupled device camera for low-angle elastic light scattering *Rev. Sci. Instrum.* **68** 2265–74
Forrest S R and Witten T A 1979 Long-range correlations in smoke-particle aggregates *J. Phys. A: Math. Gen.* **12** L109–17
Gautam P, Maughan J B, Ilavsky J and Sorensen C M 2020 Light scattering study of highly absorptive, non-fractal, hematite aggregates *J. Quant. Spectrosc. Radiat. Transf.* **246** 106919
Gautam P and Sorensen C M 2020 A light-scattering study of highly refractive, irregularly shaped MoS_2 particles *J. Quant. Spectrosc. Radiat. Transf.* **242** 106757
Heinson Y W, Chakrabarti A and Sorensen C M 2016 A light-scattering study of Al_2O_3 abrasives of various grit sizes *J. Quant. Spectrosc. Radiat. Transf.* **180** 84–91
Heinson Y W, Maughan J B, Ding J C, Chakrabarti A, Yang P and Sorensen C M 2016 Q-space analysis of light scattering by ice crystals *J. Quant. Spectrosc. Radiat. Transf.* **185** 86–94
Heinson Y W, Maughan J B, Heinson W R, Chakrabarti A and Sorensen C M 2016 Light scattering Q-space analysis of irregularly shaped particles *J. Geophys. Res. Atmos.* **121** 682–91
Jackson J D 1999 *Classical Electrodynamics.* (New York: Wiley)
Kalashnikova O V and Sokolik I N 2004 Modeling the radiative properties of nonspherical soil-derived mineral aerosols *J. Quant. Spectrosc. Radiat. Transf.* **87** 137–66
Laan E C, Volten H, Stam D M, Munoz O, Hovenier J W and Roush T L 2009 Scattering matrices and expansion coefficients of martian analogue palagonite particles *Icarus* **199** 219–30
Mackowski D W and Mishchenko M I 2011 A multiple sphere T-matrix FORTRAN code for use on parallel computer clusters *J. Quant. Spectrosc. Radiat. Transf.* **112** 2182–92
Mandelbrot B B and Mandelbrot B B 1982 *The Fractal Geometry of Nature.* (San Francisco, CA: W.H. Freeman)
Maughan J B, Chakrabarti A and Sorensen C M 2017 Rayleigh scattering and the internal coupling parameter for arbitrary particle shapes *J. Quant. Spectrosc. Radiat. Transf.* **189** 339–43
Maughan J B, Chakrabarti A and Sorensen C M 2017 Rayleigh scattering and the internal coupling parameter for arbitrary shapes *J. Quant. Spectrosc. Radiat. Transf.* **189** 339–43
Maughan J B, Sorensen C M and Chakrabarti A 2016 Q-space analysis of light scattering by Gaussian random spheres *J. Quant. Spectrosc. Radiat. Transf.* **174** 14–21
Mokhtari T, Sorensen C M and Chakrabarti A 2005 Multiple-scattering effects on static light-scattering optical structure factor measurements *Appl. Opt.* **44** 7858–61
Muinonen K, Nousiainen T, Lindqvist H, Munoz O and Videen G 2009 Light scattering by Gaussian particles with internal inclusions and roughened surfaces using ray optics *J. Quant. Spectrosc. Radiat. Transf.* **110** 1628–39

Muinonen K, Zubko E, Tyynela J, Shkuratov Y G and Videen G 2007 Light scattering by Gaussian random particles with discrete-dipole approximation *J. Quant. Spectrosc. Radiat. Transf.* **106** 360–77

Munoz O, Moreno F, Guirado D, Dabrowska D D, Volten H and Hovenier J W 2012 The Amsterdam–Granada light scattering database *J. Quant. Spectrosc. Radiat. Transf.* **113** 565–74

Munoz O, Volten H, de Haan J F, Vassen W and Hovenier J W 2000 Experimental determination of scattering matrices of olivine and Allende meteorite particles *Astron. Astrophys.* **360** 777–88

Munoz O, Volten H, de Haan J F, Vassen W and Hovenier J W 2001 Experimental determination of scattering matrices of randomly oriented fly ash and clay particles at 442 and 633 nm *J. Geophys. Res. Atmos.* **106** 22833–44

Munoz O, Volten H, de Haan J F, Vassen W and Hovenier J W 2002 Experimental determination of the phase function and degree of linear polarization of El Chichon and Pinatubo volcanic ashes *J. Geophys. Res. Atmos.* **107 ACL** 41–48

Munoz O, Volten H, Hovenier J W, Min M, Shkuratov Y G, Jalava J P, van der Zande W J and Waters L B F M 2006 Experimental and computational study of light scattering by irregular particles with extreme refractive indices: hematite and rutile *Astron. Astrophys.* **446** 525–35

Munoz O, Volten H, Hovenier J W, Nousiainen T, Muinonen K, Guirado D, Moreno F and Waters L B F M 2007 Scattering matrix of large saharan dust particles: experiments and computations *J. Geophys. Res. Atmos.* **112** D13215

Munoz O, Volten H, Hovenier J W, Veihelmann B, van der Zande W J, Waters L B F M and Rose W I 2004 Scattering matrices of volcanic ash particles of mount St. Helens, redoubt, and mount spurr volcanoes *J. Geophys. Res. Atmos.* **109** D16201

Ordal M A, Bell R J, Alexander R W, Long L L and Querry M R 1985 Optical-properties of 14 metals in the infrared and far infrared: Al, Co, Cu, Au, Fe, Pb, Mo, Ni, Pd, Pt, Ag, Ti, V, and W. *Appl. Opt.* **24** 4493–9

Purcell E M and Pennypacker C R 1973 Scattering and absorption of light by nonspherical dielectric grains *Astrophys. J.* **186** 705–14

Slepian Z 2012 The average projected area theorem—generalization to higher dimensions arXiv: 1109.054v4

Sokolik I N and Toon O B 1999 Incorporation of mineralogical composition into models of the radiative properties of mineral aerosol from UV to IR wavelengths *J. Geophys. Res. Atmos.* **104** 9423–44

Sorensen C M 2013 Q-space analysis of scattering by dusts *J. Quant. Spectrosc. Radiat. Transf.* **115** 93–5

Sorensen C M, Heinson Y W, Heinson W R, Maughan J B and Chakrabarti A 2017 Q-space analysis of the light scattering phase function of particles with any shape. *Atmosphere* **8** 68

Sorensen C M, Maughan J B and Chakrabarti A 2017 The partial light scattering cross section of spherical particles **34** 681–4

Sorensen C M and Shi D 2000 Guinier analysis for homogeneous dielectric spheres of arbitrary size *Opt. Commun.* **178** 31–6

Sorensen C M, Zubko E, Heinson W R and Chakrabarti A 2014 Q-space analysis of scattering by small irregular particles *J. Quant. Spectrosc. Radiat. Transf.* **133** 99–105

van de Hulst H C 1957 *Light Scattering by Small Particles* (New York: Wiley)

Volten H, Munoz O, Hovenier J W and Waters L B F M 2006 An update of the Amsterdam light scattering database *J. Quant. Spectrosc. Radiat. Transf.* **100** 437–43

Volten H, Munoz O, Rol E, de Haan J F, Vassen W, Hovenier J W, Muinonen K and Nousiainen T 2001 Scattering matrices of mineral aerosol particles at 441.6 nm and 632.8 nm *J. Geophys. Res.* **106** 17375–17401

Wang Y, Chakrabarti A and Sorensen C M 2015 A light-scattering study of the scattering matrix elements of Arizona Road Dust *J. Quant. Spectrosc. Radiat. Transf.* **163** 72–9

Weast R C (ed) 1976–1977 *Handbook of Chemistry and Physics* (Cleveland, OH: CRC Press)

Yang P, Bi L, Baum B A, Liou K N, Kattawar G W, Mishchenko M I and Cole B 2013 Spectrally consistent scattering absorption, and polarization properties of atmospheric ice crystals at wavelengths from 0.2 to 100 mu m *J. Atmos. Sci.* **70** 330–47

Yang P and Liou K N 1996 Geometric-optics-integral-equation method for light scattering by nonspherical ice crystals *Appl. Opt.* **35** 6568–84

Yim C Y, O'Brien M, McEvoy N, Winters S, Mirza I, Lunney J G and Duesberg G S 2014 Investigation of the optical properties of MoS_2 thin films using spectroscopic ellipsometry *Appl. Phys. Lett.* **104** 103114

Yurkin M A and Hoekstra A G 2011 The discrete-dipole-approximation code ADDA: capabilities and known limitations *J. Quant. Spectrosc. Radiat. Transf.* **112** 2234–47

Yurkin M A, Maltsev V P and Hoekstra A G 2007 The discrete dipole approximation for simulation of light scattering by particles much larger than the wavelength *J. Quant. Spectrosc. Radiat. Transf.* **106** 546–57

Zhang H, Ma Y G, Wan Y, Rong X, Xie Z, Wang W and Dai L 2015 Measuring the refractive index of highly crystalline monolayer MoS_2 with high confidence *Sci. Rep.* **5** 8440

Zubko E, Kimura H, Shkuratov Y, Muinonen K, Yamamoto T, Okamoto H and Videen G 2009 Effect of absorption on light scattering by agglomerated debris particles *J. Quant. Spectrosc. Radiat. Transf.* **110** 1741–9

Zubko E, Shkuratov Y, Kiselev N N and Videen G 2006 DDA simulations of light scattering by small irregular particles with various structure *J. Quant. Spectrosc. Radiat. Transf.* **101** 416–34

Zubko E, Shkuratov Y, Mishchenko M and Videen G 2008 Light scattering in a finite multi-particle system *J. Quant. Spectrosc. Radiat. Transf.* **109** 2195–206

Appendix A

The Dirac delta function

The Fourier transformation of $F(k)$ to $f(x)$ is

$$f(x) = 1/\sqrt{2\pi} \int_{-\infty}^{\infty} F(k)e^{-ikx}dk \tag{A.1}$$

The functions $F(k)$ and $f(x)$ are Fourier transform pairs. The inverse transform of $f(x)$ to $F(k)$ is

$$F(k) = 1/\sqrt{2\pi} \int_{-\infty}^{\infty} f(x')e^{ikx'}dx' \tag{A.2}$$

Notice the positive exponent for the $e^{ikx'}$ term in equation (A.2) versus the negative exponent in equation (A.1). Now substitute equation (A.2) into equation (A.1) to find

$$f(x) = 1/\sqrt{2\pi} \int_{-\infty}^{\infty} \left[1/\sqrt{2\pi} \int_{-\infty}^{\infty} f(x')e^{ikx'}dx' \right] e^{-ikx}dk \tag{A.3}$$

$$f(x) = 1/2\pi \int_{-\infty}^{\infty} \left[\int_{-\infty}^{\infty} f(x')e^{ikx'}dx' \right] e^{-ikx}dk \tag{A.4}$$

$$f(x) = 1/2\pi \int_{-\infty}^{\infty} f(x') \left[\int_{-\infty}^{\infty} e^{ikx'-ikx}dk \right] dx' \tag{A.5}$$

Given (A.5), we define the *Dirac delta function* as

$$\delta(x' - x) = 1/2\pi \int_{-\infty}^{\infty} e^{ik(x'-x)}dk \tag{A.6}$$

With this we have

$$f(x) = \int_{-\infty}^{\infty} \delta(x' - x)f(x')dx' \tag{A.7}$$

doi:10.1088/978-0-7503-5157-7ch11

Equation (A.7) is often used to define the Dirac delta function.

Substitution of $f(x') = 1$ into equation (A.7) yields

$$\int_{-\infty}^{\infty} \delta(x' - x)dx' = 1 \tag{A.8}$$

In three dimensions, equation (A.6) modifies to

$$\delta(\boldsymbol{r}' - \boldsymbol{r}) = (1/2\pi)^3 \int_{-\infty}^{\infty} e^{i\boldsymbol{k}\cdot(\boldsymbol{r}'-\boldsymbol{r})}d\boldsymbol{k} \tag{A.9}$$

Good mathematicians tell us that the Dirac delta function is not really a function. Nevertheless, it is heuristically useful to think of it as a function that is zero everywhere except when its argument is zero, where it is infinite. Then, equation (A.8) tells us that the area under the delta function curve is unity.

IOP Publishing

Light Scattering and Absorption by Particles
The Q-space approach
Christopher M Sorensen

Appendix B

Aggregation

The volume fraction of a system of single, monomeric particles (a sol) is

$$f_v = N_m V_{part}/V_{sys} \sim N_m R^3(1)/V_{sys} \quad \text{(B.1)}$$

where N_m is the number of monomers in the system, V_{part} is the monomer particle volume, V_{sys} is the volume of the entire system, and $R(N)$ is the radius of particle composed of N monomers.

The volume per particle is related to the mean nearest neighbor separation, R_{nn}, by

$$V_{sys}/N_m \sim R_{nn}^3 \quad \text{(B.2)}$$

Aggregation will lead the monomers to form aggregates called N-mers, i.e. N monomers per aggregate. N is a function of time, $N(t)$, and thus can represent the time dependence of the system. A statement of conservation of mass during aggregation is

$$N_m = NN_c \quad \text{(B.3)}$$

where N_c is the number of cluster aggregates.

For the aggregating sol the mean nearest neighbor aggregate (cluster) separation, R_{nc}, in analogy to relation (B.2), is

$$V_{sys}/N_c \sim R_{cn}^3 \quad \text{(B.4)}$$

Combining (B.2), (B.3) and (B.4) we find

$$R_{nc}^3 \sim NR_{nn}^3 \quad \text{(B.5)}$$

which can be rewritten as

$$R_{nc} \sim N^{1/3} \quad \text{(B.6)}$$

Relation (B.6) tells us that the aggregates get further apart on average as the average number of monomers in the aggregate increases with time.

The aggregates that form will have a mass scaling dimension D_m. Because the aggregate mass is proportional to the number of monomers N in the aggregate, we have

$$N \sim R^{D_m} \tag{B.7}$$

which can be rewritten as

$$R \sim N^{1/D_m} \tag{B.8}$$

If the aggregates are dense, perhaps coalescing liquid droplets or compact assemblies of monomers, such that the aggregate scaling dimension is the spatial dimension, i.e. $D_m = 3$, then relations (B.6) and (B.8) have the same N dependence, hence the same time dependence. This means their ratio, which compares separation to size, will stay constant with time.

If instead the aggregates are not compact, but ramified in some manner like a fractal aggregate, such that the aggregate scaling dimension is less than the spatial dimension, i.e. $D_m < 3$, then relations (B.6) and (B.8) will not have the same N dependence, hence not the same time dependence. The ratio will be

$$R_{nc}/R \sim N^{(D_m-3)/3D_m} \tag{B.9}$$

While both the mean separation and mean size increase with time, this ratio decreases with time when $D_m < 3$. This relative separation continues to decrease until the aggregates touch and the system gels (Sorensen and Chakrabarti 2011).

Reference

Sorensen C M and Chakrabarti A 2011 The sol to gel transition in irreversible particulate systems *Soft Matter* 7 2284–96

Appendix C

A theory for DLCA fractal aggregate morphology

Here, I present a simple analytic theory for the morphology of diffusion limited cluster–cluster aggregates (DLCAs): the Restricted hierarchical model (RHM) (Sorensen and Oh 1998, Heinson *et al* 2015).

The general scaling relation for a fractal aggregate, equation (9.3), is

$$N = k_0(R_g/a)^D \tag{C.1}$$

In equation (C.1), N is the number of monomers in the aggregate, k_0 is the scaling prefactor, R_g is the aggregate radius of gyration, a is the monomer radius and D is the fractal dimension. The goal of the theory is to calculate values for k_0 and D, and to quantitatively describe the aggregate shape.

The RHM assumes circular or spherical monomers. It is hierarchical in the sense that $N = 1, 2, 4, 8, \ldots$. The restriction comes with combinations of two N-mers being constrained to occur only with the joining of the shortest side of one to the longest side of the other.

Start in $d = 2$ spatial dimensions. Figure C.1 shows a series with 1-mer, 2-mer, 4-mer, 8-mer and so on. In the $N = 8$ aggregate of dimensions 3×5, the dashed lines show how it was constructed from two $N = 4$ aggregates of dimensions 2×3 joined side to end, i.e. the 3-side of one to the 2-end of the other.

Interestingly, the dimensions of circumscribing rectangles of the series of aggregates so created follow the Fibonacci series.

$$f_n = 1, 1, 2, 3, 5, 8, 13, 21, 34, 55, \ldots \tag{C.2}$$

The side-to-end aspect ratio R of the aggregates is equal to the ratio of two consecutive Fibonacci numbers, which in the limit $n \to \infty$ is the Divine proportion, also known as the Golden mean, and often designated by φ.

$$R = \lim_{n\to\infty}\left(f_{n+1}/f_n\right) = \varphi = 1.618\ldots \tag{C.3}$$

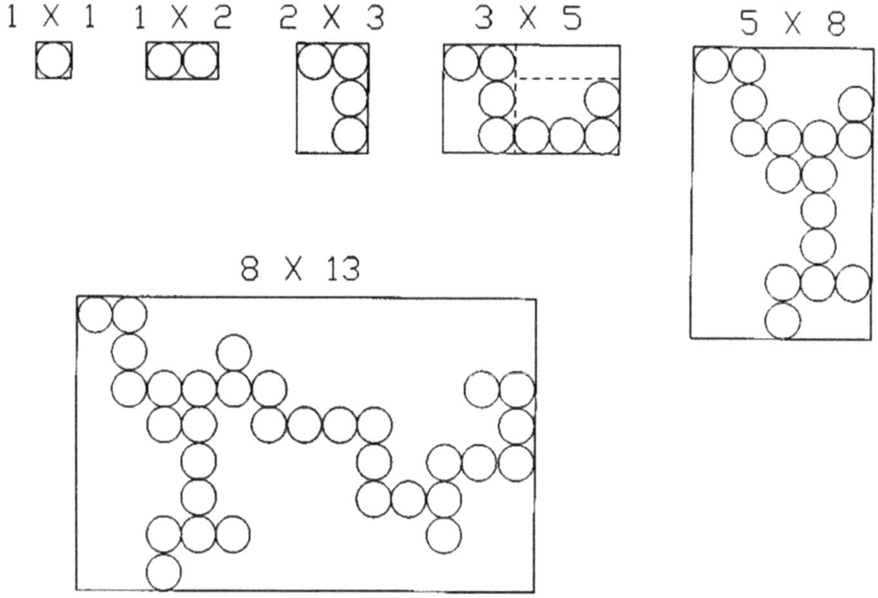

Figure C.1. Two-dimensional aggregates created with the RHM. Reprinted figure with permission from Sorensen and Oh (1998), Copyright (1998) by the American Physical Society.

Now recognize that with each new side-to-end aggregation, the aggregates grow in mass (N) by a factor of two, while the linear size (an end or a side) increases by a factor of φ. Hence, the scaling dimension of these aggregates, which is also called the fractal dimension, is given by

$$D_2 = \log 2/\log \varphi = 1.44 \tag{C.4}$$

In equation (C.4), the subscript 2 refers to the spatial dimension, d. The value $D_2 = 1.44$ is in excellent agreement with numerous $d = 2$ DLCA simulations, which find, on average, $D_2 = 1.44 \pm 0.03$.

The RHM can be generalized to higher dimensions. Then d-dimensional circumscribing rectangles surround d-dimensional spheres. The longest rectangle side is joined to the shortest side. In $d = 3$ dimensions, a progression of end lengths proceeds from the monomer box of dimension $1 \times 1 \times 1$ as follows

$$f_{n,\,3} = 1,\ 1,\ 1,\ 2,\ 3,\ 4,\ 6,\ 9,\ 13,\ 19,\ 28,\ \ldots \tag{C.5}$$

We call $f_{n,3}$ the three-dimensional Fibonacci series. Note that the rule of construction is to add next nearest neighbors. With this, we generalize the Divine proportion of equation (C.3) to the three-dimensional Divine proportion, φ_3, defined by

$$\varphi_3 = \lim_{n\to\infty}\left(f_{n+1,3}/f_{n,\,3}\right) = 1.465\,57\ldots \tag{C.6}$$

Once again recognize that with each new longest side to shortest end aggregation, the aggregates grow in mass (N) by a factor of two while the linear size (an end or a side) increases by a factor of φ_3. Hence the scaling dimension, i.e. the fractal dimension, of these aggregates is given by

$$D_3 = \log 2/\log \varphi_3 = 1.81 \tag{C.7}$$

This theoretical value of the three-dimensional fractal dimension is in good agreement with the profusion of experimental and simulation values measured through the years.

All these results can be extended to any dimension. It can be shown that the d-dimensional divine proportion is given by

$$\varphi_d^{d} - \varphi_d^{d-1} - 1 = 0 \tag{C.8}$$

In Heinson *et al* (2015) we compared the prediction of the RHM to a well-known, off-lattice DLCA aggregation simulation. The RHM was also simulated. Both $d = 2$ and 3 aggregations were studied. The details of both simulations are described in the article.

Figure C.2 shows a data comparison of the DLCA and RHM simulations. Plotted is N versus (R_g/a) to test equation (C.1). There one essentially sees a perfect overlap of the two models. The conclusion is that the RHM has successfully modeled the canonical DLCA simulation.

One should not overlook the fact that the RHM leads us to the conclusion that aggregation leads to aggregates with a self-preserving shape. Indeed, one might ask: how else could it be? This is reminiscent of the self-preserving size distributions that also occur as a result of aggregation.

Figure C.3 shows distributions of side length ratios for the smallest circumscribing d-dimensional rectangles of the aggregates that resulted from these

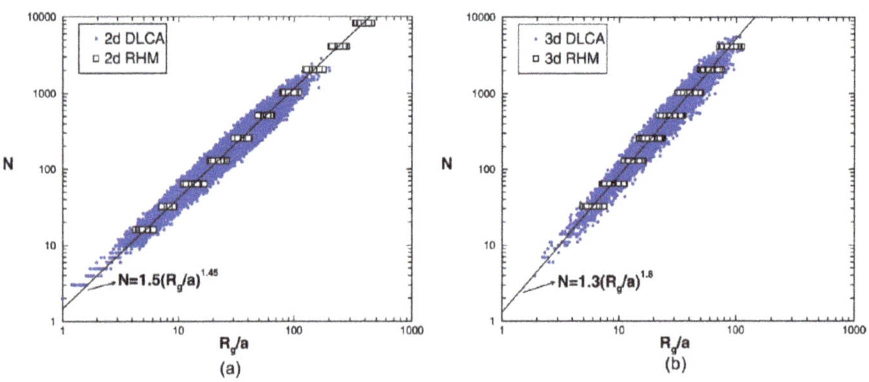

Figure C.2. Number of monomers N versus the monomer normalized radius of gyration, R_g/a, for DLCA (closed circles) and RHM (open squares) simulated aggregates for $d = 2$ and 3 spatial dimensions, (a) and (b), respectively. From Heinson *et al* (2015), reprinted by permission of the publisher (Taylor & Francis Ltd, http://www.tandfonline.com).

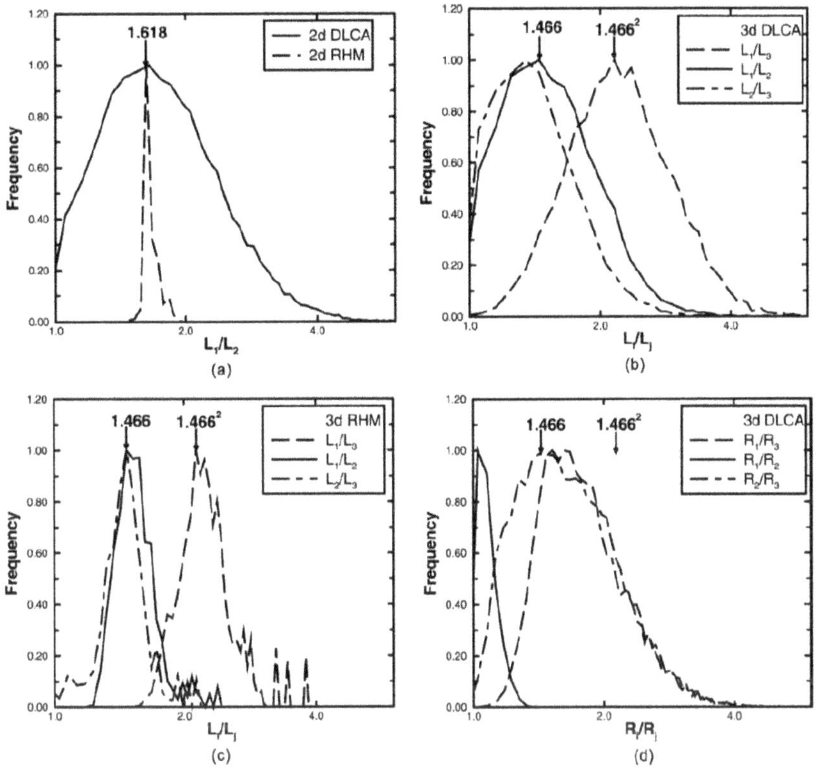

Figure C.3. Distributions of side length ratios L_i/L_j for: (a) circumscribing rectangles around $d = 2$ DLCA and RHM aggregates. Arrow indicates the Divine proportion $\varphi_2 = 1.618$. (b) Circumscribing rectangles around $d = 3$ DLCA aggregates. Arrows indicate the $d = 3$ Divine proportion $\varphi_3 = 1.466$ and its square. (c) Circumscribing rectangles around $d = 3$ RHM aggregates. Arrows indicate the $d = 3$ Divine proportion $\varphi_3 = 1.466$ and its square. (d) Distribution of principle radii of gyration ratios, R_i/R_j, for $d = 3$ DLCA aggregates. Arrows indicate the $d = 3$ Divine proportion $\varphi_3 = 1.466$ and its square. From Heinson *et al* (2015), reprinted by permission of the publisher (Taylor & Francis Ltd, http://www.tandfonline.com).

simulations. All the distributions peak at the values of the Divine proportion or its generalization to $d = 3$.

References

Heinson W R, Chakrabarti A and Sorensen C M 2015 Divine proportion shape invariance of diffusion limited cluster–cluster aggregates *Aerosol Sci. Technol.* **49** 786–92

Sorensen C M and Oh C 1998 Divine proportion shape preservation and the fractal nature of cluster–cluster aggregates *Phys. Rev.* E **58** 7545–8

IOP Publishing

Light Scattering and Absorption by Particles
The Q-space approach
Christopher M Sorensen

Appendix D

The radius of gyration

We will prove that

$$R_g^2 = \frac{1}{2}\int u^2 g(\boldsymbol{u})d\boldsymbol{u} \tag{D.1}$$

We have

$$\int u^2 g(\boldsymbol{u})d\boldsymbol{u} = \iint u^2 n(\boldsymbol{r})n(\boldsymbol{r}+\boldsymbol{u})d\mathbf{r}d\mathbf{u} \tag{D.2}$$

Rewrite the right hand side as

$$\int\left[\int u^2 n(\boldsymbol{r}+\boldsymbol{u})d\boldsymbol{u}\right]n(\boldsymbol{r})d\boldsymbol{r} \tag{D.3}$$

The quantity in the brackets is

$$[\ldots] = \int u^2 n(\boldsymbol{r}+\boldsymbol{u})d\boldsymbol{u} \tag{D.4}$$

Let $\boldsymbol{u} = \boldsymbol{x} - \boldsymbol{r}$ to obtain

$$[\ldots] = \int(\boldsymbol{x}-\boldsymbol{r})^2 n(\boldsymbol{x})d\boldsymbol{x} \tag{D.5}$$

$$[\ldots] = \int(x^2 - 2\boldsymbol{x}\bullet\boldsymbol{r} + r^2)n(\boldsymbol{x})d\boldsymbol{x} \tag{D.6}$$

Now recall that $R_g^2 = \int x^2 n(\boldsymbol{x})d\boldsymbol{x}$, $\int \boldsymbol{x}n(\boldsymbol{x})d\boldsymbol{x} = 0$ and $\int n(\boldsymbol{x})d\boldsymbol{x} = 1$ to find

$$[\ldots] = R_g^2 - 0 + r^2 \tag{D.7}$$

Substitute this back into equation (D.3)

$$\int\left[\int R_g^2 - 0 + r^2\right]n(\boldsymbol{r})d\boldsymbol{r} = R_g^2 + 0 + R_g^2 \tag{D.8}$$

Finally, remember that equation (D.3) was a rewrite of the right hand side of equation (D.2), thus

$$\int u^2 g(\boldsymbol{u}) d\boldsymbol{u} = 2R_g^2 \tag{D.9}$$

which proves equation (D.1).

Subject Index

www.ingramcontent.com/pod-product-compliance
Ingram Content Group UK Ltd.
Pitfield, Milton Keynes, MK11 3LW, UK
UKHW050000050425
457129UK00002B/6

9 780750 351553